Molecular Mechanisms of Neuropathic Pain

Molecular Mechanisms of Neuropathic Pain

Guest Editors

Hiroshi Ueda
Norimitsu Morioka

MDPI

Basel • Beijing • Wuhan • Barcelona • Belgrade • Novi Sad • Cluj • Manchester

Guest Editors
Hiroshi Ueda
Graduate Institute of Pharmacology
National Defense Medical University
Taipei
Taiwan

Norimitsu Morioka
Department of Pharmacology
Hiroshima University
Hiroshima
Japan

Editorial Office
MDPI AG
Grosspeteranlage 5
4052 Basel, Switzerland

This is a reprint of the Special Issue, published open access by the journal *Cells* (ISSN 2073-4409), freely accessible at: https://www.mdpi.com/journal/cells/special_issues/92B43N4AJW.

For citation purposes, cite each article independently as indicated on the article page online and as indicated below:

Lastname, A.A.; Lastname, B.B. Article Title. *Journal Name* **Year**, *Volume Number*, Page Range.

ISBN 978-3-7258-6620-5 (Hbk)
ISBN 978-3-7258-6621-2 (PDF)
https://doi.org/10.3390/books978-3-7258-6621-2

Contents

About the Editors . . . . **vii**

Hiroshi Ueda, Hiroyuki Neyama, Naoki Dozono, Junken Aoki and Jerold Chun
Hyperalgesia in the Psychological Stress-Induced Fibromyalgia Model Shows Sexual Dimorphism Mediated by LPA_1 and LPA_3
Reprinted from: *Cells* **2025**, *14*, 1022, https://doi.org/10.3390/cells14131022 . . . . **1**

Hanin Abdulbaset AboTaleb, Hani A. Alturkistani, Gamal S. Abd El-Aziz, Emad A. Hindi, Mervat M. Halawani, Mona Ali Al-Thepyani, et al
The Antinociceptive Effects and Sex-Specific Neurotransmitter Modulation of Metformin in a Mouse Model of Fibromyalgia
Reprinted from: *Cells* **2024**, *13*, 1986, https://doi.org/10.3390/cells13231986 . . . . **17**

Rafael Marins Rezende, Roney Santos Coimbra, Markus Kohlhoff, Lukiya Silva Campos Favarato, Hércia Stampini Duarte Martino, Luciano Bernardes Leite, et al
Effects of Tryptophan and Physical Exercise on the Modulation of Mechanical Hypersensitivity in a Fibromyalgia-like Model in Female Rats
Reprinted from: *Cells* **2024**, *13*, 1647, https://doi.org/10.3390/cells13191647 . . . . **46**

Simeng Ma, Yoki Nakamura, Suzuna Uemoto, Kenta Yamamoto, Kazue Hisaoka-Nakashima and Norimitsu Morioka
Intranasal Treatment with Cannabinoid 2 Receptor Agonist HU-308 Ameliorates Cold Sensitivity in Mice with Traumatic Trigeminal Neuropathic Pain
Reprinted from: *Cells* **2024**, *13*, 1943, https://doi.org/10.3390/cells13231943 . . . . **57**

Fumihiro Saika, Tetsuya Sato, Takeru Nakabayashi, Yohji Fukazawa, Shinjiro Hino, Kentaro Suzuki, et al
Male-Dominant Spinal Microglia Contribute to Neuropathic Pain by Producing CC-Chemokine Ligand 4 Following Peripheral Nerve Injury
Reprinted from: *Cells* **2025**, *14*, 484, https://doi.org/10.3390/cells14070484 . . . . **69**

Yuto Shibata, Yuki Matsumoto, Keita Kohno, Yasuharu Nakashima and Makoto Tsuda
Microgliosis in the Spinal Dorsal Horn Early After Peripheral Nerve Injury Is Associated with Damage to Primary Afferent Aβ-Fibers
Reprinted from: *Cells* **2025**, *14*, 666, https://doi.org/10.3390/cells14090666 . . . . **84**

Ayumi Tsukada, Yui Uekusa, Etsuro Ohta, Akito Hattori, Manabu Mukai, Dai Iwase, et al
Association Between Synovial NTN4 Expression and Pain Scores, and Its Effects on Fibroblasts and Sensory Neurons in End-Stage Knee Osteoarthritis
Reprinted from: *Cells* **2025**, *14*, 395, https://doi.org/10.3390/cells14060395 . . . . **98**

Rita Martín-Ramírez, María Ángeles González-Nicolás, Karen Álvarez-Tosco, Félix Machín, Julio Ávila, Manuel Morales, et al
Cilastatin Modulates DPEP1- and IQGAP1-Associated Neuro-Glio-Vascular Inflammation in Oxaliplatin-Induced Peripheral Neurotoxicity
Reprinted from: *Cells* **2025**, *14*, 1294, https://doi.org/10.3390/cells14161294 . . . . **113**

Leah Chang, Zala Čok and Lei Yu
Protein Kinases as Mediators for miRNA Modulation of Neuropathic Pain
Reprinted from: *Cells* **2025**, *14*, 577, https://doi.org/10.3390/cells14080577 . . . . **135**

Qiqi Zhou and George Nicholas Verne
Molecular Mechanisms and Pathways in Visceral Pain
Reprinted from: *Cells* **2025**, *14*, 1146, https://doi.org/10.3390/cells14151146 . . . . **152**

Mariacristina Mazzitelli, Takaki Kiritoshi, Peyton Presto, Zachary Hurtado, Nico Antenucci, Guangchen Ji, et al
BDNF Signaling and Pain Modulation
Reprinted from: *Cells* **2025**, *14*, 476, https://doi.org/10.3390/cells14070476 . . . . **168**

About the Editors

Hiroshi Ueda

Hiroshi Ueda is a professor at the Graduate Institute of Pharmacology, National Defense Medical University, Taipei, Taiwan. He is a neuropharmacologist renowned for his research in the neurobiology of chronic pain and neuroprotective mechanisms, particularly those involving prothymosin α. Over his career, he has held prominent academic and leadership positions at esteemed institutions. His work has significantly advanced the understanding of pain mechanisms, resulting in numerous high-impact publications and influential contributions to therapeutic development. He completed his academic training at Kyoto University, earning his B.Sc., M.Sc., and Ph.D. in pharmacology, and has held a series of influential academic and research positions over several decades, as Assistant and Associate Professor at Kyoto University and Yokohama City University, Chairman and Professor at Nagasaki University Graduate School of Biomedical Sciences, and Visiting Professor at McGill University. He currently serves as Chair Professor at the National Defense Medical University and Principal Investigator at the Research Institute for Production Development. He has focused on the neurobiology of chronic pain and the role of prothymosin α as a neuroprotective alarmin molecule. He has contributed to a better understanding of the molecular and cellular mechanisms underlying fibromyalgia and neuropathic pain, identifying novel therapeutic targets such as lysophosphatidic acid receptors. He has received several prestigious awards recognizing his exceptional contributions to pharmacology and neuroscience. Over the years, his sustained research contribution has been acknowledged through awards such as the Pharmaceutical Society of Japan Award and the Gayle A. Olson & Richard D. Olson Prize.

Norimitsu Morioka

Norimitsu Morioka is a professor in the Department of Pharmacology, Graduate School of Biomedical and Health Sciences, Hiroshima University, Hiroshima, Japan. He specializes in neuropharmacology, with a particular focus on pain and depression. For many years, he has analyzed the mechanisms involved in the onset and maintenance of chronic pain conditions in the spinal dorsal horn using animal models and primary cultures of neurons and glial cells. He has identified numerous drug target molecules and contributed to the advancement of pharmacological research in the field of pain. After obtaining his bachelor's, master's, and doctoral degrees at Hiroshima University, he served as an assistant professor, lecturer, and associate professor at the Graduate School of Biomedical and Health Sciences, Hiroshima University. He was appointed professor in 2016 and currently heads his own laboratory. Currently, he is expanding his research field to focus on changes in microglial activity in the brain, particularly in areas such as the hippocampus and prefrontal cortex, in addition to the spinal cord, which are involved not only in abnormal pain sensitivity but also in emotional and cognitive dysfunction associated with chronic pain conditions.

cells

MDPI

Article

Hyperalgesia in the Psychological Stress-Induced Fibromyalgia Model Shows Sexual Dimorphism Mediated by LPA_1 and LPA_3

Hiroshi Ueda [1,2,3,4,*], Hiroyuki Neyama [1,5], Naoki Dozono [1,2], Junken Aoki [6] and Jerold Chun [7]

1 Department of Pharmacology and Therapeutic Innovation, Nagasaki University Institute of Biomedical Sciences, Nagasaki 852-8521, Japan; neyama.hiroyuki.5y@kyoto-u.ac.jp (H.N.); 44naoki7@gmail.com (N.D.)
2 Department of Molecular Pharmacology, Graduate School of Pharmaceutical Sciences, Kyoto University, Kyoto 606-8501, Japan
3 Laboratory for the Study of Pain, Research Institute for Production Development, Kyoto 606-0805, Japan
4 Graduate Institute of Pharmacology, National Defense Medical Center, Nei-hu, Taipei 114201, Taiwan
5 Center for Cancer Immunotherapy and Immunobiology, Graduate School of Medicine, Kyoto University, Kyoto 606-8501, Japan
6 Department of Health Chemistry, Graduate School of Pharmaceutical Sciences, The University of Tokyo, Tokyo 113-0033, Japan; jaoki@mol.f.u-tokyo.ac.jp
7 Center for Neurological Diseases, Sanford Burnham Prebys Medical Discovery Institute, La Jolla, CA 92037, USA; jchun@sbpdiscovery.org
* Correspondence: ueda1hiroshi@icloud.com; Tel.: +81-075-781-1107

Abstract: Since the initial report indicating that LPA_1 signaling plays a key role in initiating nerve injury-induced neuropathic pain (NeuP), subsequent studies using knockout mice and $LPA_{1/3}$ antagonists have demonstrated that LPA_1 and LPA_3 signaling impact NeuP and fibromyalgia (FM) models. In the present study, we identified hyperalgesia sexual dimorphism involving $LPA_{1/3}$ signaling in the intermittent psychological stress induced-related FM-like model called intermittent psychological stress (IPS)-induced generalized pain (IPGP) model where the hyperalgesia in IPGP mice was abolished in LPA_1- and LPA_3-knock-out mice. Pharmacological intervention by intraperitoneal (i.p.) treatments with the $LPA_{1/3}$ antagonist Ki16425 consistently prevented hyperalgesia. However, intracerebroventricular treatments with Ki16425 abolished hyperalgesia in male, but not female, mice. Notably, intrathecal treatments of Ki16425 did not prevent hyperalgesia. Further studies revealed that splenocytes derived from female IPGP mice could initiate hyperalgesia via adoptive transfer in naïve mice, and this effect was abolished when donor mice were pre-treated with Ki16425 (i.p.). Thus, these studies identify male-specific $LPA_{1/3}$-mediated mechanisms in the brain underlying IPGP, as well as distinct LPA-$LPA_{1/3}$-mediated peripheral immune mechanisms.

Keywords: LPA_1; LPA_3; fibromyalgia; empathy; psychological stress; Neurometer; splenocytes; clodronate liposome; knock-out mouse; Ki16425

1. Introduction

Sexual dimorphism in chronic pain has increasingly been discussed in recent years [1–4]. Fibromyalgia (FM) is overwhelmingly female-dominant and sex differences remain an important issue. However, the full picture of FM mechanisms is still not fully clear. Here, we considered the involvement of lysophosphatidic acid (LPA, a bioactive lipid mediator) receptors. Previous studies using LPA receptor 1 (LPA_1)-KO mice have revealed that pain-related phenotypes are completely abolished in animal models of pain [5–7], including the following: hyperalgesia in neuropathic pain (NeuP), such as partial sciatic nerve ligation (pSNL); chemotherapeutic agent paclitaxel-induced models; type-1 and type-2 diabetes-related models; the central poststroke pain model. Furthermore, the role of LPA and LPA_1 signaling in NeuP, has also been reported in animal models of lumbar spinal canal stenosis [8], trigeminal ganglionic compression [9], and joint neuropathy [10,11]. In the NeuP model, intense and non-selective stress produced LPA and LPA synthesis was subsequently amplified in a feed-forward manner through the activation of LPA_1 and LPA receptor 3 (LPA_3) on microglia [6,7]. This mechanism provides evidence that LPA plays an important role in the formation of chronic pain. Interestingly, LPA production was found to increase even 2–3 weeks after pSNL [12] and continuous administration of $LPA_{1/3}$ antagonists completely attenuated established chronic pain. These findings support $LPA_{1/3}$ signaling as a significant component of pain memory. Thus, the inhibitors of LPA-LPA_1 signaling have been considered as potential targets for the treatment of NeuP [13,14].

Multiple animal models for FM have been developed for the purpose of developing diagnostics and treatments, based on symptomatology and possible pathogenesis. These include the vagotomized rat model [15]; twice repeated muscular acid injection-induced generalized pain model (AcGP) [16]; intermittent cold stress (ICS)-induced generalized pain (ICGP) model [17]; intermittent sound stress model [18]; reserpine administration-induced model [19]; intermittent psychological stress (IPS)-induced generalized pain (IPGP) model [20,21]. Recently, we reported that both IPGP and ICGP models share similar pathophysiological and pharmacotherapeutic features, which are clinically found in FM patients [21]. Among these models, we found evidence that LPA_1 signaling is involved in the mechanisms of IPGP, ICGP, and AcGP-type FM models, by using LPA_1-KO mice, and treatment with LPA_1-antagonists [20,21].

Although details of molecular and cellular mechanisms of NeuP and FM models through LPA_1 signaling remain elusive, there are reports that microglial activation through LPA-LPA_1 signaling plays a key role in NeuP, FM, and neuroinflammation [7,22–27]. Pioneering studies have also described a sexual dimorphism in the mechanisms of NeuP. They found that microglial cells are required for NeuP in male but not female mice, whereas T-cells are likely to be involved in female mice [28–33]. In the present study, we report the sexual dimorphism that LPA_1 and LPA_3 mechanisms are differentially involved in the brain and periphery, including splenic immunity, using FM models.

2. Materials and Methods

2.1. Animal Experiments

Male and female C57BL/6J mice (6–12 weeks old) from Japan SLC (Shizuoka, Japan) or TEXAM Corporation (Nagasaki, Japan), and homozygous mutant mice for the LPA_1 receptor gene (LPA_1-KO), and LPA_3 receptor gene (LPA_3-KO) weighing 20–25 g were used. Male and female homozygous mutant LPA_1- and LPA_3-KO mice were kindly provided by Prof. Jerold Chun, co-author. LPA_1- and LPA_3-KO mice had been backcrossed for 10 generations with C57BL/6J, which were used for wild-type mice to match genetic background. Whilst we backcrossed, homozygous mutant litter was confirmed by genotyping, as previously reported [34,35].

They were housed at 22 ± 2 °C and 55 ± 5% relative humidity with a 12 h light/dark cycle with a standard laboratory diet and water ad libitum. All experiments were conducted in accordance with the ethical guidelines of the Kyoto University Animal Research Committees (Approval number: 19-38), Nagasaki University Animal Research Committees (Approval number: 1607201325-8), and complied with the fundamental guidelines for the proper conduct of animal experiments and related activities in academic research institutions under the jurisdiction of the Ministry of Education, Culture, Sports, Science and Technology, Japan. In addition, the present study has been performed to comply with ARRIVE guidelines [36]. For randomization, the mechanical pain thresholds of each animal were measured before experiments, and animals were assigned to the control and treatment groups, which had approximately equal pain thresholds. Investigators who oversaw the project, the experimental design, and the conduct of experiments were independent to achieve blinding. The total number of animals used in this study was 172.

2.2. IPGP and ICGP Model

For the IPGP model, mice were exposed to intermittent psychological stress (IPS) by using the communication box (CBX-9M, Muromachi-Kikai, Tokyo, Japan) that has nine compartments (10 cm × 10 cm) with divided transparent plastic walls. Physical electrical stress (120 times) was given to 5 mice through the grid floor by a shock generator (CSG-001, Muromachi-Kikai, Tokyo, Japan) and cycler timer (CBX-CT, Muromachi-Kikai, Tokyo, Japan). These mice were individually put in the compartment located at the center and 4 corners without cover, as previously reported [20,21]. For IPS, mice which were in the remaining 4 compartments with insulating plastic cover, were exposed to psychological stress through seeing, hearing, and smelling the foot-shocked mice. In all experimental paradigms, stress was given once per day for 5 days. The post-stress day 1 stands as P1. For ICGP, mice were exposed to intermittent cold stress (ICS) as previously described [17,21]. In this model, mice were kept in a room under the condition of 22 ± 2 °C and humidity of 60 ± 5%. During the stress period, two mice were kept in each cage (12 × 15 × 10.5 cm), and fed ad libitum with a standard laboratory diet and tap water. On the 1st day (or day 0), mice were kept in a cold room at 4 ± 2 °C from 16:30 PM. In the cold room, mice were placed on a stainless-steel floor to induce a rapid temperature change and covered with plexiglass cage. The next day, mice were removed at 10:00 AM (day 1), and kept at 24 ± 2 °C, leaving the stainless floor in the cold room. Thirty min later, mice were put back in the cold room for another 30 min. These processes were repeated until 16:30 PM, followed by the placement of mice in the cold room overnight. On the next day (day 2), the treatment of mice with alternate temperature change processes was performed again. Finally, on day 3 (or post-stress day 1/P1), mice were removed from the cold room at 10:00 AM, kept at room temperature for at least 1 h, and used for tests. When constant cold stress (CCS) was given for 3 days, mice also acquired hyperalgesia, but the hyperalgesia completely recovered within 5 days. This suggests that generalized and chronic pain induced by ICS is not simply attributed to the cold stress. There was no apparent skin damage after ICS or CCS.

2.3. Partial Sciatic Nerve Ligation (pSNL) Model

For the pSNL model, mice received a small skin incision exposing the right-side sciatic nerve. The dorsal half of the sciatic nerve at the upper thigh level was tightly ligated with a sterile silk suture (virgin silk thread-K, USP 9-0, Kono Seisakusho, Chiba, Japan), as previously described [12]. All procedures should be smoothly but carefully performed, minimizing systemic neuroinflammatory influences.

2.4. Pain Tests

In the mechanical paw pressure test, mice were acclimatized on a 6 mm × 6 mm wire-mesh grid floor for over 1 h. A mechanical stimulus was then applied to the middle of the hind paw plantar surface using a digital von Frey anesthesiometer, with a rigid tip of 0.8 mm in outer diameter (Model 2390, 90 g probe, IITC Inc., Woodland Hills, CA, USA), and a transducer indicator (Model 1601; IITC Inc., Woodland Hills, CA, USA). The mechanical paw withdrawal threshold was determined from the average of three trials of paw pressure. A cut-off pressure of 20 g was set to avoid tissue damage. In the electrical stimulation-induced paw withdrawal (EPW) test, on the other hand, electrodes of Neurometer® Current Perception Threshold/C (CPT/C; Neurotron Inc., Baltimore, MD, USA) were attached to the planter and insteps of the hind paw, and transcutaneous nerve stimuli with sine-wave pulses of 2000 Hz or 250 Hz were applied. The minimum intensity at which each mouse withdrew its paw (cutoff time: 3 s) was evaluated by nociceptive current threshold (μA), as described previously [21]. The in vivo patch-clamp recording studies using acutely isolated rat spinal cord slices characterized that transcutaneous stimuli at 2000 Hz and 250 Hz caused excitatory synaptic responses in the substantia gelatinosa neurons, which correspond to Aβ- and Aδ-fiber stimulation, respectively [37]. This method has an advantage that no acclimatization time is necessary since the experimenter holds the mouse.

2.5. Clodronate Liposome Treatments

To deplete peripheral macrophages, dichloromethylene diphosphonate (clodronate) encapsulated liposomes (0.05 mg/10 μL; HYGIEIA BIOSCIENCE, Osaka, Japan) or PBS vehicle control liposomes were administrated by i.p. injection at a volume of 200 μL/mouse, as previously described [38].

2.6. Isolation and Adoptive Transfer of Splenocytes

To investigate peripheral immunity contribution to hyperalgesia in female mice with IPGP, the adoptive transfer of splenocytes from a donor mouse into a naïve mouse was attempted on the analogy of previous studies [27,39]. The spleen was isolated from IPS-treated female mice at P19 and placed in 3 mL of ice-cold RPMI 1640 medium (Gibco, Grand Island, NY, USA) containing 2% fetal bovine serum (FBS), followed by mincing using the plunger of a 1 mL injection syringe. The cell suspension was filtrated through a 70 μm mesh-cell strainer, washed with 5 mL of ice-cold PBS containing 2% FBS, and incubated in 3 mL of red blood cell lysis buffer (Abcam, San Diego, CA, USA) for 3 min at room temperature. The reaction was stopped by adding 5 mL of ice-cold PBS containing 2% FBS, washed, and resuspended in 3 mL of the same buffer. An aliquot of the suspension (200 μL, 1×10^6 cells) was then injected into the right retro-orbital sinus (retro-orbital injection) of the naïve male mouse under isoflurane (4%) anesthesia [40,41].

2.7. Drug Treatments

$LPA_{1/3}$ antagonist Ki16425 was dissolved in saline for the intraperitoneal (i.p.) injection (100 μL/10 g of body weight) or in the aCSF (125 mM NaCl, 3.8 mM KCl, 1.2 mM KH_2PO_4, 26 mM $NaHCO_3$, 10 mM glucose, pH 7.4) for the intracerebroventricular (i.c.v., 5 μL) or intrathecal (i.t., 5 μL) injection. The i.c.v. injection was performed using the method from Haley and McCormick [42]. The i.t. injection was given into the space between spinal L5 and L6 segments according to the method described by Hylden and Wilcox [43].

2.8. Statistical Analysis

All data are shown as mean ± S.E.M. Statistical value was calculated by GraphPad Prism 8 (GraphPad Software Inc., La Jolla, CA, USA) with p value set at under 0.05. Sample size was decided by our previous reports [7]. Statistical differences between the group were analyzed using two-way repeated measures ANOVA followed by Bonferroni's or Tukey's multiple comparisons test. Individual statistics were presented in figure legends.

3. Results

3.1. Involvement of LPA_3 Signaling in IPGP and ICGP Models in Male Mice

It has already been reported that LPA_1 plays a key role in IPGP and ICGP models using male LPA_1-KO mice. As shown in Figure 1A,B, male wild-type (WT) mice exhibited stable mechanical hyperalgesia from P1 to P22, after IPS induction from D1 to D5. Mechanical hyperalgesia was completely absent in male LPA_3-KO mice with IPS treatment and LPA_3-KO mice without IPS treatment. In addition to the equivalent pathophysiological and pharmacotherapeutic features between IPGP and ICGP models, a similar anti-hyperalgesia role of LPA_1 [20] and LPA_3 was found in the ICGP model (Figure 1C,D). Mechanical hyperalgesia was observed in male WT mice from P1 to P19 after ICS treatment. Male LPA_3-KO mice, in contrast, did not show mechanical hyperalgesia, either with ICS treatment or without ICS treatment.

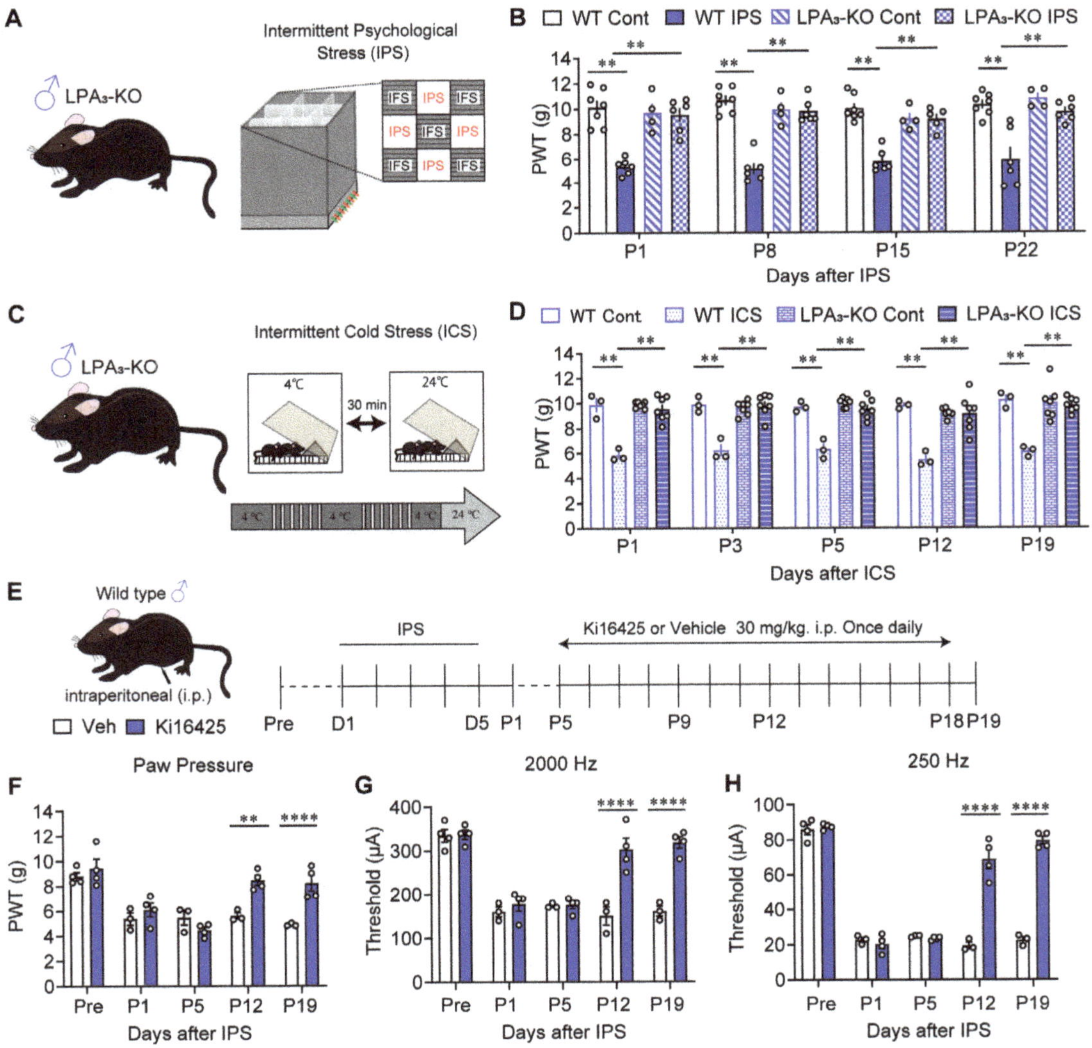

Figure 1. Reversal of male IPGP and ICGP by genetic deletion of LPA_3 or repeated i.p. treatments with $LPA_{1/3}$ antagonist. (**A**) Illustrative diagram of intermittent psychological stress (IPS)-induced

generalized pain (IPGP) model. (**B**) Reversal of IPS-induced mechanical hyperalgesia by genetic deficiency of LPA_3 in male mice. Mechanical hyperalgesia at P1–P22 was observed in the paw withdrawal test. (**C**) Illustrative diagram of intermittent cold stress (ICS)-induced generalized pain (ICGP) model. (**D**) Reversal of ICS-induced mechanical hyperalgesia (P1–P19) by genetic deficiency of LPA_3 in male mice. (**E**) Schedule of IPS-treatments and systemic (i.p.)-treatments with Ki16425. (**F–H**) Time-dependent change in threshold in the IPGP model treated with vehicle or Ki16425 in the paw pressure test (**F**), EPW 2000 Hz (**G**), and EPW 250 Hz (**H**). (**B**) ** $p < 0.01$, two-way ANOVA followed by Tukey's multiple comparisons test (WT-Cont, $n = 6$; WT-IPS, $n = 6$; LPA_3-KO-Cont, $n = 4$; LPA_3-KO-IPS, $n = 5$). (**D**) ** $p < 0.01$, two-way ANOVA, followed by Tukey's multiple comparisons test (WT-Cont, $n = 3$; WT-ICS, $n = 3$; LPA_3-KO-Cont, $n = 7$; LPA_3-KO-ICS, $n = 7$). (**F–H**) ** $p < 0.01$, **** $p < 0.0001$, two-way ANOVA followed by Bonferroni's multiple comparisons test (Veh, $n = 3$–4; Ki16425, $n = 4$).

Next, male mice were administered the $LPAR_{1/3}$ antagonist Ki16425 30 mg/kg (i.p.), once daily from P5 to P18 (Figure 1E). The IPGP-induced mechanical hyperalgesia reversed from P12 to P19 (Figure 1F). To test the noxious paw-withdrawal response (hypersensitivity), mice received a 2000 Hz current for non-noxious Aβ fibers, and 250 Hz current for noxious Aδ fibers generated by a Neurometer® [44]. The IPS-induced hypersensitivity at P1 and P5 was significantly reversed at P12 and P19 after repeated Ki16425 treatment (Figure 1G,H).

3.2. *Involvement of LPA_1 and LPA_3 Signaling in the IPGP Model in Female Mice*

IPS-induced pain remained stable until at least P19 in female WT mice (Figure 2B–D). In LPA_1-KO mice with IPS, mechanical hyperalgesia was not observed at P1 and P5. In LPA_3-KO mice, mechanical hyperalgesia was not observed at P1, but a reduction in paw withdrawal threshold occurred at P5 suggesting partial hyperalgesia suppression may occur, although without significance. When Ki16425 was administered continuously in WT mice from P5 to P18 (Figure 2A), mechanical hyperalgesia was significantly blocked at P12 and P19 (Figure 2B). Similar observations were also observed with Neurometer® stimulation experiments. LPA_1-KO mice did not develop hypersensitivity at P1 and P5, evaluated by 2000 Hz and 250 Hz stimulation (Figure 2C,D). LPA_3-KO mice did not show hypersensitivity at P1 with 2000 Hz stimulation but did display hypersensitivity at P5. At 250 Hz the anti-hypersensitivity effect was partial but significant at P1 and P5. Thus, the blocking of mechanical hyperalgesia and Neurometer hypersensitivity in LPA_3-KO mice appears to be less evident than in LPA_1-KO mice. Therefore, LPA_3 may be more important in the early or developing stage of abnormal pain, but less so in the maintenance stage. When Ki16425 was i.p. administered, hypersensitivity to the stimulation with 2000 Hz and 250 Hz at P12 and P19 significantly decreased. The reversal of hypersensitivity was more evident at the later time point, P19.

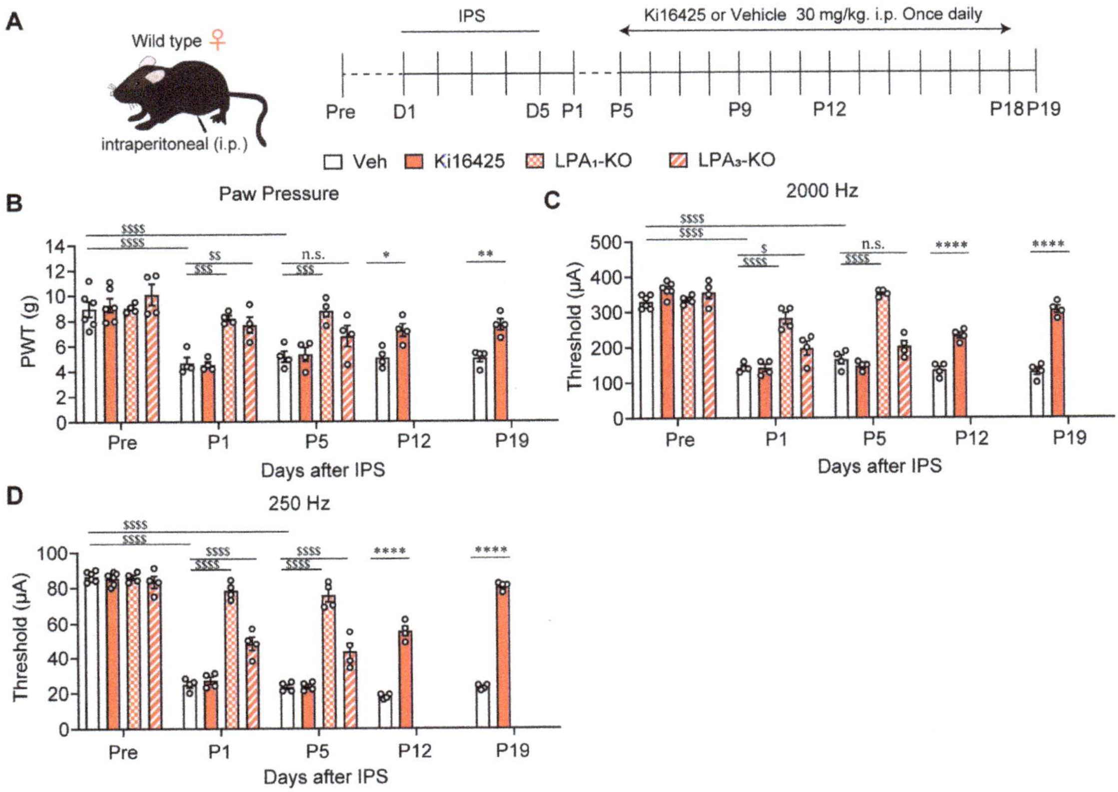

Figure 2. Reversal of female IPGP by genetic deletion of $LPA_{1/3}$ or repeated i.p. treatments with Ki16425. (**A**) Schedule of IPS-treatments and systemic (i.p.)-treatments with Ki16425. (**B–D**) Time-dependent reversal of IPS-induced mechanical hyperalgesia (**B**) and hypersensitivity in EPW tests with 2000 Hz (**C**) or 250 Hz (**D**) by genetic deficiency of LPA_1 or LPA_3, or by Ki16425 (i.p.) in female mice. (**B–D**) * $p < 0.05$, ** $p < 0.01$, **** $p < 0.0001$, Veh vs. Ki16425 (0–19), and two-way ANOVA followed by Bonferroni's multiple comparisons test (Veh, n = 4–6; Ki16425, n = 4–6). $ $p < 0.05$, $$ $p < 0.01$, $$$ $p < 0.001$, $$$$ $p < 0.0001$, n.s.; not significant, Veh vs. LPA_1-KO or LPA_3-KO, nd two-way ANOVA followed by Tukey's multiple comparisons test (0–5; Veh, n = 4–6; LPA_1-KO, n = 4; LPA_3-KO, n = 4).

3.3. Sex Differences in the Effect of Intracerebroventricular Administration of Ki16425

Next, Ki16425 at 10 nmol was administered intracerebroventricularly (i.c.v.) to male mice once daily from P5 to P11 (Figure 3A). At P12, IPS-induced mechanical hyperalgesia, and Neurometer hypersensitivity at 2000 Hz and 250 Hz were significantly blocked (Figure 3B–D). Surprisingly, Ki16425 (i.c.v.) administration to female mice showed no anti-hyperalgesia or anti-hypersensitivity effect (Figure 3E–G).

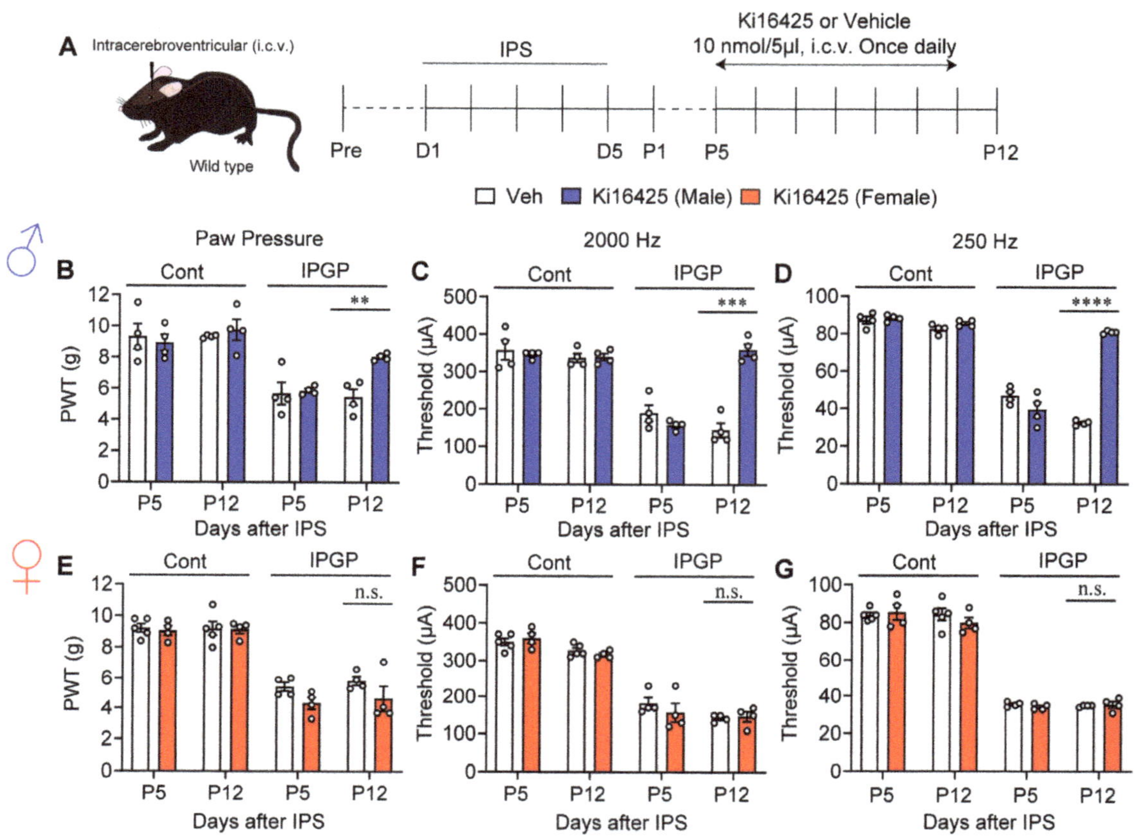

Figure 3. Male-specific reversal of hyperalgesia and hypersensitivity by repeated i.c.v. treatments with Ki16425. (**A**) Schedule of IPS-treatments and repeated i.c.v. treatments with Ki16425. (**B–D**) Reversal of IPS-induced mechanical hyperalgesia (**B**) and hypersensitivity in EPW tests with 2000 Hz (**C**) or 250 Hz (**D**) by Ki16425 (i.c.v.)-treatments in male mice. (**E–G**) Lack of reversal of IPS-induced mechanical hyperalgesia (**E**) and hypersensitivity in EPW tests with 2000 Hz (**F**) or 250 Hz (**G**) by Ki16425 (i.c.v.)-treatments in female mice. (**B–G**) ** $p < 0.01$, *** $p < 0.001$, **** $p < 0.0001$, n.s.; not significant, and two-way ANOVA followed by Bonferroni's multiple comparisons test (male Veh, $n = 4$; female Veh, $n = 5$; Ki16425, $n = 4$).

3.4. Lack of Blocking Effects on Hyperalgesia and Hypersensitivity by Intrathecal Administration of Ki16425

Ki16425 at 10 nmol was then intrathecally (i.t.) administered to male and female mice daily from P5 to P11 (Figure 4A). No effect in response to treatment was observed in mechanical hyperalgesia, or Neurometer hypersensitivity at 2000 Hz and 250 Hz, up until P12 (Figure 4B–G).

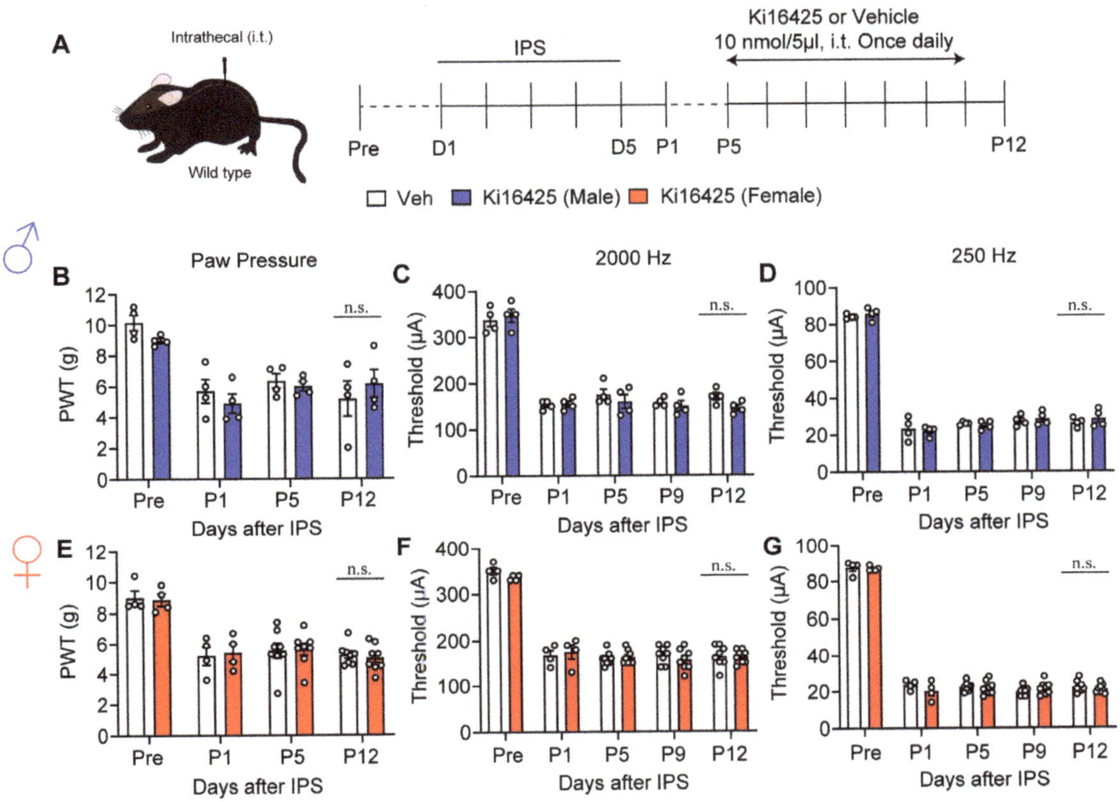

Figure 4. Lack of reversal of hyperalgesia and hypersensitivity by i.t. treatments with Ki16425. (**A**) Schedule of IPS-treatments and repeated i.t. treatments with Ki16425. (**B–G**) Lack of reversal of IPS-induced mechanical hyperalgesia (**B**,**E**) and hypersensitivity in EPW tests with 2000 Hz (**C**,**F**) or 250 Hz (**D**,**G**) by Ki16425 (i.t.)-treatments in male (**B–D**) and female mice (**E–G**). (**B–G**) n.s.; not significant and two-way ANOVA followed by Bonferroni's multiple comparisons test (male Veh, $n = 4$; female Veh, $n = 4$–8; Ki16425, $n = 4$–8).

3.5. Lack of Blocking Effect of Clodronate Liposome on Female IPGP Model

It is known that i.p. treatment with clodronate liposome depletes peripheral macrophage to block neuropathic pain [38]. To evaluate effects of clodronate liposome treatment in the present study, liposomes containing 25 mg/kg of clodronate were given 30 min before and 6 days after the pSNL (Figure 5A,B). The mechanical hyperalgesia at the ipsilateral paw of male mice at P1 and P7 was abolished by clodronate liposome treatment (Figure 5C). The hypersensitivity evaluated by 2000 Hz and 250 Hz stimulation was also significantly reversed by clodronate liposome treatment (Figure 5D,E).

In the IPGP model in female mice, clodronate liposomes were given 30 min before the first day of IPS and after day 5 of IPS (Figure 5F,G). The mechanical hyperalgesia or Neurometer hypersensitivity evaluated by 2000 Hz and 250 Hz stimulation at P1 and P5 after IPS was not affected by clodronate liposomes (Figure 5H–J).

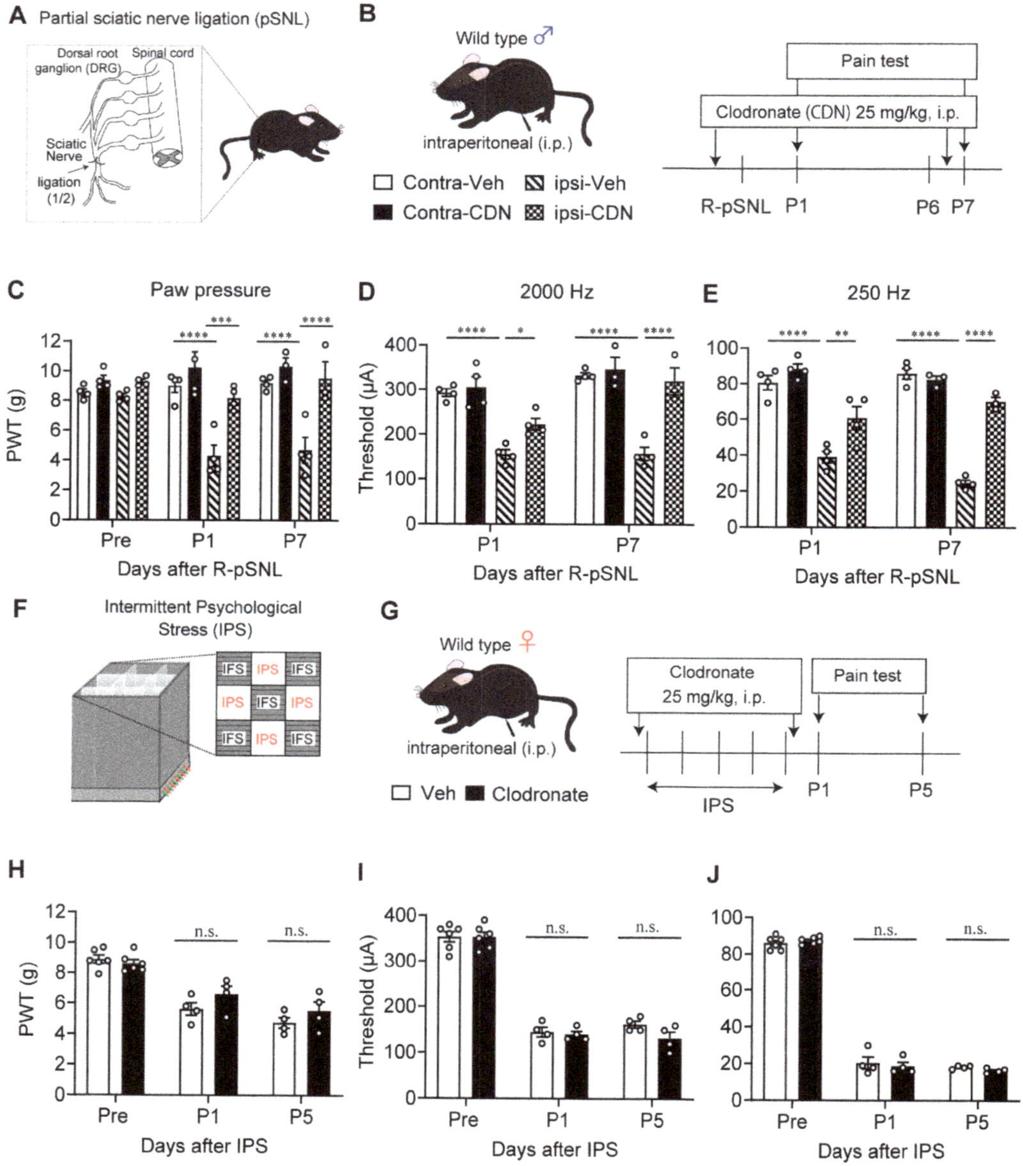

Figure 5. Lack of reversal of IPS-induced hyperalgesia and hypersensitivity by clodronate liposome in the IPGP model. (**A**) Illustrative diagram of partial sciatic nerve injury (pSNL) model. (**B**) Schedule of clodronate treatments and pain tests in the partial sciatic nerve injury (pSNL) model in male mice. (**C–E**) Reversal of pSNL-induced mechanical hyperalgesia (**C**) or hypersensitivity in the EPW 2000 Hz (**D**) or 250 Hz (**E**) tests by clodronate liposome treatments. (**F**) Illustrative diagram of IPGP model. (**G**) Schedule of clodronate treatments and pain tests in the IPGP model in female mice. (**H–J**) Lack of reversal of the IPS-induced hyperalgesia (**H**) or hypersensitivity (**I**,**J**) by clodronate liposome treatments. (**C–E**) * $p < 0.05$, ** $p < 0.01$, *** $p < 0.001$, **** $p < 0.0001$, and two-way ANOVA followed by Tukey's multiple comparisons test (contra-Veh, $n = 4$; ipsi-Veh, $n = 4$; contra-Clodronate, $n = 3$–4; ipsi-Clodronate, $n = 3$–4). (**H–J**) n.s.; not significant and two-way ANOVA followed by Bonferroni's multiple comparisons test (Veh, $n = 4$–6; Clodronate, $n = 4$–6).

3.6. Blocking Effect of Ki16425 on the Induction of Pain Hypersensitivity by Splenocytes Derived from Female IPGP Model Mice

Next, splenocytes were prepared at P19 after the IPS in female mice and were then administered (i.v.) to naïve mice (Figure 6A,B). This induced significant mechanical hyperalgesia or electrical hypersensitivity evaluated by 2000 Hz and 250 Hz stimulation at P1 (Figure 6C–E). The hyperalgesia and hypersensitivity were significantly reversed by repeated Ki16425-treatments (i.p.), from P5 to P18 in donor mice (Figure 6C–E).

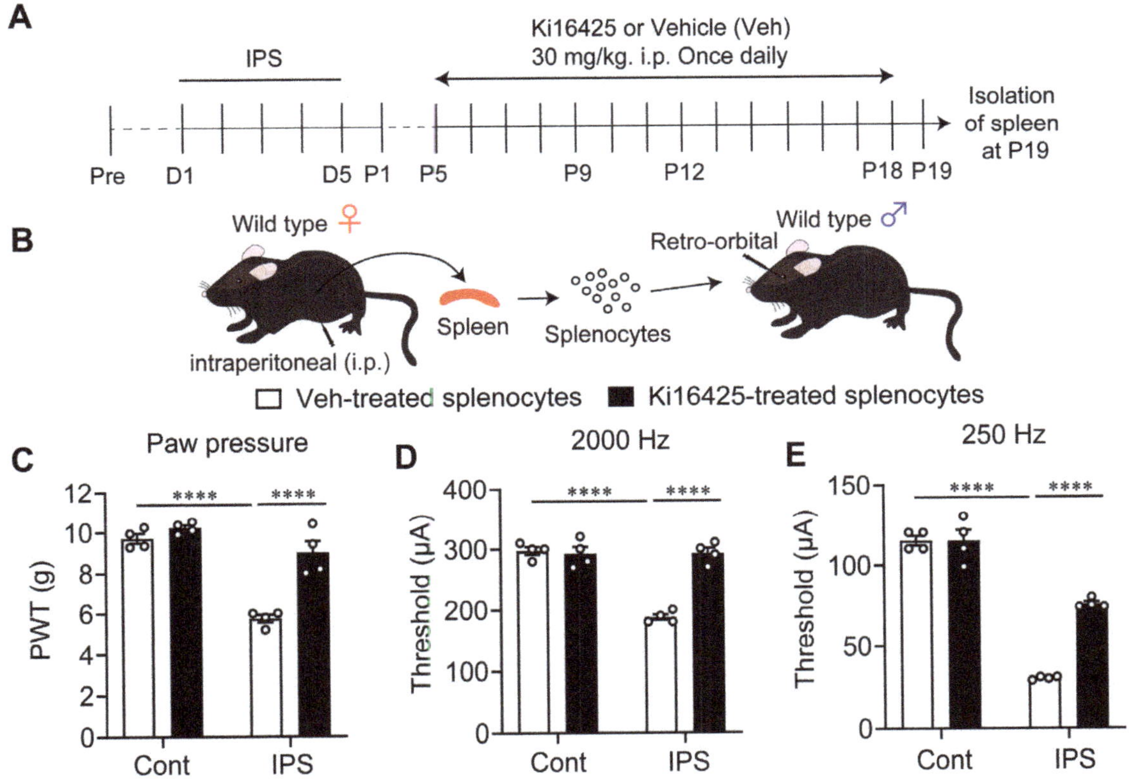

Figure 6. Abnormal pain by splenocytes derived from IPS-treated mice. (**A**) Schedule of IPS treatments and repeated i.p. treatments with Ki16425 in female donor mice. (**B**) Experimental procedure for splenocytes injection. (**C–E**) Ki16425-reversible hyperalgesia (**B**) and hypersensitivity (**C**,**D**) in naïve female mice by adoptive transfer of splenocytes from IPS-treated donor female mice. Ki16425 (i.p.) was treated in donor mice, as described in the legend of Figure 3. (**C–D**) **** $p < 0.0001$ and two-way ANOVA followed by Tukey's multiple comparisons test ($n = 4$).

4. Discussion

Previous reports have shown that thermal hyperalgesia in IPGP and ICGP models are completely abolished in male LPA_1-KO mice [20]. The present study evaluated the sex-dependent involvements of LPA_1 and LPA_3 in the IPGP model. Mechanical paw withdrawal threshold was tested using an electronic digital von Frey anesthesiometer and the electrical stimulation-induced paw withdrawal test (EPW) was performed using a Neurometer, as briefly summarized in Table 1. In the IPGP model, male LPA_1- or LPA_3-KO mice did not experience mechanical hyperalgesia and EPW hypersensitivity. Similar effects were observed in male LPA_1- or LPA_3-KO mice after systemic administration (i.p.) of Ki16425 (an $LPA_{1/3}$ antagonist). In female LPA_1-KO mice, mechanical hyperalgesia and EPW hypersensitivity were also absent. Whilst in LPA_3-KO mice, there was no, or a very weak blocking effect, even when treated with Ki16425 (i.p.). Sexual dimorphism of $LPA_{1/3}$ receptor signaling was more evident when Ki16425 was administered into the brain (i.c.v.). Repeated i.c.v. treatments with Ki16425 completely reversed the mechanical hyperalgesia and EPW hypersensitivity in male mice, but no reversal was observed in female mice. As Ki16425 (i.t.) had no effect on male and female mice, brain LPA_1 and LPA_3 signaling likely play key roles in the mechanisms underlying IPGP in male mice. Female LPA_3 mice, in contrast, show no, or very limited contribution of LPA_3 signaling. There are reports that LPA_1 and LPA_3 signaling are associated with activating microglia, namely for the self-amplification of LPA and production of brain-derived neurotrophic factor in the NeuP mechanisms [7,45]. All these findings are reminiscent of previous reports that NeuP is associated with microglial activation in male, but not female mice [32,33].

Table 1. Summary of LPA_1 and LPA_3-mediated hyperalgesia and hypersensitivity.

IPS Model	Sex	Hyperalgesia (Paw Pressure)	Hypersensitivity (2000 Hz)	Hypersensitivity (250 Hz)
LPA_1-KO	Male	– (1)	No data	No data
	Female	–	–	–
LPA_3-KO	Male	– (2)	No data	No data
	Female	–	–	–
WT Ki 16425, i.p.	Male	–	–	–
	Female	–	–	–
WT Ki 1642, i.c.v.	Male	–	–	–
	Female	+	+	+
WT Ki 16425, i.t.	Male	+	+	+
	Female	+	+	+

(1): Data have been reported elsewhere [20]. (2): The blockade of hyperalgesia was also observed in the ICS model.

There have been many reports on the role of macrophages in NeuP models [46]. One experimental evidence for this can be seen in a study showing the disappearance of NeuP by using clodronate liposomes, which phagocytose activated peripheral macrophages and cause cell death via metabolic toxin clodronate [38]. In the present study, we verified the validity of this technique using the pSNL model. As reported before [47], the present study confirmed that clodronate liposome treatment abolished pSNL-induced hyperalgesia and hypersensitivity. However, clodronate liposome treatment had no effect in the IPGP model at P1 and P5. Therefore, it is evident that peripheral macrophages are not involved in IPGP mechanisms.

Accumulating studies have reported that the peripheral immune system is involved in the sexual dimorphism of NeuP mechanisms. These suggest brain microglia are involved in male, but not female mice. In female mice, the peripheral immune system, but not central, likely plays a key role in the NeuP mechanism [30,33]. To further investigate a possible involvement of the peripheral immune system, we explored the involvement of splenocytes. This was based on the analogy of our previous study, in which splenocytes derived from donor mice with chronic muscular pain induced by twice repeated acid injections, caused hyperalgesia in naïve mice [27]. The present study confirmed this speculation by finding that systemic (retro-orbital) injection of splenocytes from donor mice in the IPGP model, into naïve mice, caused significant hyperalgesia and hypersensitivity. As systemic Ki16425-treatment of female donor mice reversed the abnormal pain activity of splenocytes, peripheral $LPA_{1/3}$ mechanisms are likely involved in this reversal. However, it remains unclear whether $LPA_{1/3}$ signaling in splenocytes is responsible for such mechanisms. For this discussion, it is necessary to perform several key experiments in female mice treated with IPS, such as whether $LPA_{1/3}$ or biosynthesis enzyme autotaxin are upregulated in splenic T-cells or B-cells. Current reports suggest that plasma cytokine levels (possibly from T-cells) in FM patients may be associated with FM development [48–50]. Furthermore, pain-related immunoglobulin G (IgGs) may be derived plasma cells differentiated from activated B-cells [51]. A recent study also revealed that IgGs from FM patients were accumulated in satellite glial cells, possibly leading to DRG neuron stimulation [52–54].

It remains unclear what mechanisms are involved in male brain-specific LPA_1 and LPA_3 mechanisms in the IPGP model. Future studies on male-specific microglial roles, as seen in the case of AcGP-type FM model [27], may be noteworthy. Indeed, recent studies demonstrated the male-specific expression of microglia-derived molecules, which are associated with sexually dimorphic NeuP [55–57]. Thus, it should be interesting to investigate whether LPA_1 or LPA_3 signaling is associated with the actions on these specific molecules.

5. Conclusions

In conclusion, accumulating studies demonstrated that $LPA_{1/3}$ signaling plays key roles in the mechanisms underlying several NeuP and FM models [7]. The present study firstly demonstrated that there is sexual dimorphism in $LPA_{1/3}$ signaling in the IPGP model. Upon i.c.v. treatment with an $LPA_{1/3}$-antagonist, hyperalgesia and hypersensitivity in male, but not female mice, were abolished, although systemic treatment reversed the abnormal pain in both male and female mice. Furthermore, studies revealed that peripheral macrophages were not involved, whilst splenocytes derived from female IPGP model mice caused hyperalgesia in naïve mice. Thus, these findings advocate the necessity to identify the male-specific $LPA_{1/3}$-mediated mechanisms in the brain underlying IPGP, as well as molecular-based peripheral immune mechanisms, which may contribute to the translational potential of the research, such as diagnostics and treatments by targeting involved molecules.

Author Contributions: Conceptualization, H.U.; methodology, H.U.; validation, H.N. and N.D.; formal analysis, H.N. and N.D.; investigation, H.N. and N.D.; resources, H.U.; writing—original draft preparation, H.U., H.N., J.A., and J.C.; writing—review and editing, H.U., H.N., and J.C.; supervision, H.U.; project administration, H.U.; funding acquisition, H.U. All authors have read and agreed to the published version of the manuscript.

Funding: This study was supported by KAKENHI JP17H01586, JP19K21592, and JP21H03024 (HU) from the Japan Society for the Promotion of Science (JSPS), and Platform for Drug Discovery, Informatics, and Structural Life Science [16am0101012j0005] (HU) from Japan Agency for Medical Research and Development (AMED), and NSTC 113-2320-B-016-004 (HU) from NSTC (R.O.C. Taiwan).

Institutional Review Board Statement: All experiments were conducted in accordance with the ethical guidelines of the Kyoto University Animal Research Committees (Approval number: 19–38, approval date: 2019-04-01), Nagasaki University Animal Research Committees (Approval number: 1607201325-8, approval date: 20 July 2016) and complied with the fundamental guidelines for the proper conduct of animal experiments and related activities in academic research institutions under the jurisdiction of the Ministry of Education, Culture, Sports, Science and Technology, Japan.

Informed Consent Statement: Not applicable.

Data Availability Statement: Data are available upon request to the authors.

Acknowledgments: We thank Shuji Kaneko and Hisashi Shirakawa for the help in animal studies in Kyoto University and Myles Sant-Cassia for English editing of the manuscript.

Conflicts of Interest: Author Jerold Chun has an employment relationship with Neurocrine Biosciences, a company that may potentially benefit from the research results. Jerold Chun's relationship with Neurocrine Biosciences has been reviewed and approved by Sanford Burnham Prebys Medical Discovery Institute in accordance with its Conflict of Interest Policies. The remaining authors declare that the research was conducted in the absence of any commercial or financial relationships that could be construed as a potential conflict of interest.

Abbreviations

The following abbreviations are used in this manuscript:

FM	Fibromyalgia
ICGP	Intermittent Cold Stress-Induced Generalized Pain
IPGP	Intermittent Psychological Stress-Induced Generalized Pain
LPA	Lysophosphatidic acid
NeuP	Neuropathic Pain
pSNL	Partial Sciatic Nerve ligation
IGGs	Immunoglobulin G

References

1. Fillingim, R.B.; King, C.D.; Ribeiro-Dasilva, M.C.; Rahim-Williams, B.; Riley, J.L., III. Sex, gender, and pain: A review of recent clinical and experimental findings. *J. Pain* **2009**, *10*, 447–485. [CrossRef]
2. Todd, A.; McNamara, C.L.; Balaj, M.; Huijts, T.; Akhter, N.; Thomson, K.; Kasim, A.; Eikemo, T.A.; Bambra, C. The European epidemic: Pain prevalence and socioeconomic inequalities in pain across 19 European countries. *Eur. J. Pain* **2019**, *23*, 1425–1436. [CrossRef]
3. Tsang, A.; Von Korff, M.; Lee, S.; Alonso, J.; Karam, E.; Angermeyer, M.C.; Borges, G.L.; Bromet, E.J.; Demytteneare, K.; de Girolamo, G.; et al. Common chronic pain conditions in developed and developing countries: Gender and age differences and comorbidity with depression-anxiety disorders. *J. Pain* **2008**, *9*, 883–891. [CrossRef]
4. Zimmer, Z.; Fraser, K.; Grol-Prokopczyk, H.; Zajacova, A. A global study of pain prevalence across 52 countries: Examining the role of country-level contextual factors. *Pain* **2022**, *163*, 1740–1750. [CrossRef]
5. Inoue, M.; Rashid, M.H.; Fujita, R.; Contos, J.J.; Chun, J.; Ueda, H. Initiation of neuropathic pain requires lysophosphatidic acid receptor signaling. *Nat. Med.* **2004**, *10*, 712–718. [CrossRef]
6. Ueda, H. Lysophosphatidic acid signaling is the definitive mechanism underlying neuropathic pain. *Pain* **2017**, *158* (Suppl. S1), S55–S65. [CrossRef]
7. Ueda, H. Pathogenic mechanisms of lipid mediator lysophosphatidic acid in chronic pain. *Prog. Lipid Res.* **2021**, *81*, 101079. [CrossRef]
8. Uranbileg, B.; Ito, N.; Kurano, M.; Saigusa, D.; Saito, R.; Uruno, A.; Kano, K.; Ikeda, H.; Yamada, Y.; Sumitani, M.; et al. Alteration of the lysophosphatidic acid and its precursor lysophosphatidylcholine levels in spinal cord stenosis: A study using a rat cauda equina compression model. *Sci. Rep.* **2019**, *9*, 16578. [CrossRef]
9. Ahn, D.K.; Lee, S.Y.; Han, S.R.; Ju, J.S.; Yang, G.Y.; Lee, M.K.; Youn, D.H.; Bae, Y.C. Intratrigeminal ganglionic injection of LPA causes neuropathic pain-like behavior and demyelination in rats. *Pain* **2009**, *146*, 114–120. [CrossRef]
10. McDougall, J.J.; Albacete, S.; Schuelert, N.; Mitchell, P.G.; Lin, C.; Oskins, J.L.; Bui, H.H.; Chambers, M.G. Lysophosphatidic acid provides a missing link between osteoarthritis and joint neuropathic pain. *Osteoarthr. Cartil.* **2017**, *25*, 926–934. [CrossRef]
11. O'Brien, M.S.; Philpott, H.T.A.; McDougall, J.J. Targeting the Nav1.8 ion channel engenders sex-specific responses in lysophosphatidic acid-induced joint neuropathy. *Pain* **2019**, *160*, 269–278. [CrossRef]
12. Ueda, H.; Neyama, H.; Nagai, J.; Matsushita, Y.; Tsukahara, T.; Tsukahara, R. Involvement of lysophosphatidic acid-induced astrocyte activation underlying the maintenance of partial sciatic nerve injury-induced neuropathic pain. *Pain* **2018**, *159*, 2170–2178. [CrossRef]

13. Edamura, T.; Sumitani, M.; Hayakawa, K.; Inoue, R.; Abe, H.; Tsuchida, R.; Chikuda, H.; Ogata, T.; Kurano, M.; Aoki, J.; et al. Different Profiles of the Triad of Lysophosphatidylcholine, Lysophosphatidic Acid, and Autotaxin in Patients with Neuropathic Pain Diseases: A Preliminary Observational Study. *Pain Ther.* **2022**, *11*, 1439–1449. [CrossRef]
14. Velasco, M.; O'Sullivan, C.; Sheridan, G.K. Lysophosphatidic acid receptors (LPARs): Potential targets for the treatment of neuropathic pain. *Neuropharmacology* **2017**, *113*, 608–617. [CrossRef]
15. Khasar, S.G.; Miao, F.J.; Janig, W.; Levine, J.D. Vagotomy-induced enhancement of mechanical hyperalgesia in the rat is sympathoadrenal-mediated. *J. Neurosci.* **1998**, *18*, 3043–3049. [CrossRef]
16. Sluka, K.A.; Kalra, A.; Moore, S.A. Unilateral intramuscular injections of acidic saline produce a bilateral, long-lasting hyperalgesia. *Muscle Nerve* **2001**, *24*, 37–46. [CrossRef]
17. Nishiyori, M.; Ueda, H. Prolonged gabapentin analgesia in an experimental mouse model of fibromyalgia. *Mol. Pain* **2008**, *4*, 52. [CrossRef]
18. Khasar, S.G.; Dina, O.A.; Green, P.G.; Levine, J.D. Sound stress-induced long-term enhancement of mechanical hyperalgesia in rats is maintained by sympathoadrenal catecholamines. *J. Pain* **2009**, *10*, 1073–1077. [CrossRef]
19. Nagakura, Y.; Oe, T.; Aoki, T.; Matsuoka, N. Biogenic amine depletion causes chronic muscular pain and tactile allodynia accompanied by depression: A putative animal model of fibromyalgia. *Pain* **2009**, *146*, 26–33. [CrossRef]
20. Ueda, H.; Neyama, H. LPA1 receptor involvement in fibromyalgia-like pain induced by intermittent psychological stress, empathy. *Neurobiol. Pain* **2017**, *1*, 16–25. [CrossRef]
21. Ueda, H.; Neyama, H. Fibromyalgia Animal Models Using Intermittent Cold and Psychological Stress. *Biomedicines* **2023**, *12*, 56. [CrossRef]
22. Amaral, R.F.; Geraldo, L.H.M.; Einicker-Lamas, M.; TCLS, E.S.; Mendes, F.; Lima, F.R.S. Microglial lysophosphatidic acid promotes glioblastoma proliferation and migration via LPA_1 receptor. *J. Neurochem.* **2021**, *156*, 499–512. [CrossRef]
23. Gaire, B.P.; Choi, J.W. Critical Roles of Lysophospholipid Receptors in Activation of Neuroglia and Their Neuroinflammatory Responses. *Int. J. Mol. Sci.* **2021**, *22*, 7864. [CrossRef]
24. Gaire, B.P.; Sapkota, A.; Song, M.R.; Choi, J.W. Lysophosphatidic acid receptor 1 (LPA_1) plays critical roles in microglial activation and brain damage after transient focal cerebral ischemia. *J. Neuroinflamm.* **2019**, *16*, 170. [CrossRef]
25. Kwon, J.H.; Gaire, B.P.; Park, S.J.; Shin, D.Y.; Choi, J.W. Identifying lysophosphatidic acid receptor subtype 1 (LPA_1) as a novel factor to modulate microglial activation and their TNF-α production by activating ERK1/2. *Biochim. Biophys. Acta. Mol. Cell Biol. Lipids* **2018**, *1863*, 1237–1245. [CrossRef]
26. Plastira, I.; Bernhart, E.; Joshi, L.; Koyani, C.N.; Strohmaier, H.; Reicher, H.; Malle, E.; Sattler, W. MAPK signaling determines lysophosphatidic acid (LPA)-induced inflammation in microglia. *J. Neuroinflamm.* **2020**, *17*, 127. [CrossRef]
27. Ueda, H.; Dozono, N.; Tanaka, K.; Kaneko, S.; Neyama, H.; Uchida, H. Allodynia by Splenocytes from Mice with Acid-Induced Fibromyalgia-Like Generalized Pain and Its Sexual Dimorphic Regulation by Brain Microglia. *Front. Neurosci.* **2020**, *14*, 600166. [CrossRef]
28. Boerner, K.E.; Chambers, C.T.; Gahagan, J.; Keogh, E.; Fillingim, R.B.; Mogil, J.S. Conceptual complexity of gender and its relevance to pain. *Pain* **2018**, *159*, 2137–2141. [CrossRef]
29. Mapplebeck, J.C.; Beggs, S.; Salter, M.W. Molecules in pain and sex: A developing story. *Mol. Brain* **2017**, *10*, 9. [CrossRef]
30. Mapplebeck, J.C.S.; Beggs, S.; Salter, M.W. Sex differences in pain: A tale of two immune cells. *Pain* **2016**, *157* (Suppl. S1), S2–S6. [CrossRef]
31. Mogil, J.S. Qualitative sex differences in pain processing: Emerging evidence of a biased literature. *Nat. Rev. Neurosci.* **2020**, *21*, 353–365. [CrossRef]
32. Sorge, R.E.; LaCroix-Fralish, M.L.; Tuttle, A.H.; Sotocinal, S.G.; Austin, J.S.; Ritchie, J.; Chanda, M.L.; Graham, A.C.; Topham, L.; Beggs, S.; et al. Spinal cord Toll-like receptor 4 mediates inflammatory and neuropathic hypersensitivity in male but not female mice. *J. Neurosci.* **2011**, *31*, 15450–15454. [CrossRef]
33. Sorge, R.E.; Mapplebeck, J.C.; Rosen, S.; Beggs, S.; Taves, S.; Alexander, J.K.; Martin, L.J.; Austin, J.S.; Sotocinal, S.G.; Chen, D.; et al. Different immune cells mediate mechanical pain hypersensitivity in male and female mice. *Nat. Neurosci.* **2015**, *18*, 1081–1083. [CrossRef]
34. Contos, J.J.; Fukushima, N.; Weiner, J.A.; Kaushal, D.; Chun, J. Requirement for the lpA1 lysophosphatidic acid receptor gene in normal suckling behavior. *Proc. Natl. Acad. Sci. USA* **2000**, *97*, 13384–13389. [CrossRef]
35. Ye, X.; Hama, K.; Contos, J.J.; Anliker, B.; Inoue, A.; Skinner, M.K.; Suzuki, H.; Amano, T.; Kennedy, G.; Arai, H.; et al. LPA3-mediated lysophosphatidic acid signalling in embryo implantation and spacing. *Nature* **2005**, *435*, 104–108. [CrossRef]
36. du Sert, N.P.; Hurst, V.; Ahluwalia, A.; Alam, S.; Avey, M.T.; Baker, M.; Browne, W.J.; Clark, A.; Cuthill, I.C.; Dirnagl, U.; et al. The ARRIVE guidelines 2.0: Updated guidelines for reporting animal research. *Br. J. Pharmacol.* **2020**, *177*, 3617–3624. [CrossRef]

37. Koga, K.; Furue, H.; Rashid, M.H.; Takaki, A.; Katafuchi, T.; Yoshimura, M. Selective activation of primary afferent fibers evaluated by sine-wave electrical stimulation. *Mol. Pain* **2005**, *1*, 13. [CrossRef]
38. Nguyen, T.; Du, J.; Li, Y.C. A protocol for macrophage depletion and reconstitution in a mouse model of sepsis. *STAR Protoc.* **2021**, *2*, 101004. [CrossRef]
39. Yamashita, S.; Dozono, N.; Tobori, S.; Nagayasu, K.; Kaneko, S.; Shirakawa, H.; Ueda, H. Peripheral Beta-2 Adrenergic Receptors Mediate the Sympathetic Efferent Activation from Central Nervous System to Splenocytes in a Mouse Model of Fibromyalgia. *Int. J. Mol. Sci.* **2023**, *24*, 3465. [CrossRef]
40. Yardeni, T.; Eckhaus, M.; Morris, H.D.; Huizing, M.; Hoogstraten-Miller, S. Retro-orbital injections in mice. *Lab Anim.* **2011**, *40*, 155–160. [CrossRef]
41. University Animal Care Committee Standard Operating Procedure. Available online: https://www.queensu.ca/animals-in-science/policies-procedures/sop/mice/7-11 (accessed on 3 July 2025).
42. Haley, T.J.; McCormick, W.G. Pharmacological effects produced by intracerebral injection of drugs in the conscious mouse. *Br. J. Pharmacol. Chemother.* **1957**, *12*, 12–15. [CrossRef]
43. Hylden, J.L.; Wilcox, G.L. Intrathecal morphine in mice: A new technique. *Eur. J. Pharmacol.* **1980**, *67*, 313–316. [CrossRef]
44. Matsumoto, M.; Xie, W.; Ma, L.; Ueda, H. Pharmacological switch in Aβ-fiber stimulation-induced spinal transmission in mice with partial sciatic nerve injury. *Mol. Pain* **2008**, *4*, 25. [CrossRef]
45. Ma, L.; Nagai, J.; Ueda, H. Microglial activation mediates de novo lysophosphatidic acid production in a model of neuropathic pain. *J. Neurochem.* **2010**, *115*, 643–653. [CrossRef]
46. Kiguchi, N.; Kobayashi, D.; Saika, F.; Matsuzaki, S.; Kishioka, S. Pharmacological Regulation of Neuropathic Pain Driven by Inflammatory Macrophages. *Int. J. Mol. Sci.* **2017**, *18*, 2296. [CrossRef]
47. Liu, T.; van Rooijen, N.; Tracey, D.J. Depletion of macrophages reduces axonal degeneration and hyperalgesia following nerve injury. *Pain* **2000**, *86*, 25–32. [CrossRef]
48. Ellergezen, P.; Alp, A.; Cavun, S.; Celebi, M.; Macunluoglu, A.C. Pregabalin inhibits proinflammatory cytokine release in patients with fibromyalgia syndrome. *Arch. Rheumatol.* **2023**, *38*, 307–314. [CrossRef]
49. Gerdle, B.; Leinhard, O.D.; Lund, E.; Lundberg, P.; Forsgren, M.F.; Ghafouri, B. Pain and the biochemistry of fibromyalgia: Patterns of peripheral cytokines and chemokines contribute to the differentiation between fibromyalgia and controls and are associated with pain, fat infiltration and content. *Front. Pain Res.* **2024**, *5*, 1288024. [CrossRef]
50. Peck, M.M.; Maram, R.; Mohamed, A.; Crespo, D.O.; Kaur, G.; Ashraf, I.; Malik, B.H. The Influence of Pro-inflammatory Cytokines and Genetic Variants in the Development of Fibromyalgia: A Traditional Review. *Cureus* **2020**, *12*, e10276. [CrossRef]
51. Goebel, A.; Krock, E.; Gentry, C.; Israel, M.R.; Jurczak, A.; Urbina, C.M.; Sandor, K.; Vastani, N.; Maurer, M.; Cuhadar, U.; et al. Passive transfer of fibromyalgia symptoms from patients to mice. *J. Clin. Investig.* **2021**, *131*, e144201. [CrossRef]
52. Fanton, S.; Menezes, J.; Krock, E.; Sandstrom, A.; Tour, J.; Sandor, K.; Jurczak, A.; Hunt, M.; Baharpoor, A.; Kadetoff, D.; et al. Anti-satellite glia cell IgG antibodies in fibromyalgia patients are related to symptom severity and to metabolite concentrations in thalamus and rostral anterior cingulate cortex. *Brain Behav. Immun.* **2023**, *114*, 371–382. [CrossRef]
53. Jakobsson, J.E.; Menezes, J.; Krock, E.; Hunt, M.A.; Carlsson, H.; Vaivade, A.; Khoonsari, P.E.; Agalave, N.M.; Sandstrom, A.; Kadetoff, D.; et al. Fibromyalgia patients have altered lipid concentrations associated with disease symptom severity and anti-satellite glial cell IgG antibodies. *J. Pain* **2025**, *29*, 105331. [CrossRef]
54. Krock, E.; Morado-Urbina, C.E.; Menezes, J.; Hunt, M.A.; Sandstrom, A.; Kadetoff, D.; Tour, J.; Verma, V.; Kultima, K.; Haglund, L.; et al. Fibromyalgia patients with elevated levels of anti-satellite glia cell immunoglobulin G antibodies present with more severe symptoms. *Pain* **2023**, *164*, 1828–1840. [CrossRef]
55. Fan, C.Y.; McAllister, B.B.; Stokes-Heck, S.; Harding, E.K.; de Vasconcelos, A.P.; Mah, L.K.; Lima, L.V.; van den Hoogen, N.J.; Rosen, S.F.; Ham, B.; et al. Divergent sex-specific pannexin-1 mechanisms in microglia and T cells underlie neuropathic pain. *Neuron* **2025**, *113*, 896–911.e899. [CrossRef]
56. Lister, K.C.; Wong, C.; Cai, W.; Uttam, S.; Stecum, P.; Rodrigues, R.; Hooshmandi, M.; Brown, N.; Fan, J.; Francois-Saint-Cyr, N.; et al. 4E-BP1-dependent translation in microglia controls mechanical hypersensitivity in male and female mice. *J. Clin. Investig.* **2025**, *135*, e180190. [CrossRef]
57. Saika, F.; Sato, T.; Nakabayashi, T.; Fukazawa, Y.; Hino, S.; Suzuki, K.; Kiguchi, N. Male-Dominant Spinal Microglia Contribute to Neuropathic Pain by Producing CC-Chemokine Ligand 4 Following Peripheral Nerve Injury. *Cells* **2025**, *14*, 484. [CrossRef]

cells

MDPI

Article

The Antinociceptive Effects and Sex-Specific Neurotransmitter Modulation of Metformin in a Mouse Model of Fibromyalgia

Hanin Abdulbaset AboTaleb [1,2,*], Hani A. Alturkistani [3], Gamal S. Abd El-Aziz [3], Emad A. Hindi [2,3], Mervat M. Halawani [2,3], Mona Ali Al-Thepyani [2,4] and Badrah S. Alghamdi [1,2,*]

1 Department of Physiology, Faculty of Medicine, King Abdulaziz University, Jeddah 21589, Saudi Arabia
2 Neuroscience and Geroscience Research Unit, King Fahd Medical Research Center, King Abdulaziz University, Jeddah 21589, Saudi Arabia; eahindi@kau.edu.sa (E.A.H.); mhalwani@kau.edu.sa (M.M.H.); mahalthepyani@kau.edu.sa (M.A.A.-T.)
3 Department of Clinical Anatomy, Faculty of Medicine, King Abdulaziz University, Jeddah 22252, Saudi Arabia; hturkustani@kau.edu.sa (H.A.A.); dr_gamal_said@yahoo.com (G.S.A.E.-A.)
4 Department of Chemistry, College of Sciences & Arts, King Abdulaziz University, Rabigh 21911, Saudi Arabia
* Correspondence: htaleb0004@stu.kau.edu.sa (H.A.A.); basalghamdi@kau.edu.sa (B.S.A.)

Abstract: Fibromyalgia (FM) is a chronic and debilitating condition characterized by diffuse pain, often associated with symptoms such as fatigue, cognitive disturbances, and mood disorders. Metformin, an oral hypoglycemic agent, has recently gained attention for its potential benefits beyond glucose regulation. It has shown promise in alleviating neuropathic and inflammatory pain, suggesting that it could offer a novel approach to managing chronic pain conditions like FM. This study aimed to further explore metformin's analgesic potential by evaluating its effects in an experimental FM model induced by reserpine in both male and female mice. After the administration of 200 mg/kg metformin to male and female mice, the FM-related symptoms were assessed, including mechanical allodynia, thermal hyperalgesia, and depressive-like behaviors. A histological examination of the thalamus, hippocampus, and spinal cord was conducted using haematoxylin and eosin staining. The neurotransmitter and proinflammatory cytokines levels were measured in the brains and spinal cords. Our results have shown that metformin treatment for seven days significantly reversed these FM-like symptoms, reducing pain sensitivity and improving mood-related behaviors in both the male and female mice. Additionally, metformin exhibited neuroprotective effects, mitigating reserpine-induced damage in the hippocampus, thalamus, and spinal cord. It also significantly lowered the levels of the proinflammatory cytokine interleukin 1-beta (IL-1β) in the brain and spinal cord. Notably, metformin modulated the neurotransmitter levels differently between the sexes, decreasing glutamate and increasing serotonin and norepinephrine in the male mice, but not in the females. These findings underscore metformin's potential as an alternative therapy for FM, with sex-specific differences suggesting distinct mechanisms of action.

Keywords: fibromyalgia; metformin; chronic pain; antinociceptive; neurotransmitters; histological changes; IL-1β; serotonin; norepinephrine

1. Introduction

Fibromyalgia (FM) is a chronic condition characterized by widespread pain, often accompanied by symptoms such as depression, sleep disturbances, fatigue, and cognitive dysfunction [1]. FM affects approximately 3–9% of the global population and is significantly more prevalent in females than in males, with a female-to-male ratio of 3:1 [2]. This condition is a major cause of disability, impacting patients' quality of life and daily functioning [3,4]. Currently, there are no universally effective treatments for FM, largely due to the lack of consensus on its etiopathogenesis [3,4]. Since the 1990s, scientists have recognized a potential link between FM and neurohormonal alterations [5]. Central pain processing dysfunctions and neuroinflammation are also reported as plausible contributors to generalized pain sensitization in FM patients [6]. Additionally, FM patients exhibit reduced activity in their descending pain modulatory system across several brain regions, including the rostral anterior cingulate cortex, thalamus, periaqueductal gray, and rostral ventromedial medulla [7]. Animal studies further suggest that an imbalance—characterized by increased glutamatergic and decreased Gamma-aminobutyric acid (GABA)-ergic neurotransmission in the insular cortex or anterior cingulate cortex—may drive heightened pain sensitivity by intensifying central pain processing in FM [8]. Consequently, disruptions in neurotransmitter levels are now recognized as key elements in the pathophysiology of FM [9]. Despite these insights, the Food and Drug Administration (FDA) has endorsed three drugs for FM treatment: the antiseizure drug pregabalin and the antidepressants duloxetine and milnacipran [10]. Unfortunately, these drugs often achieve minimal pain reduction, and not all patients experience substantial pain relief [11]. Moreover, the side effects of these drugs sometimes outweigh their benefits [12–15]. Although lethal side effects such as heart failure, hepatic failure, or serotonin syndrome are rare, they must still be considered [11]. For these reasons, preclinical studies are crucial in explaining the pathophysiology of FM and identifying new therapeutic targets. These studies are essential for exploring innovative treatment options, particularly given the limitations in efficacy associated with current therapeutic approaches.

Metformin is a hypoglycemic agent that has been widely used as a first-line treatment for type 2 diabetes [16]. Recent research has unveiled a novel relationship between metformin and pain management, suggesting that metformin may serve as a promising antinociceptive agent [17]. It has been found to have valuable impacts on various persistent pain pathologies, such as complex regional pain syndrome, neuropathic pain, diabetic neuropathy, and FM-type pain [18–21]. Interestingly, metformin has also been shown to alleviate common associative symptoms of chronic pathological pain, including depression, anxiety, and cognitive impairment, in both rodents and humans [20]. Notably, recent studies suggest that metformin can improve serotonin and norepinephrine levels in the brain [22,23]. Additionally, metformin administration has been shown to normalize the brain glutamate levels in diabetic epileptic rodent models [24]. Moreover, FM is frequently associated with overweight, obesity, and related metabolic disturbances, which may exacerbate symptom severity and adversely affect one's quality of life. Given metformin's known effects on these parameters, it presents intriguing potential as a therapeutic agent for FM [25].

While preclinical studies have shown the promising benefits of metformin in some chronic pain conditions, research on its antinociceptive effects in FM-like pain remains limited. This study aims to fill this gap by examining the antinociceptive effects of metformin on an experimental FM model. We hypothesize that metformin exhibits antinociceptive effects in a FM mouse model by alleviating pain sensitivity and improving mood-related behaviors through modulating neurotransmitter levels, reducing neuroinflammation, and protecting the neuronal structure in the brain and spinal cord. Therefore, this research aims to assess the influence of metformin on FM symptoms and explore its therapeutic potential in addressing certain aspects of FM-related pathophysiology. This study included male and female mice to identify any potential sex differences in response to treatment that could help guide more personalized treatments for FM in the future.

2. Materials and Methods

2.1. Animals

This study involved a total of 104 Swiss albino mice (30–40 g, 9–10 weeks old), consisting of 52 males and 52 females. All the animals were sourced from the King Fahad Medical Research Center in Jeddah, Saudi Arabia. A maximum of five mice were housed in clear, transparent polycarbonate cages with straw bedding, which was regularly changed. The mice had constant access to food and water and were kept on a standard 12 h light/dark cycle at a room temperature of 23 ± 2 °C. The animals were acclimatized to the laboratory environment for four days before the start of the experiments. Behavioral evaluations were conducted between 9:00 a.m. and 2:00 p.m. This study followed the ethical guidelines set by the Biomedical Ethics Committee of King Abdulaziz University (approval no. 236-24) and the Animal Care and Use Committee (ACUC) of the Animal House Unit, King Fahad Medical Research Center, Jeddah, Saudi Arabia.

2.2. Drugs

Reserpine (98 %; Acros Organs—Thermo Fisher Scientific, Waltham, MA, USA), metformin hydrochloride (Solarbio Science & Technology Co., Ltd., Beijing, China), and pregabalin (Saudi Pharmaceutical Industries, Jeddah, Saudi Arabia) were used in this study. The reserpine was diluted to its final concentration in 0.5% glacial acetic acid (v/v in distilled water) and administered subcutaneously to the mice at a volume of 6.3 mL/kg. The FM-like state was induced in the mice by injecting reserpine (0.5 mg/kg/day) for three consecutive days, following the method described in the literature [26,27]. We selected this dosing regimen based on a preliminary study to determine the most appropriate dose of reserpine for inducing FM-like symptoms in our specific mouse strain [28]. Solutions of metformin and pregabalin were prepared daily in sterile saline and administered to the mice intraperitoneally at a volume of 10 mL/kg. Pregabalin was used as a positive control drug, and it was administered to the mice at the same schedule of treatment. The doses of metformin (200 mg/kg) and pregabalin (30 mg/kg) were based on established protocols from previous research [18,29].

2.3. Treatment and Assessment Strategy

Animals were randomly assigned into four groups: the vehicle control group, the RES + saline group, the RES + metformin group, and the RES + pregabalin group. For the first 3 days of the experiment, the vehicle control group received subcutaneous injections of vehicle solution (0.5% glacial acetic acid in distilled water), and the other groups received subcutaneous injections of reserpine. Beginning on day 4 and continuing through day 10 of the experiment, the mice received daily intraperitoneal injections of saline, metformin, or pregabalin.

A sensory behavioral analysis was performed on the 4th and 10th days of the study to evaluate the effects of single and 7-day treatments, respectively. The drug was injected first, and sensory tests were conducted 3 h post-injection. Motor behavioral analyses were carried out on the 7th and 9th days, while depression-related behaviors were assessed on the 7th, 8th, and 9th days. To minimize any potential influence of drug injections on the results, injections were administered after the completion of the motor and depression tests. We conducted behavioral tests in the middle of the study, ensuring that each group of mice was tested only once per day for each type of assessment. On day 11, the animals were dissected, and brain and spinal cord samples were collected for biochemical and histological evaluations. Figure 1 displays the timeline of this study.

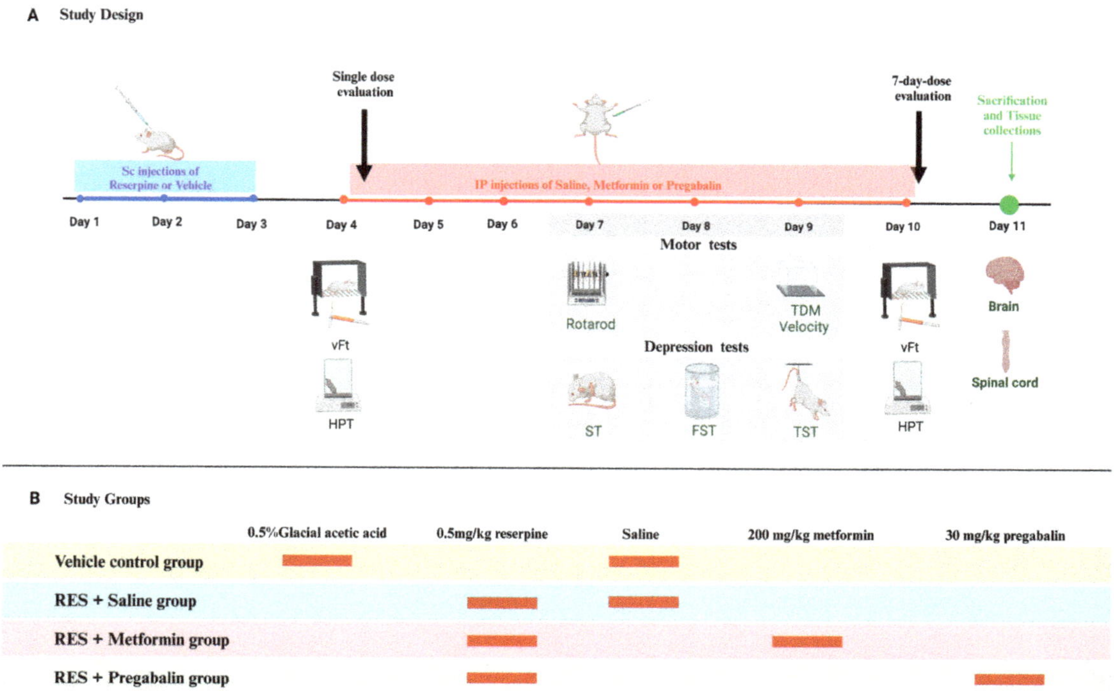

Figure 1. Schematic representation of the overall design and timeline of the study. (**A**) Total duration of the study was 11 days. Reserpine (RES) was administered subcutaneously (Sc) to the mice during the first 3 days. Behavioral tests were performed on days 4, 7, 8, 9, and 10. On day 11, the mice were sacrificed, and brain and spinal cord samples were collected for histopathological and biochemical analysis. (**B**) Study groups and the agents they received. Abbreviations: vFt, von Frey test; HPT, hot plate test; TDM, total distance moved; FST, forced swimming test; TST, tail suspension test; ST, splash test; IP, intraperitoneal injection; Sc, subcutaneous injection. Created in BioRender. AboTaleb, H. https://BioRender.com/z08v394 (1 January 2024).

2.4. *Assessment of the Pain Threshold*

2.4.1. Von Frey Test (vFt)

The mechanical thresholds were measured by using von Frey filaments following the up-down paradigm, as described in the literature [30]. The mice were positioned individually in clear plexiglass boxes on the elevated mesh platform measuring 90 × 38 cm, with a grid of square holes approximately 5 × 5 mm. The filaments (the Aesthesio® set, UGO-37450-275) with increasing stiffness (0.04–4 g) were employed to the plantar surface of the left hind paw, and the responses were recorded in an XO pattern. Paw lifting was recorded as (X), indicating a positive response, while no response within 5 s was recorded as (O). The stimulation began with 0.6 g and continued until the completion of five succeeding positive responses (assigned a score of 0.04), five consecutive negative responses (assigned a score of 4), or four readings following the first different response. The threshold for paw withdrawal was measured in grams (g) and calculated using the threshold calculator website designed by Christensen SL and his groups [31]. A noticeable drop in the paw withdrawal threshold relative to the control values was considered mechanical allodynia [32]. The mice were habituated to the mesh platform for at least 45 min before testing.

2.4.2. Hot Plate Test

A hot plate apparatus (Ugo Basile, Gemonio, Italy) was used to assess the thermal threshold in the mice as described previously [33]. The temperature of the hot plate's surface was kept constant at 55 ± 0.1 °C [34]. The mice then were positioned on the heated metal plate that enclosed via a clear acrylic cylinder. To avoid causing any tissue damage, the latency to heat response, indicated by flicking, hind paw licking, or jumping, was recorded with a 30 s cut-off period [35].

2.5. Assessment of Motor Behavior

2.5.1. Open Field Test (OFT)

On the 9th day of the experiment, each mouse was positioned in the center of an acrylic box arena (45 cm × 45 cm × 34 cm) and given 3 min to move around freely in a sound-attenuated room, under low-intensity light. The movement of the mice, including their total distance moved (TDM) in centimeters and the velocity in centimeters per sec (cm/s), was tracked and recorded using the EthoVision XT8A system (Noldus Information Technology, Wageningen, The Netherlands) [36].

2.5.2. Rotarod Test

This study aimed to examine the effects of different interventions on fatigue and muscle coordination using a rotarod test, which is also recommended for assessing nociception in rodents [37]. The mice were trained on a rotarod at a speed of 9 rpm until they could remain on the rotating device for 30 s without falling. A total of 1 h later, their ability to stay on the rotarod at a fixed speed of 20 rpm was recorded for up to 240 s [38,39]. This procedure was repeated for three trials at 10 min intervals, and then the average performance was calculated. The rotarod test was conducted on the 7th day of the study.

2.6. Measurement of Depressive-like Behavior

2.6.1. Splash Test (ST)

An ST was performed to measure anti-anhedonia-like behavior, which is one of the major signs of depression [40]. In this test, self-cleaning behavior was assessed by squirting a sucrose solution (300 µL of 10%) onto each mouse's dorsal coat and recording its grooming behavior for 5 min [7]. The grooming behavior was recorded by either licking or scratching the mice at their body or limbs, particularly in the areas affected by the splash solution. The ST was conducted on the 7th day of the study.

2.6.2. Forced Swimming Test (FST)

An FST was used to identify potential depressive-like behavior in the experimental animals [41]. The test was performed following the method described in the literature [42]. In this test, each mouse was placed in a separate glass cylinder (10 cm in diameter and 20 cm in height) filled with water maintained at 23–25 °C, and allowed to swim for 6 min. The immobility of the mice during the last 4 min was recorded in sec [34]. The absence of paw movements other than those required to maintain the mouse's head above water was defined as immobility [27]. The FST was performed on the 8th day of the study.

2.6.3. Tail Suspension Test (TST)

To evaluate depression-like behavior, TSTs have been commonly utilized [43]. In this test, the mice were suspended by their tails approximately 50 cm above the bench, with their tails adhered to the wall using adhesive tape and a barrier to obstruct the mouse's view. The suspension lasted for 6 min, and the total immobility time was evaluated during the last 4 min. The mice were considered immobile when they stopped struggling to escape the suspended position [44]. The TST was performed on the 9th day of the study.

2.7. Tissue Harvesting and Processing

At the end of the study, the mice were killed via cervical dislocation, and their brains and spinal cords were harvested. Four samples from each group were immediately fixed in 10% formalin and then processed for histopathological analysis within 48 h. The remaining samples were stored at −80 °C for further biochemical analysis.

2.7.1. Histopathological Analysis

The standard protocol described in Alqurashi et al. 2022 was adhered to for performing hematoxylin and eosin (H&E) staining [44]. Briefly, after 16 h of tissue processing using the Spin Tissue Processor STP120 (Especialidades Médicas Myr, S.L., Tarragona, Spain), the midsagittal hemisected brains were immersed in paraffin wax and sectioned into 4 μm-thick slices using microtomes (LEICA RM 2255, Leica Microsystems, Germany). To characterize the histopathological changes, the slides were stained with H&E using the Myr AutoStainer (Especialidades Médicas Myr, S.L., Tarragona, Spain) and examined under an Olympus BX53 light microscope to assess the alterations in the hippocampus, thalamus, and spinal cord. The representative areas were captured at ×100, ×200, and ×400 magnifications using an Olympus DP73 camera and the images were processed using Olympus CellSens Entry software (Olympus Corporation, Tokyo, Japan).

2.7.2. Biochemical Analysis

The minced brain and spinal cord tissues were weighed and added to a homogenization buffer containing 0.25 M sucrose, 1 mM EDTA, 5 mM MOPS, and 0.1% ethanol, with the pH adjusted to 7.2 using 1 M NaOH. The tissues were prepared at a ratio of 0.1 of tissue per 1 mL of solution. The tissue was thoroughly homogenized and then centrifuged at 1000 *g* for 10 min at 4 °C. The resulting supernatant was collected for the subsequent estimation of the neurotransmitter and proinflammatory cytokine levels.

2.7.3. Measurement of Serotonin and Norepinephrine Levels

The serotonin and norepinephrine levels were determined using ELISA kits (SEKSM-0016 and SEKSM-0019, respectively) obtained from Solarbio, in accordance with the manufacturer's instructions (Solarbio Science & Technology Co., Ltd., Beijing, China). Briefly, the test procedure involved adding 50 μL of samples and standards to each well of a microplate, followed immediately by 50 μL of a working solution of a biotin-conjugated anti-ST/5-HT or anti-NA/NE antibody. At 37 °C, the plate was incubated for 45 min. Upon cleaning the plate, 100 μL of a streptavidin-labeled detection antibody was introduced and incubated for another half hour at 37 °C. After aspirating and washing it three times, 90 μL of substrate solution was added, followed 30 min later by 50 μL of stop solution. The optical absorbance of each well was then immediately measured at 450 nm using a microplate reader (BioTek Instruments, Inc., Winooski, VT, USA), and the absorbance measurements were compared to the calibration plot for standard solutions to determine the readings.

2.7.4. Measurement of Glutamic Acid Levels

The glutamate levels were assessed using the Glutamic Acid Content Assay kit (WST-1 chromogenic method) obtained from Solarbio, as per the manufacturer's instructions (Solarbio Science & Technology Co., Ltd., Beijing, China). The absorbance values were measured at 450 nm using a microplate reader (BioTek Instruments, Inc., Winooski, VT, USA), and the results were established by comparing these values with the calibration plot for standard solutions.

2.7.5. Measurement of TNF-α and IL-1β Levels

The levels of IL-1β and TNF-α were determined using ELISA kits (SEKM-0002 and SEKM-0034, respectively) from Solarbio following the manufacturer's instructions (Solarbio Science & Technology Co., Ltd., Beijing, China). Briefly, the assay procedure involved adding 100 μL of the samples and standards to the microplate wells and incubating them at 37 °C for 90 min. After washing the plate, 100 μL of biotin-conjugated anti-mouse IL-1β or TNF-α antibody working solution was added to each well. The plate was incubated at 37 °C for 60 min. After washing the plate again, 100 μL of the streptavidin-labeled detection antibody was added to the microplate wells and incubated again at 37 °C for 30 min. After aspirating and washing it three times, 100 μL of substrate solution was added. After 15 min, 50 μL of stop solution was added, and the optical absorbance of each well was immediately read at 450 nm using a microplate reader (BioTek Instruments, Inc., Winooski, VT, USA). The concentrations were determined by comparing the absorbance values with the calibration chart for standard solutions.

2.8. Statistical Analysis

Statistical analyses were performed using GraphPad PRISM software for Windows (version 10, GraphPad Software, San Diego, CA, USA). The data were displayed as the mean ± standard error of the mean (SEM), and the figure legends included the number of mice or samples used in each analysis. The normality was assumed using a Shapiro–Wilk's test. The differences between the groups were analyzed using an ordinary one-way ANOVA followed by Tukey's multiple comparison test, or Kruskal–Wallis tests as appropriate. A p-value of less than 0.05 was considered statistically significant.

3. Results

3.1. Metformin Treatment Reversed Reserpine-Induced Mechanical Allodynia and Thermal Hypersensitivity in Male and Female Mice After 7 Days of Treatment

The vFt is a valuable tool for understanding pain mechanisms and evaluating potential therapies in research on pain disorders [35]. To evaluate the effects of single and continuous 7-day metformin injections on established reserpine-induced mechanical allodynia in male and female mice, the responses were assessed using the vFt. Starting on day 4 of the study, the mice received intraperitoneal injections of metformin (200 mg/kg), pregabalin (30 mg/kg), or saline.

In the male mice, reserpine administration significantly decreased the mechanical nociceptive thresholds at days 4 and 10 of the study compared to the vehicle control group (p = 0.0030 and p = 0.0021, respectively). Seven days of metformin injections reversed the reserpine-induced mechanical allodynia (p = 0.0061), whereas a single metformin injection did not produce a significant acute effect (Figure 2A,B).

Similar results were observed in the female mice, where subcutaneous injections of reserpine significantly reduced the paw withdrawal threshold compared to the control mice receiving the vehicle on days 4 and 10 (p = 0.0027 and p = 0.0482, respectively). Seven days of metformin treatment significantly reduced reserpine-induced mechanical allodynia compared to the saline-treated mice (p = 0.0044). However, it did not produce a significant acute effect after a single dose (Figure 2C,D).

In contrast to the effects of metformin, a single pregabalin injection significantly increased the paw withdrawal threshold compared to the saline group in both the male (p = 0.0271) and female mice (p = 0.0013).

To assess the effects of both single and continuous 7-day metformin injections on the established reserpine-induced thermal hypersensitivity in the male and female mice, we measured their responses using a hot plate test. The hot plate test evaluates an animal's response to a painful thermal stimulus, allowing researchers to quantify their sensitivity to heat pain [35]. In the male mice, reserpine injections significantly reduced their latency to respond to heat stimulus compared to the vehicle control group in both the tested days (p = 0.0113 and p = 0.0131, respectively). Although a single dose of metformin did not produce an acute positive effect on their thermal hypersensitivity, metformin treatment for seven consecutive days significantly increased their latency to respond to heat stimulus compared to the saline-treated mice (p = 0.0486; Figure 3A,B).

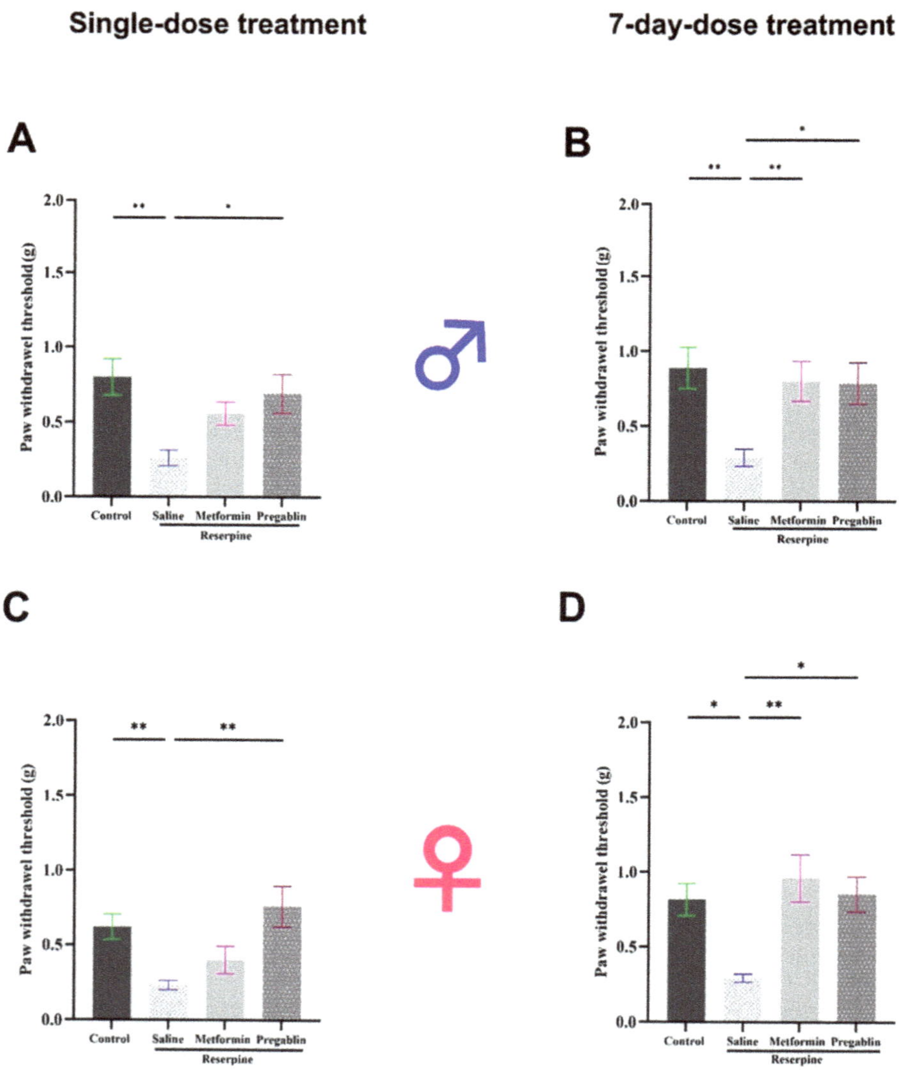

Figure 2. Effect of different interventions on mechanical allodynia in male and female mice tested after single and 7-day treatment. (**A**,**B**) Paw withdrawal threshold in von Frey test in male mice. Reserpine administration (0.5 mg/kg) reduced paw withdrawal threshold on all tested days. A single dose of metformin (200 mg/kg) did not reverse the effects of reserpine on mechanical threshold, whereas 7-day dosing significantly alleviated mechanical allodynia. (**C**,**D**) Paw withdrawal threshold in von Frey test in female mice. Single dose of metformin had no effect, while 7-day dosing significantly alleviated mechanical allodynia. Pregabalin (30 mg/kg), used as a positive control, effectively restored paw withdrawal threshold toward control levels across all testing days in both male and female mice. Each bar represents mean, and vertical lines indicate standard error mean (SEM) for 7–12 mice/group. Asterisks above lines indicate significant difference between groups where * $p < 0.05$ and ** $p < 0.01$, otherwise, a non-significant difference is recorded.

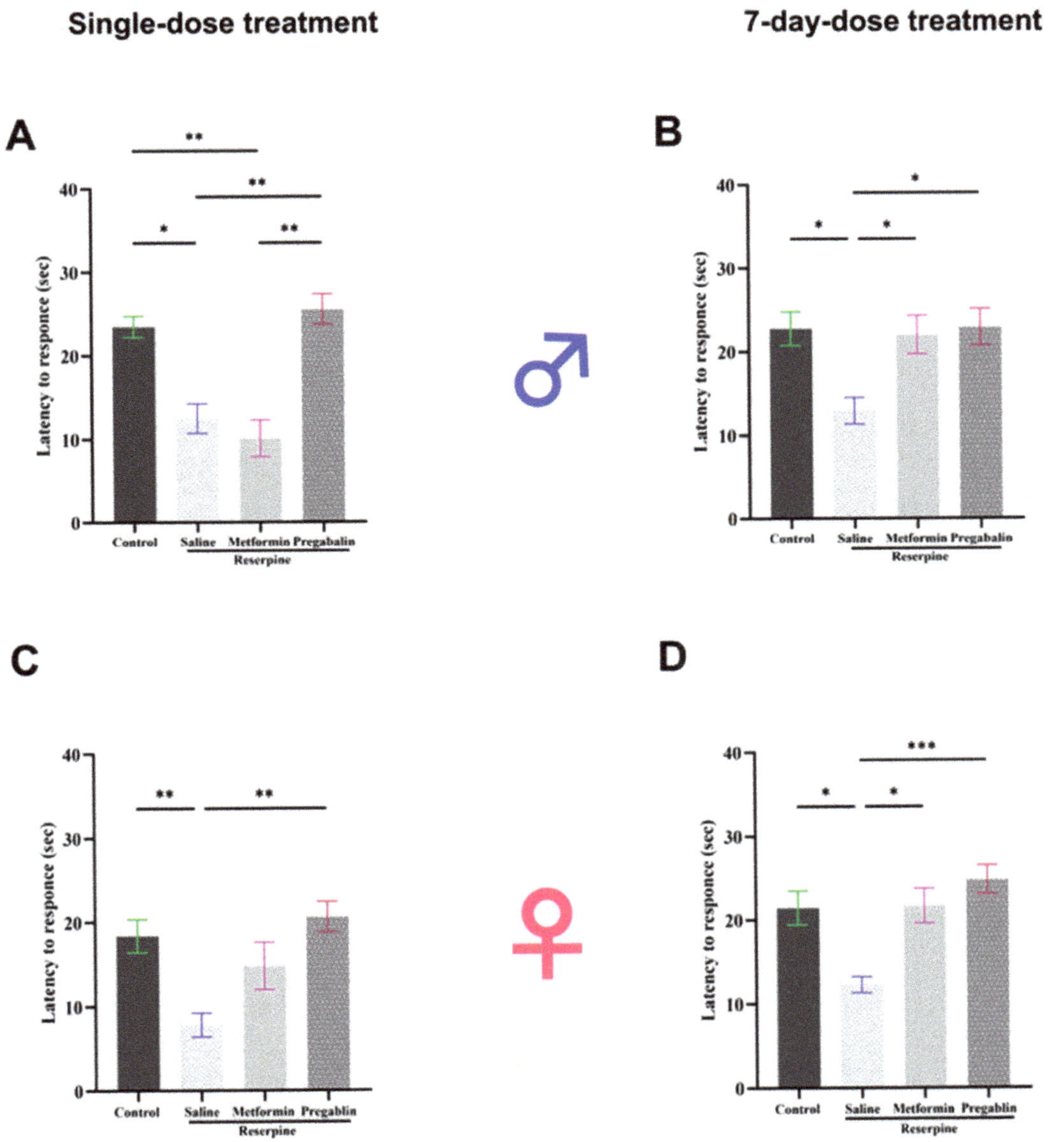

Figure 3. Effect of different interventions on thermal hypersensitivity in male and female mice tested after single and 7-day treatment. (**A**,**B**) Latency to respond in hot plate test in male mice. Reserpine administration (0.5 mg/kg) reduced latency of response to hot stimuli on all tested days. Single dose of metformin (200 mg/kg) did not reverse effects of reserpine on hot threshold, while 7-day dosing significantly alleviated thermal hypersensitivity. (**C**,**D**) Latency to respond to hot stimulus in female mice. Single metformin dose had no effect, while 7-day dosing significantly alleviated thermal hypersensitivity. Pregabalin (30 mg/kg), used as a positive control, effectively restored latency to respond to hot stimulus toward control levels across all testing days in both male and female mice. Each bar represents mean, and vertical lines indicate standard error mean (SEM) for 8–13 mice/group. Significant difference between groups is indicated by asterisks above the lines where * $p < 0.05$, ** $p < 0.01$, and *** $p < 0.001$; otherwise, a non-significant difference is recorded.

Similar results were observed in the female mice as subcutaneous injections of reserpine significantly reduced their latency to respond to hot stimulus on days 4 and 10 compared to the vehicle control group ($p = 0.0096$ and $p = 0.0158$, respectively). Although a single injection of metformin did not affect the thermal hypersensitivity, continuous metformin treatment for seven consecutive days significantly increased their latency to respond to hot stimulus compared to saline-treated mice ($p = 0.0134$; Figure 3C,D).

In contrast to the effects of metformin, a single pregabalin injection significantly increased their latency to respond to hot stimuli compared to the saline group in both the male ($p = 0.0075$) and female mice ($p = 0.0020$).

3.2. Metformin Treatment Reversed Motor Defects in Male Mice Only

While FM primarily manifests symptoms related to pain, some patients may also experience motor defects or impairments. One of the prevalent associated symptoms of FM is fatigue, which affects about 76% of FM patients [45]. Through motor tests, we aimed to assess fatigue in the mice. TDM and velocity tests evaluated spontaneous motor behavior, which can indirectly reflect fatigue-like symptoms. The rotarod test, on the other hand, measures forced motor activity, providing a more targeted assessment of fatigue.

Therefore, the effects of reserpine and metformin on motor performance were evaluated through various behavioral tests in the male and female mice. We used the OFT to assess the TDM and velocity of the mice, therefore evaluating their spontaneous motor activity. Reserpine administration significantly reduced the TDM in both the male and female mice compared to the controls ($p = 0.0308$ and $p = 0.0007$, respectively). Metformin treatment reversed the reserpine-induced reduction in their TDM, but this effect was significant only in the male mice ($p = 0.0149$; Figure 4A,B).

Reserpine administration also led to a significant decrease in the velocity in both the male and female mice during the OFT ($p = 0.0092$ and $p = 0.0008$, respectively). Metformin treatment successfully reversed the reduction in velocity caused by reserpine in male mice ($p = 0.0285$; Figure 4C), whereas no significant effect was observed in the female mice (Figure 4D).

In addition, the forced motor behavior was assessed in the mice using a rotarod test. Reserpine administration significantly reduced the rotating time only in the male mice compared to the vehicle control group ($p = 0.0287$), indicating fatigue and impaired motor coordination. Metformin treatment effectively reversed this reduction, restoring their normal motor performance ($p = 0.0141$; Figure 4E). However, the female mice showed no differences across all the groups in the rotarod test (Figure 4F).

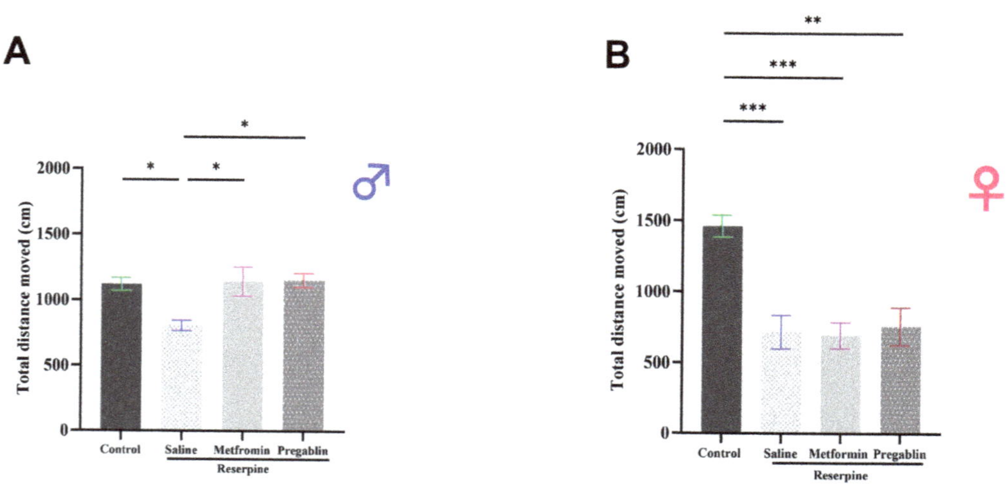

Figure 4. *Cont.*

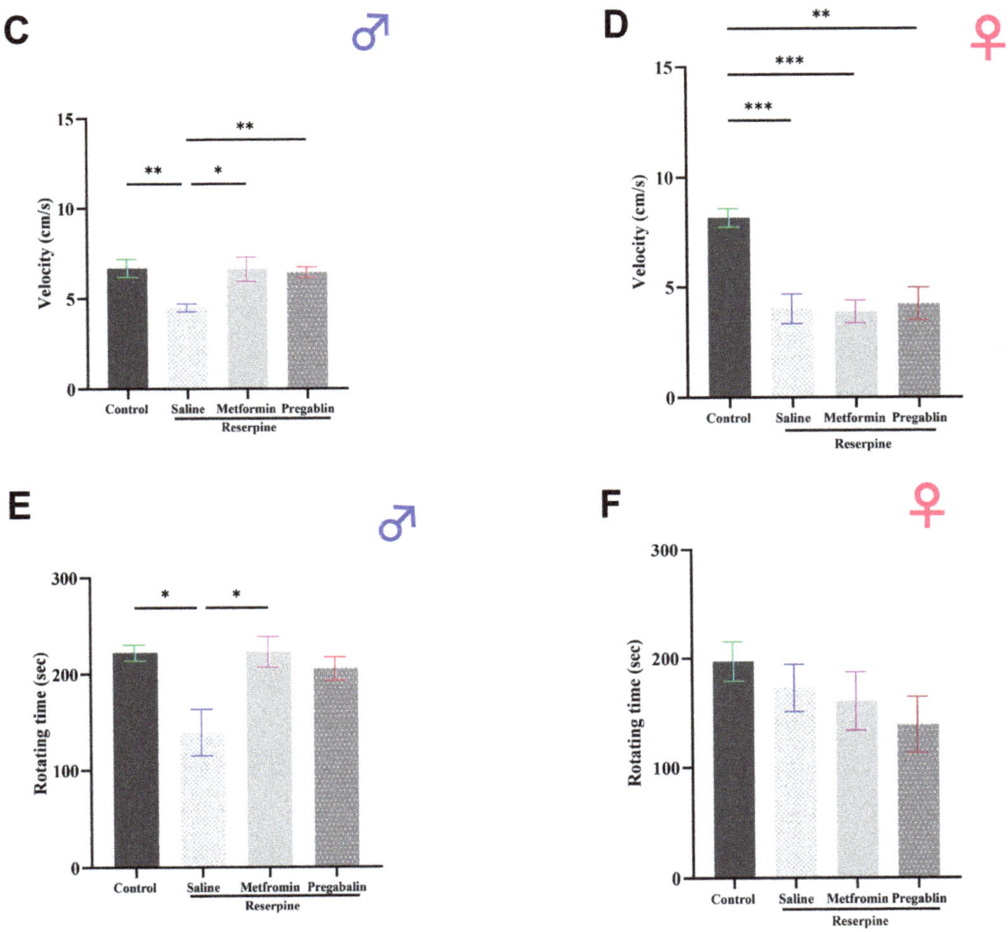

Figure 4. Effect of different interventions on motor performances in male and female mice. (**A**,**B**) Total distance moved (TDM) and (**C**,**D**) velocity of mice in open field test, performed on day 9 of study. Reserpine administration reduced TDM and velocity in both male and female mice. Metformin administration reversed effects of reserpine on TDM and velocity in male mice only. (**E**,**F**) Rotating time in rotarod test, conducted on day 7 of study. Reserpine administration reduced rotating time only in male mice, and metformin administration reversed that effect. Each bar represents mean, and vertical lines indicate standard error mean (SEM) for 6–12 mice/group. Asterisks above lines indicate a significant difference between groups where * $p < 0.05$, ** $p < 0.01$, and *** $p < 0.001$; otherwise, non-significant difference is recorded.

3.3. Metformin Demonstrated Antidepressant-Like Effects in Male and Female Mice

Next, to explore the potential antidepressant-like effects of metformin, we assessed the depressive-like behaviors in the male and female mice using the ST, FST, and TST. Both the male and female mice displayed pronounced depression-like behavior after three days of reserpine injections, as shown in Figure 5. Metformin treatments significantly increased the grooming time in the ST compared to the saline group in both the male ($p = 0.0002$; Figure 5A) and female mice ($p = 0.0036$; Figure 5B). Moreover, metformin administration significantly reduced the total immobility time in the FST compared to the saline group in both the male ($p = 0.0014$; Figure 5C) and female mice ($p = 0.0047$; Figure 5D). Similarly, metformin administration significantly reduced the total immobility time in the

TST compared to the saline group in both the male (p = 0.0017; Figure 5E) and female mice (p = 0.0059; Figure 5F). Interestingly, metformin's effects surpassed those of pregabalin, as pregabalin treatment did not improve the depression-like behavior in the male and female mice across all the tests, except for its impacts on the grooming time during the ST in the male mice.

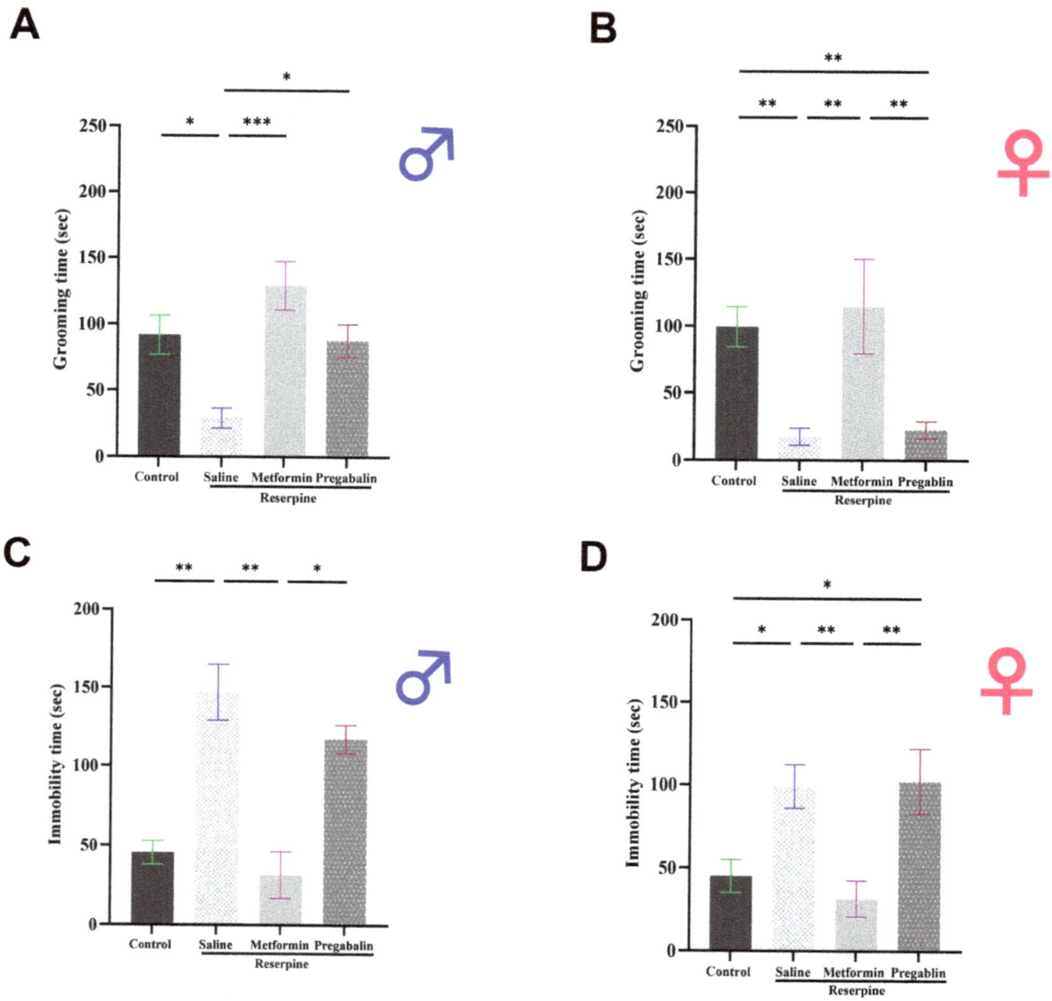

Figure 5. *Cont.*

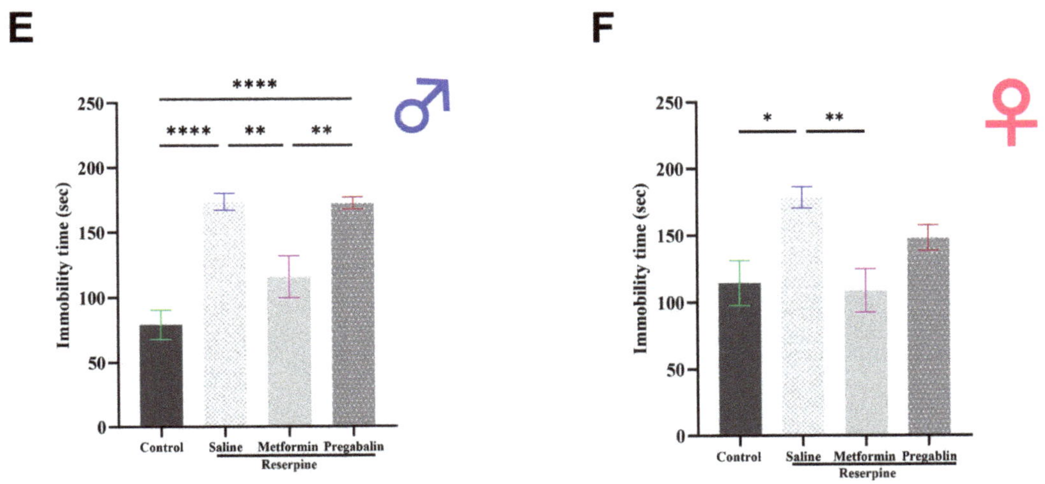

Figure 5. Effect of different interventions on depressive-like behavior in male and female mice. (**A**,**B**) Grooming time in splash test, conducted on day 7 of study. Reserpine injections significantly decreased grooming time in both male and female mice. Metformin treatment reversed decrease in both sexes. (**C**,**D**) Immobility time in forced swimming test, performed on day 8 of study. Reserpine injections significantly increased immobility time in both male and female mice. Metformin reversed increase in both sexes. (**E**,**F**) Immobility time in tail suspension test, performed on day 9 of study. Reserpine injections significantly increased immobility time in both male and female mice. Metformin treatment successfully reversed increase in both sexes. Notably, pregabalin administration had no significant effect in any tests except for splash test, where it showed effect only in male mice. Each bar represents mean, and vertical lines indicate standard error mean (SEM) for 8–12 mice/group. Asterisks above lines indicate significant difference between groups where * $p < 0.05$, ** $p < 0.01$, *** $p < 0.001$, and **** $p < 0.0001$; otherwise, non-significant difference is recorded.

3.4. Metformin Modulated Neurotransmitter Levels in the Brain and Spinal Cord of Male Mice, but Not Females

Following behavioral tests, we dissected the animals and collected their brains and spinal cords to measure their neurotransmitter levels (day 11). In the male mice, the serotonin and norepinephrine levels significantly decreased while the glutamate levels significantly increased following reserpine administration compared to the vehicle control group ($p = 0.0003$, $p = 0.0002$, and $p = 0.0488$, respectively). Metformin treatment restored the levels of serotonin and norepinephrine toward the control, therefore revealing significant differences compared to the RES + saline group ($p < 0.0001$; Figure 6A,C). In addition, metformin significantly reduced the glutamate levels compared to the RES + saline group ($p = 0.0059$; Figure 6E).

The results in the female mice were markedly different. Reserpine administration significantly decreased the serotonin and norepinephrine levels in the female mice compared to the vehicle control group ($p < 0.0001$, and $p = 0.0002$, respectively). However, the metformin and pregabalin treatments failed to improve the serotonin and norepinephrine levels in the brains of the female mice (Figure 6B,D). No significant differences were observed in the glutamate levels among the female mice groups (Figure 6F).

The results observed in the spinal cord closely mirrored those found in the brain. In the male mice, the serotonin and norepinephrine levels significantly decreased in the RES + saline group compared to the vehicle control group, while the glutamate levels significantly increased ($p = 0.0003$, $p = 0.0046$, and $p = 0.0393$, respectively). Metformin treatment significantly alleviated the reductions in the serotonin and norepinephrine levels and reduced the elevated glutamate levels compared to the RES + saline group ($p = 0.0029$, $p < 0.0001$, and $p = 0.0267$, respectively; Figure 7A,C,E).

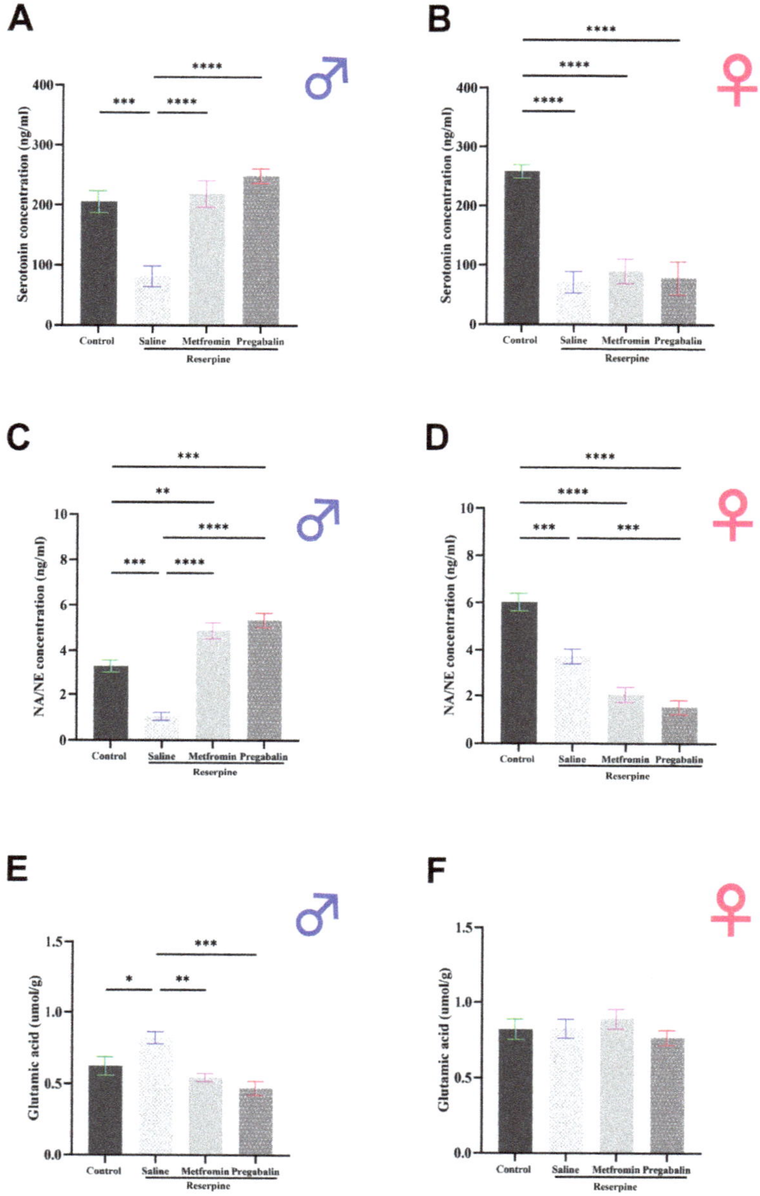

Figure 6. Effect of different interventions on neurotransmitter levels in total brains of male and female mice. (**A**,**B**) Serotonin levels: reserpine injections significantly decreased serotonin levels in both male and female mice. Metformin treatment reversed decrease in male mice only. (**C**,**D**) Norepinephrine levels: reserpine injections significantly decreased norepinephrine levels in both male and female mice. Metformin reversed decrease in male mice only. (**E**,**F**) Glutamate levels: reserpine injections significantly increased glutamate levels in brains of male mice, and metformin treatment reversed increase. In female mice, no significant differences were observed between any groups. Each bar represents mean, and vertical lines indicate standard error mean (SEM) for 6–8 mice/group. Asterisks above lines indicate significant difference between groups where * $p < 0.05$, ** $p < 0.01$, *** $p < 0.001$, and **** $p < 0.0001$; otherwise, a non-significant difference is recorded.

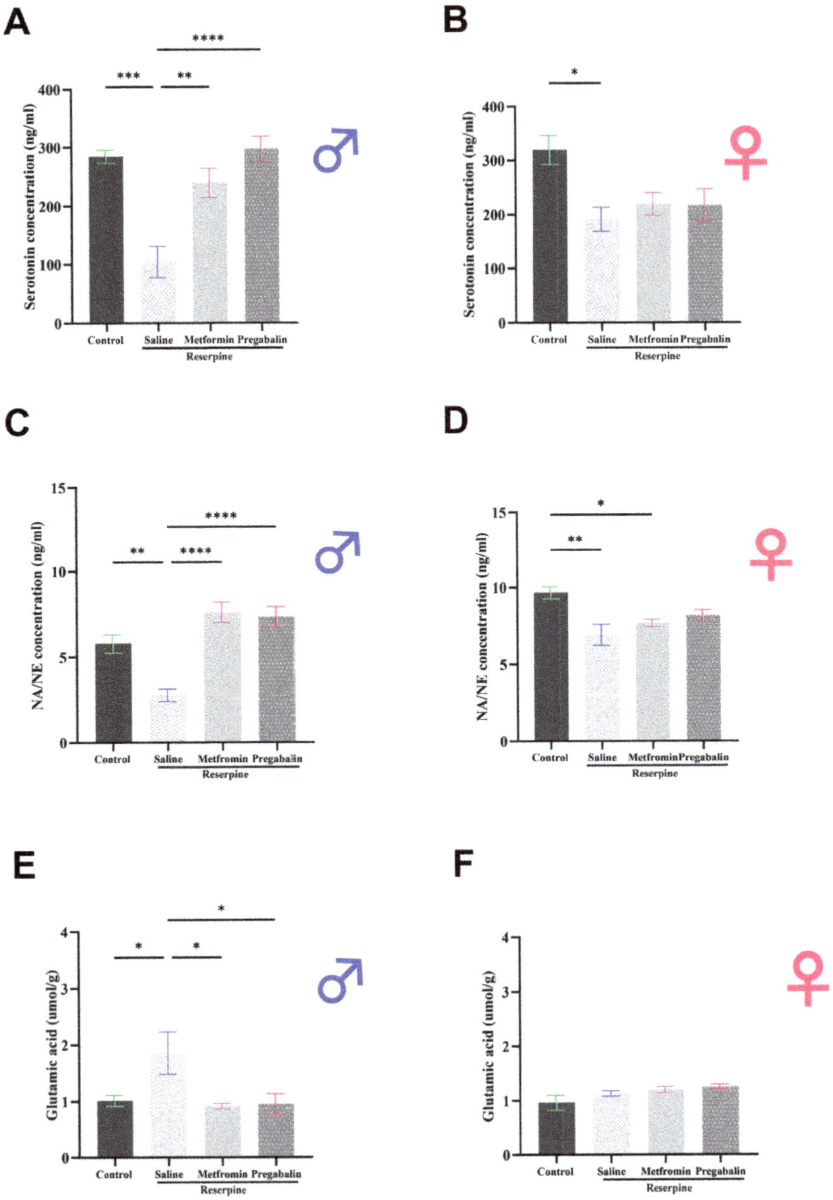

Figure 7. Effect of different interventions on neurotransmitter levels in spinal cord of male and female mice. (**A**,**B**) Serotonin levels: reserpine injections significantly decreased serotonin levels in both male and female mice. Metformin treatment reversed decrease in male mice only. (**C**,**D**) Norepinephrine levels: reserpine injections significantly decreased norepinephrine levels in both male and female mice. Metformin treatment reversed decrease in male mice only. (**E**,**F**) Glutamate levels: reserpine injections significantly increased glutamate levels in spinal cords of male mice, and metformin treatment reversed increase. No differences were observed between any groups in female mice. Each bar represents mean, and vertical lines indicate standard error mean (SEM) for 5–6 mice/group. Asterisks above lines indicate a significant difference between groups where * $p < 0.05$, ** $p < 0.01$, *** $p < 0.001$, and **** $p < 0.0001$; otherwise, a non-significant difference is recorded.

Consistent with the brain findings, the spinal cord results in the female mice were markedly different from those in the male mice. Both the serotonin and norepinephrine levels were reduced significantly in the reserpine-injected mice compared to the vehicle control group ($p = 0.0160$ and $p = 0.0023$, respectively). Metformin administration had no effect on the neurotransmitter levels in the spinal cord of the female mice (Figure 7B,D,F).

3.5. Metformin Modulated IL-1β Levels in the Brain and Spinal Cord of Both Male and Female Mice

In addition to measuring neurotransmitter levels, the levels of the proinflammatory cytokines in the male and female mice were also measured. It was observed that the IL-1β levels were significantly elevated in the brain and spinal cord of both the male and female RES + saline groups compared to the vehicle control group. Metformin treatment significantly decreased the IL-1β levels in the brain of both the male and female mice compared to the RES + saline group ($p = 0.0063$ and $p = 0.0259$, respectively; Figure 8A,B). Similarly, metformin treatment significantly decreased the IL-1β levels in the spinal cord of both the male and female mice compared to the RES + saline group ($p < 0.0001$ and $p = 0.0033$, respectively) (Figure 8C,D). The TNF-α levels also increased significantly in the brain (Figure 8E) and spinal cord (Figure 8G) of the male RES + saline groups compared to the vehicle control group ($p = 0.0146$ and $p = 0.0101$, respectively). Although, metformin treatment did not affect the TNF-α levels in the male mice in either the brain or spinal cord. In the female mice, no difference was detected between any of the groups in the TNF-α levels in either region (Figure 8F,H).

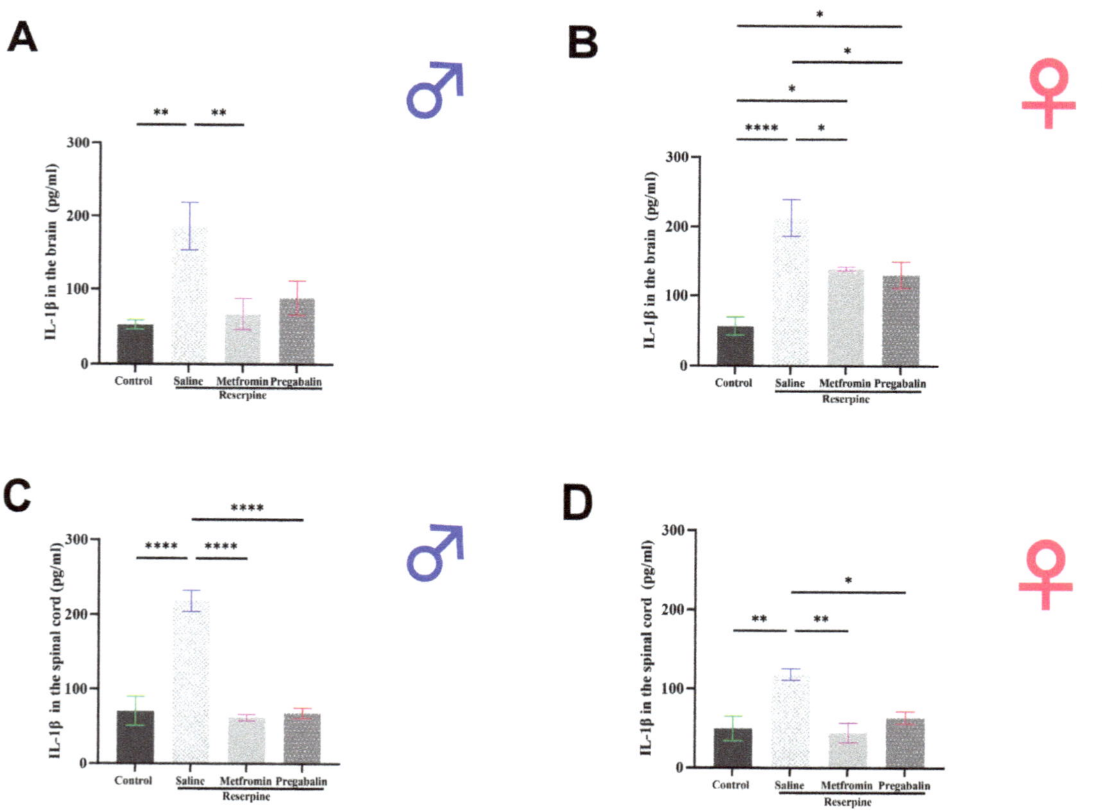

Figure 8. *Cont.*

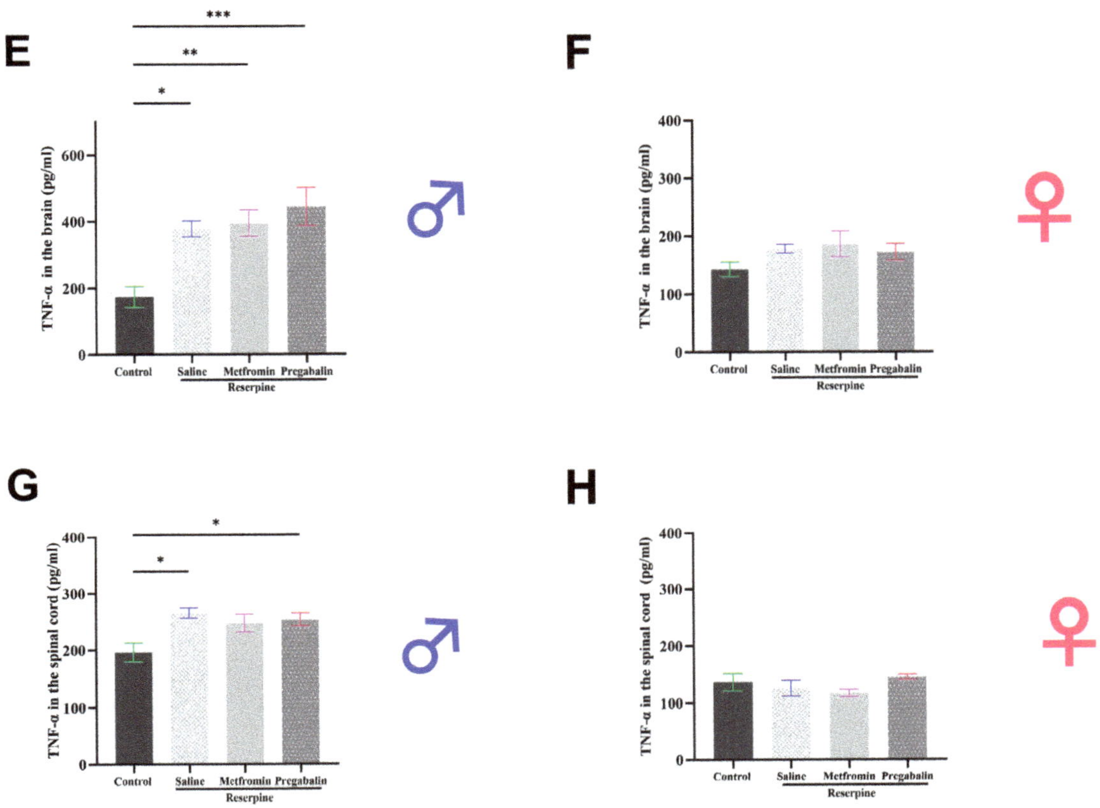

Figure 8. Effect of different interventions on proinflammatory cytokine levels in the total brain and spinal cord of male and female mice. (**A**,**B**) IL-1β levels in the brain: reserpine injections significantly increased IL-1β levels in both the male and female mice. Metformin treatment reversed this decrease in both the male and female mice. (**C**,**D**) IL-1β levels in the spinal cord: reserpine injections significantly increased the IL-1β levels in both the male and female mice. Metformin treatment reversed this decrease in both the male and female mice. (**E**,**F**) TNF-α levels in the brain: reserpine injections significantly increased the TNF-α levels in the male mice only. Metformin treatment reversed this increase in the male mice. (**G**,**H**) TNF-α levels in the spinal cord: reserpine injections significantly increased the TNF-α levels in the male mice only. However, metformin did not reverse this increase. Each bar represents the mean, and the vertical lines indicate the standard error mean (SEM) for the 4–6 mice/group. Asterisks above the lines indicate a significant difference between the groups where * $p < 0.05$, ** $p < 0.01$, *** $p < 0.001$, and **** $p < 0.0001$; otherwise, a non-significant difference is recorded.

3.6. Metformin Alleviated Reserpine Induced-Histopathological Changes in the Hippocampus, Thalamus, and Spinal Cord

To further assess the impact of metformin on the regions involved in pain perception, histological examinations of the hippocampus, thalamus, and spinal cord were conducted using H&E staining. The hippocampus is typically structured with the Cornu Ammonis (composed of CA1, CA2, CA3, and CA4) and the dentate gyrus. In the control group of both the male and female mice, the CA1, CA3, and dentate gyrus regions displayed normal morphology (Figures 9 and 10). In the RES + saline group of the male mice, the hippocampus exhibited alterations across the CA1, CA3, and dentate gyrus regions, including degenerative changes in the pyramidal and granular cell layers. Pregabalin treatment ameliorated these histopathological alterations, with most pyramidal cells preserving their morphology and displaying vesicular nuclei. Additionally, the granular cell layer demonstrated improvements. Interestingly, the metformin treatment resulted in significant histopathological improvements compared to the RES + saline group, with the CA1, CA3, and dentate gyrus regions appearing comparable to the control group (Figure 9).

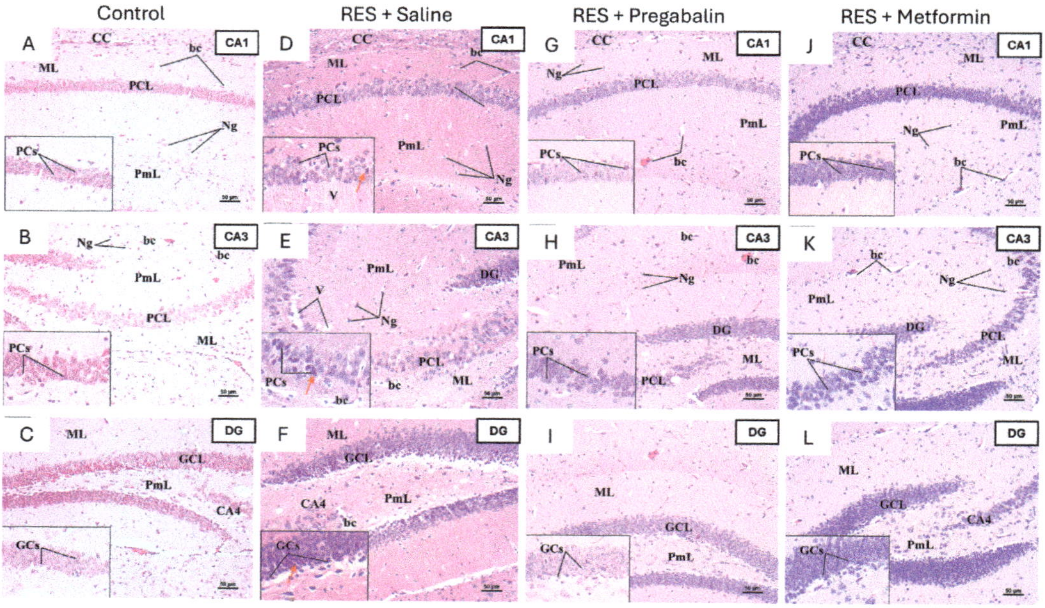

Figure 9. Representative photomicrographs of H&E-stained hippocampus sections in male mice following metformin treatment. (**A**,**B**) Control group: CA1 and CA3 regions of control group displayed all layers—outer molecular layer (ML), middle pyramidal cell layer (PCL), and inner polymorphic layer (PmL)—with normal morphology. PCL comprised well-organized pyramidal cells (PCs) containing large vesicular nuclei and pale basophilic cytoplasm. (**C**) Dentate gyrus (DG) was formed by upper and lower limbs, each consisting of three layers: ML, granular cell layer (GCL), and PmL. Inset of upper limb showed that GCL was composed of densely packed, rounded granule cells (GCs). (**D**,**E**) RES + saline group: CA1 and CA3 regions exhibited disorganized PCL compared to control group. Most PCs appeared shrunken, with dark-stained cytoplasm, ill-defined nuclei, and pericellular halos (red arrow). Both ML and PmL contained an increased number of neuroglial cells (Ng), variable-sized vacuoles (V), and dilated blood capillaries (bc). (**F**) DG exhibited several degenerated and shrunken GCs (red arrow). (**G**,**H**) RES + pregabalin group: CA1 and CA3 regions showed improvement, appearing more similar to control group. Many PCs displayed a normal appearance with vesicular nuclei, although some PCs appeared condensed with dark basophilic cytoplasm. Both ML and PmL contained more Ng cells and dilated bc. (**I**) Architectural improvements were observed in GCL, although some granular cells still appeared shrunken with condensed nuclei. (**J**,**K**) RES + metformin group: CA1 and CA3 regions demonstrated improved appearance, closely resembling control group. However, some PCs with dark basophilic cytoplasm and unclear nuclei were still present. (**L**) DG structure showed significant improvement, appearing nearly identical to control group. Inset of GCL contained densely packed, rounded-to-oval GCs without signs of degeneration. CC, Corpus Callosum. Scale bar corresponds to 50 μm (H&E × 200, Inset × 400).

Similar degenerative changes were noticed in the hippocampus of female mice following reserpine administration, including pyramidal cell degeneration, shrunken neurons with dark basophilic cytoplasm, and poorly defined nuclei in the pyramidal cell layers of CA1 and CA3. The degeneration of granular cells was also evident in the granular cell layer of the dentate gyrus. However, metformin and pregabalin treatments did not demonstrate significant improvements compared to the RES + saline group across all the regions (Figure 10).

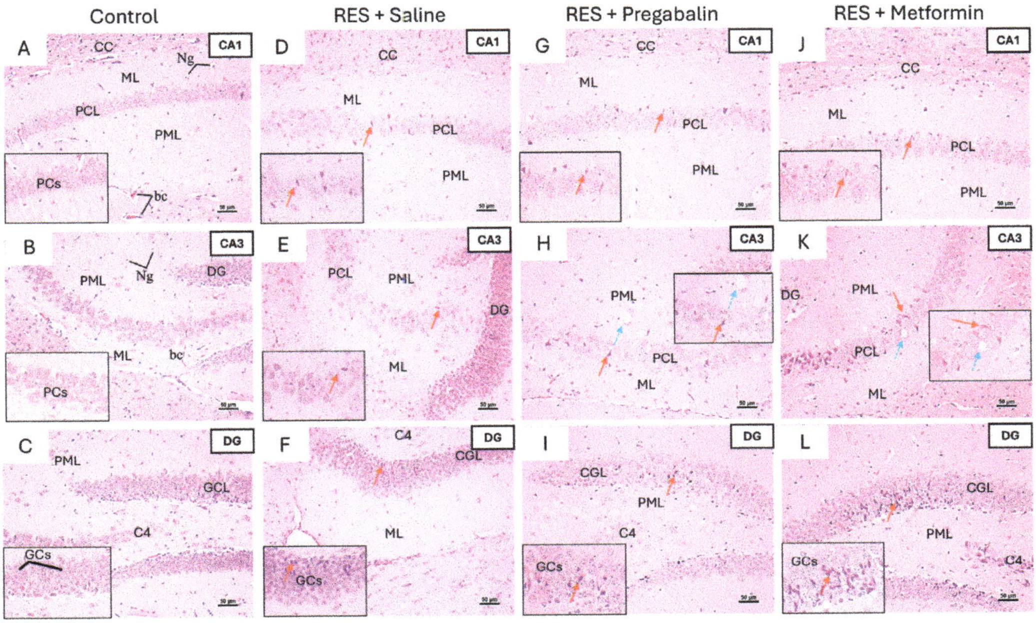

Figure 10. Representative photomicrographs of H&E-stained hippocampus sections in female mice following metformin treatment. (**A**,**B**) Control group: CA1 and CA3 regions displayed normal pyramidal cells (PCs) with large, pale vesicular nuclei. (**C**) Dentate gyrus (DG) of control group exhibited normal granular cells (GCs) with vesicular nuclei in granular cell layer (GCL). (**D**,**E**) RES + saline group: CA1 and CA3 regions in group showed several degenerated, dark, shrunken PCs with dark, ill-defined shaped nuclei (red arrow). (**F**) DG exhibited several degenerated and shrunken GCs (red arrow). (**G–I**) RES + pregabalin group: pregabalin treatment did not alleviate toxic effects of reserpine, as several degenerative changes were still evident following treatment. (**J–L**) RES + metformin group: metformin treatment also failed to mitigate toxic effects of reserpine across all areas, as degenerative changes persisted and tissue vesiculation was observed in polymorphic layer (PML) (light blue arrow). bc, Blood Capillary; CC, Corpus Callosum; ML, molecular layer; PCL, pyramidal cell layer. Scale bar corresponds to 50 µm (H&E × 200, Inset × 400).

Figure 11A,B presents the histological structure of the thalamus in the control group for both the male and female mice, which appears normal. The neuropil exhibited a multitude of healthy neurons of various types within a pinkish background. The predominant cell type observed was the principal cell, characterized by large, pale, spherical nuclei. Additionally, small cells with dark nuclei were also observed. Numerous microglial cells with small, dark nuclei and surrounding halos, along with abundant nerve fibers and tiny capillaries, were present within the neuropil. In the RES + saline-treated group, as depicted in Figure 11C,D, many principal cells and some small cells appeared degenerated and condensed compared to the control group. Figure 11E,F illustrates the thalamus from the pregabalin-treated group, showing a structure that closely resembled that of the control group, with fewer degenerated principal cells and some dilated capillaries. Metformin administration decreased the appearance of the degenerated principal cells compared to the reserpine-treated group in the male mice (Figure 11G), while the neuroprotective effects of metformin were lesser in the female mice than those seen in male mice (Figure 11H).

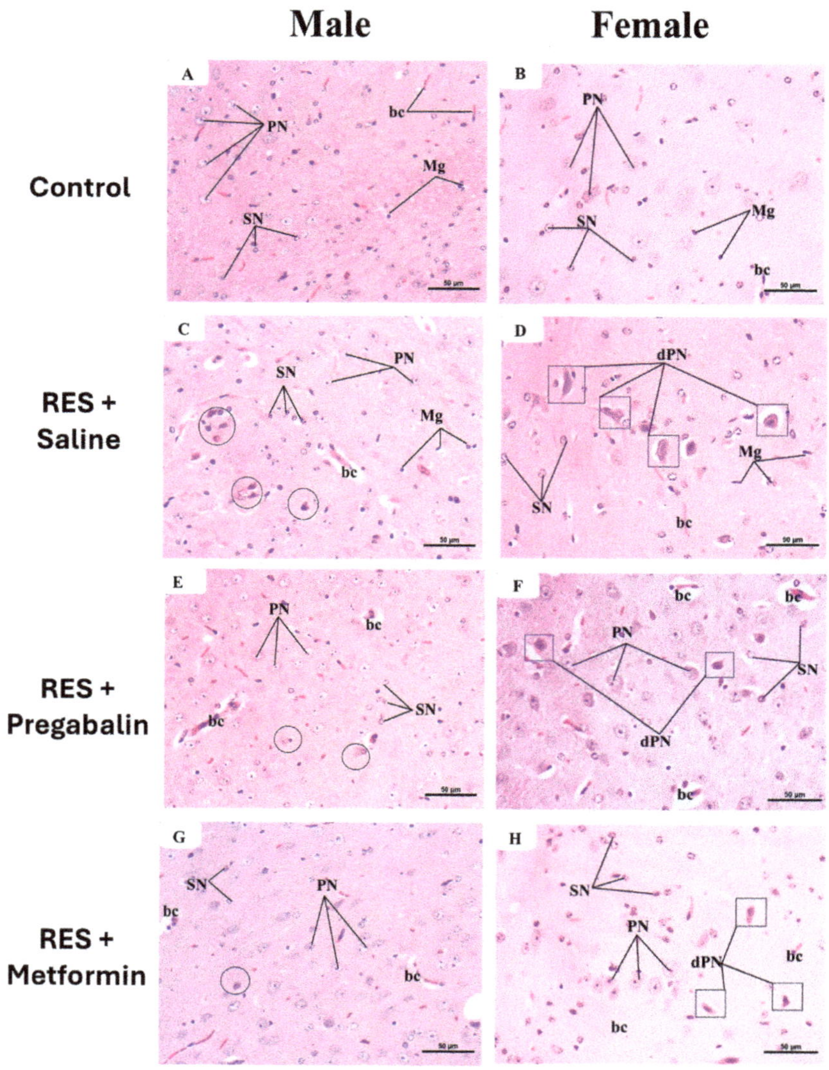

Figure 11. Representative photomicrographs of H&E-stained thalamus sections in male and female mice following metformin treatment. (**A**,**B**) Control group exhibited normal thalamic structure, featuring large principal neurons (PNs) and small neurons (SN). Numerous microglial cells (Mg) and tiny capillaries (bc) were also observed. (**C**,**D**) Reserpine (RES) + saline group displayed numerous degenerated principal neurons (dPNs), marked by black circles in male group and black squares in female group. (**E**,**F**) RES + pregabalin group also exhibited structure similar to control group but with some degenerated PNs, marked by black circles in male group and black squares in female group. (**G**,**H**) RES + metformin group showed a structure that closely resembled control group, though fewer degenerated PNs were still observed, marked by black circles in the male group and black squares in the female group. The scale bar corresponds to 50 μm (H&E × 400).

In the spinal cord (Figure 12), the examination of the H&E-stained sections at the mid-thoracic region from the control group of both the male and female mice revealed a normal histological structure. The dorsal and ventral horns of the gray matter contained various types of neurons and microglial cells within an eosinophilic neuropil matrix, along with tiny blood capillaries. The dorsal horn neurons were predominantly small multipolar neurons with small nuclei and dark cytoplasm, while the ventral horn neurons were mainly large multipolar neurons with large vesicular nuclei, basophilic cytoplasm, and prominent nucleoli (Figure 12A–D). In the RES + saline-treated male and female groups, the dorsal horn exhibited several degenerated small neurons with dark eosinophilic cytoplasm and condensed nuclei. Many large neurons in the ventral horn also appeared degenerated, characterized by dark, acidophilic cytoplasm and condensed nuclei. Both the horns exhibited various-sized vacuoles and dilated capillaries (Figure 12E–H). Pregabalin administration resulted in fewer degenerated neurons in both the dorsal and ventral horns compared to the saline-treated group in both the male and female mice. However, some vacuoles and dilated capillaries were still observed in the neuropil (Figure 12I–L). Furthermore, metformin administration to the male and female mice resulted in nearly normal neurons in both the dorsal and ventral horns, with only the irregular appearance of some congested capillaries (Figure 12M–P).

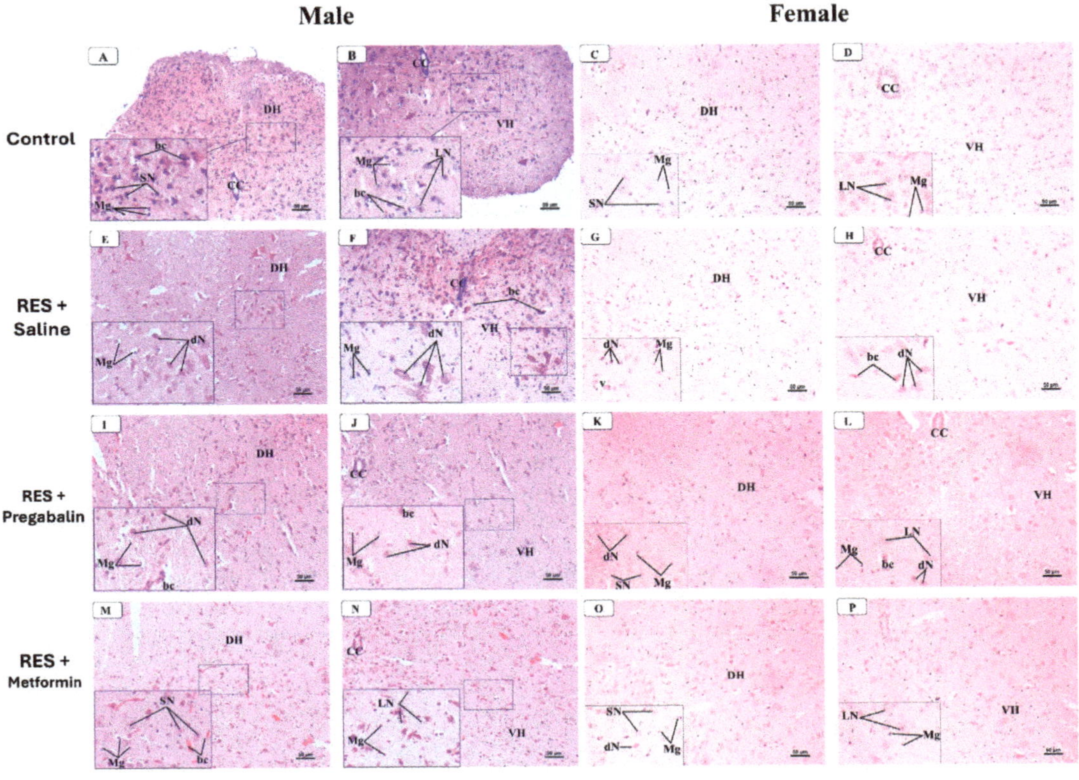

Figure 12. Representative photomicrographs of H&E-stained spinal cord sections in male and female mice following metformin treatment. (**A–D**) Control group exhibited dorsal horn of gray matter (DH) containing small multipolar neurons (SN) and ventral horn (VH) containing large multipolar neurons (LN). Microglial cells (Mg) and tiny capillaries were observed dispersed in both horns. (**E–H**) RES + saline group showed numerous degenerated neurons (dNs) in both dorsal and ventral horns, characterized by dark eosinophilic cytoplasm and condensed nuclei. Various-sized vacuoles and dilated capillaries were observed in neuropil. (**I–L**) RES + pregabalin group displayed fewer dNs in both dorsal and ventral horns. However, some vacuoles and dilated capillaries remained in neuropil. (**M–P**) RES + metformin-treated group showed nearly normal neurons in both dorsal and ventral horns, except for some congested capillaries. CC, central canal. Scale bar corresponds to 50 μm (H&E × 200, Inset × 400).

4. Discussion

The current pharmacotherapy for FM has limited efficacy and varies among individuals, with responses to medications often being subjective [46]. Additionally, the chronic nature of FM pain necessitates continuous long-term treatment, sometimes extending up to 176–205 days, while pregabalin, a first-line approved drug for FM, has a potential for abuse [47]. These factors underscore the critical need for research aimed at identifying novel therapeutic strategies for FM. In this study, the effects of metformin administration on reserpine-induced FM in male and female mice were assessed. Reserpine, known for its ability to deplete monoamine neurotransmitters, led to a significant decrease in the serotonin and norepinephrine levels in both the brain and spinal cord [26]. Furthermore, reserpine administration increased the glutamate levels in these regions, indicating potential excitotoxicity and neuronal damage [27]. These effects established an FM-like model in the rodents, as observed in the present study [48].

While the pain-relieving effects of metformin have been shown in other pain models, this study is, to the best available knowledge, the first to demonstrate its potential benefits in a model of FM in mice. The key finding is that metformin demonstrates a promising effect on FM-like pain after seven days of continuous administration in both male and female mice, as evidenced by its ability to increase the paw withdrawal threshold in a vFt test and increase the latency to respond to hot stimulus in a hot plate test. It has been proven that metformin interacts with crucial molecules that significantly contribute to nociceptive processing. Previous studies have described metformin's ability to ease pain in animal models through its ability to activate AMP-activated protein kinase (AMPK), which plays a fundamental role in controlling neural excitability [17,49]. Metformin has been found to activate the opioidergic pathways in models of nociceptive and nerve pain [50]. Metformin also prevents incision-evoked mechanical hypersensitivity and hyperalgesia [51]. It also reverses and prevents the complete occurrence of neuropathic pain, and this has a negative correlation with microglial activity in the spared nerve injury model [19].

However, a single metformin injection did not enhance reserpine-induced mechanical or thermal hyperalgesia, either in the male or female mice. According to the existing literature, the effective dose and duration of metformin in modulating the key molecules involved in nociceptive processing remain subject to debate. For instance, the oral delivery of metformin at doses of 180 mg/kg and 250 mg/kg has been shown to reduce nociceptive responses within half an hour of administration [52]. Other studies support the findings of this study, indicating that a single dose of metformin fails to produce acute analgesia, and the antinociceptive effects of metformin take several days to express [53,54].

The behavioral assessments of the motor activity in male mice indicated that reserpine-induced motor impairment notably improved after metformin administration. Metformin's neuroprotective effects and influence on neurotransmitter levels, as we will discuss later, may contribute to an improvement in motor activity, enhancing synaptic transmission and neural plasticity. In parallel to this study's findings, George Jitica and his colleagues demonstrated metformin's ability to improve motor coordination in the rotarod test following impairment induced by haloperidol, corroborating our findings [55]. Moreover, metformin increases the expression of brain-derived neurotrophic factor (BDNF), a protein that aids in the development, maintenance, and survival of neurons [56]. BDNF is also involved in neuroplasticity, which is vital for learning and memory. Improved insulin sensitivity, another benefit of metformin, has been correlated with better mental health outcomes. Insulin resistance has been linked to the etiology of depression, and metformin's potential to improve insulin sensitivity might contribute to its antidepressant effects [57]. In addition, the anti-inflammatory properties of metformin may help to improve mood and cognitive function [58]. The findings of this study align with these results, showing a significant reduction in the depressive-related behavior in the mice following metformin administration. Additionally, pregabalin was used as a positive control drug, given its FDA approval for treating FM in this study. While clinical trials have shown pregabalin's effectiveness in alleviating FM-related pain, it does not appear to exhibit antidepressant effects in individuals with FM [59]. In our study, pregabalin administration did not improve the depressive-like behavior in the mice during the TST and FST. These results highlight metformin's potential superiority over pregabalin for treating FM, as depression is a strongly associated symptom with chronic pain in FM [60].

Serotonin, norepinephrine, and glutamate neurotransmitters are critical for pain modulation, mood regulation, and sleep, and their dysregulation likely contributes to heightened pain perception and other symptoms experienced by individuals with FM [61,62]. Serotonin plays a key role in modulating pain perception and mood; its dysregulation may contribute to the sensory hypersensitivity and depressive symptoms commonly observed in FM patients [63,64]. Norepinephrine is crucial for descending pain modulation pathways, and its alterations can exacerbate pain severity and fatigue in FM [65]. Glutamate, the primary excitatory neurotransmitter in the central nervous system, is associated with amplifying pain signals and promoting neuroinflammation, potentially contributing to the chronic pain and hypersensitivity characteristic of FM [66]. One of the critical findings in our experiments is that metformin administration restored the serotonin and norepinephrine levels to near-normal values in the brain and spinal cord of male mice. Moreover, metformin reduced the glutamate levels, suggesting its role in mitigating excitotoxic processes. Consistent with our findings, metformin supplementation significantly ameliorated the serotonin and norepinephrine levels in the brains of rodents [22,23]. Furthermore, metformin administration reduced the brain glutamate to normal levels in a diabetic epileptic model [24]. While preclinical studies have highlighted metformin's ability to modify neurotransmitter levels, most research has focused on its effects on depression, particularly as a comorbid symptom of diabetes. This makes our study a pioneer in uncovering metformin's possible impact on neurotransmitter alterations in FM.

Besides neurotransmitter imbalance, nociceptive activation in FM has been linked to increasing levels of proinflammatory cytokines, including IL-1β and TNF-α [67]. Given metformin's known anti-inflammatory properties, its potential for modulating IL-1β and TNF-α levels was assessed in this study, which could be another promising mechanism for metformin to alleviate FM-related nociceptive activation [68]. In this study, metformin significantly reduced the IL-1β levels in both the brain and spinal cord to near-normal levels in both male and female mice. AMPK activation can enhance cellular metabolism and reduce oxidative stress. It also inhibits the NLRP3 inflammasome and the nuclear factor-kappa B (NF-κB) signaling pathway, thereby reducing IL-1β production [69,70]. However, it is noteworthy that metformin did not affect the TNF-α levels in the brain or the spinal cord in either the male or female mice. The lack of effect on TNF-α suggests that other regulatory mechanisms may control the production of this cytokine or that strong doses or prolonged treatment duration may be required to observe changes in TNF-α levels.

This study also explored the prophylactic effects of metformin on the histopathological changes induced by a reserpine model of FM. To achieve this, the hippocampus, thalamus, and spinal cord of the male and female mice were examined using H&E staining. Changes in hippocampal structure and function are often observed in chronic pain conditions, underscoring their importance in understanding and treating pain [71]. This study revealed significant degenerative changes in multiple hippocampal layers following reserpine administration. These histological alterations align with previous findings from an FM rat model induced by acidic saline [72]. Clinical studies on FM patients reveal hippocampal changes, including metabolic dysfunction and volume reduction, which may explain sleep disturbances and cognitive dysfunction [73,74]. In this study, metformin administration effectively reversed the histopathological changes in the hippocampus associated with reserpine administration in male mice. Consistent with present study's findings, metformin has been shown to correct the morphological state of neurons under specific pathological conditions. It enhances neuroplasticity, increases neuronal cell numbers, improves neuron survival, and reduces neuroinflammation in the CA1 and DG of rodents, thereby improving cognitive function and overall hippocampal health [75–77].

Histological changes in the thalamus resulting from reserpine administration are infrequently documented in the literature. In this study, reserpine administration led to significant changes in the thalamus of both the male and female mice, characterized by neuronal degeneration, vacuolation, and dilated capillaries. These histopathological alterations reflect the severe impact of reserpine on neural tissue integrity, contributing to the observed symptoms of FM in this experimental model, as documented in previous studies [78]. Interestingly, the findings of this study highlight metformin's potential to prevent the structural damage induced by reserpine in the thalamus, which was more obvious in male mice.

Moreover, the spinal cord is pivotal in the modulation and transmission of pain signals, acting as a primary relay center that processes nociceptive information from peripheral nerves and transmits it to the brain [79]. This study observed neuronal degeneration in the dorsal and ventral horns of the spinal cord, along with the presence of vacuoles of various sizes and dilated capillaries in the neuropil following reserpine administration. These alterations may be linked to glutamatergic excitotoxicity following reserpine exposure [80]. Remarkably, metformin administration preserved the integrity of the spinal cord, highlighting its protective effects at this level. A previous study noted that the FM induction model using reserpine resulted in morphological changes in the motor neurons within the ventral horn of the spinal cord, whereas such changes were not observed in the dorsal horn [78]. Additionally, other studies suggest that reserpine can influence spinal cord function and alter neurotransmitter levels which may indirectly affect its histological features [81].

A crucial observation in this study is that metformin has intriguing sex-dependent effects on the neurotransmitter levels in both the brain and the spinal cord. In contrast to its effect on the male mice, metformin injections were unable to equalize the serotonin and norepinephrine levels in the female mice. Metformin is a water-soluble drug that crosses cell membranes using active transporters such as an organic cation transporter (OCT-2) [82]. It has been found that testosterone can influence the expression and activity of the OCT-2 receptor [83]. Higher androgen levels in males can lead to increased OCT-2 expression [84]. Therefore, the differences in OCT-2 expression between males and females may impact the pharmacokinetics of drugs that rely on this transporter, leading to differences in drug efficacy between the sexes, as found in this study. Sex-specific differences in the effect of metformin have been described by Kufreobong and colleagues. In their study, metformin was able to prevent neuropathic pain in the male mice but not in the female mice [19]. In another study, metformin administration led to distinct changes in the signaling pathways related to AMPK and autophagy, which are critical for cellular energy balance and inflammatory responses. These changes differed significantly between the male and female subjects, contributing to differences in the therapeutic outcomes [85]. Moreover, accumulated evidence indicates significant sex differences in the expression and activity of blood–brain barrier (BBB) transporters, with gonadal hormones playing a further modulatory role [86]. Therefore, the ability of metformin to cross the BBB may differ between males and females. These differences may significantly influence the drug's effects on central nervous system processes, including the regulation of neurotransmitters. Hence, the observed inability of metformin to equalize serotonin and norepinephrine levels in female mice, despite its effects in male mice, may be partly due to these differences in BBB permeability and transporter expression. It is also important to mention that the symptoms of FM are influenced by the menstrual cycle and significant hormonal changes, such as those experienced during pregnancy and menopause [87]. These hormonal fluctuations can exacerbate symptoms, highlighting the importance of considering hormonal factors when studying FM, especially in female populations. This aspect may also provide insight into the limited efficacy of metformin in inducing full benefits in female mice, as hormonal variations could play a role in the response to treatment.

5. Strengths and Limitations

In regard to the strength of this study, this study offers valuable insights into the potential of metformin as a treatment for FM, with a detailed assessment of the neurotransmitter levels and histopathological changes in a well-established FM mouse model. The inclusion of both male and female mice allowed for an exploration of the sex-specific differences in the treatment effects.

Despite these strengths, this study has certain limitations. It relies on a single animal model and features a relatively short treatment duration, which may not fully capture the long-term therapeutic potential of metformin. Additionally, while a histopathological analysis was conducted, more detailed tissue examination, particularly at the cellular and molecular levels, is necessary to further elucidate metformin's neuroprotective effects. Future studies should incorporate extended treatment periods, the use of multiple FM models, and a more thorough investigation of tissue integrity, including advanced histological techniques, to provide a deeper understanding of the drug's effects on neural and musculoskeletal tissues.

6. Conclusions

In this study, metformin effectively alleviated mechanical and thermal hyperalgesia, improved depressive-like behaviors, reduced IL-1β levels, and prevented structural damage in the spinal cord in both male and female mice. These findings suggest that metformin could be a possible supportive agent candidate for future clinical trials aimed at alleviating hyperalgesia and depressive symptoms in FM. Importantly, the sex-specific differences observed in the neurotransmitter regulation of metformin underscore the necessity of including sex-specific analyses in its mechanism of action in the conditions associated with neurotransmitter disturbances such as FM. However, while metformin appeared to modulate monoamine levels, it is important to consider that these changes may be region-specific within the brain, potentially leading to varied behavioral outcomes, meriting further research and clinical evaluation.

Author Contributions: Conceptualization, H.A.A. (Hanin Abdulbaset AboTaleb) and B.S.A.; methodology, H.A.A. (Hanin Abdulbaset AboTaleb); H.A.A. (Hani A. Alturkistani) and M.A.A.-T.; validation, H.A.A. (Hanin Abdulbaset AboTaleb) and B.S.A.; formal analysis, H.A.A. (Hanin Abdulbaset AboTaleb); investigation, H.A.A. (Hanin Abdulbaset AboTaleb), E.A.H., M.M.H. and G.S.A.E.-A.; writing—original draft preparation, H.A.A. (Hanin Abdulbaset AboTaleb); writing—review and editing, B.S.A.; supervision, E.A.H. and B.S.A.; funding acquisition, B.S.A. All authors have read and agreed to the published version of the manuscript.

Funding: This project was funded by the Deanship of Scientific Research (DSR) at King Abdulaziz University, Jeddah, under grant no. (GPIP: 1727-828-2024). The authors, therefore, acknowledge with thanks DSR for technical and financial support.

Institutional Review Board Statement: This entire study was conducted in strict adherence to the ethical standards set by the Biomedical Ethics Committee of King Abdulaziz University (approval number: 236-24). The research protocol was developed following the Animal Care and Use Committee (ACUC) guidelines at the Animal House Unit, King Fahad Medical Research Center, which also approved all the animal procedures.

Informed Consent Statement: Not applicable.

Data Availability Statement: The data will be available upon request.

Acknowledgments: The authors would like to thank the King Fahad Medical Research Center at King Abdulaziz University for providing the opportunity to use for their facilities. We extend a special thanks to the Neuroscience and Geroscience Research Unit and Animal House Unit at the King Fahad Medical Research Center and the Histology Technical Unit, Department of Clinical Anatomy, Faculty of Medicine, King Abdulaziz University, Jeddah, Saudi Arabia.

Conflicts of Interest: The authors declare no conflicts of interest.

References

1. Sarzi-Puttini, P.; Giorgi, V.; Marotto, D.; Atzeni, F. Fibromyalgia: An update on clinical characteristics, aetiopathogenesis and treatment. *Nat. Rev. Rheumatol.* **2020**, *16*, 645–660. [CrossRef] [PubMed]
2. Queiroz, L.P. Worldwide Epidemiology of Fibromyalgia. *Curr. Pain Headache Rep.* **2013**, *17*, 356. [CrossRef] [PubMed]
3. Siracusa, R.; Di Paola, R.; Cuzzocrea, S.; Impellizzeri, D. Fibromyalgia: Pathogenesis, Mechanisms, Diagnosis and Treatment Options Update. *Int. J. Mol. Sci.* **2021**, *22*, 3891. [CrossRef] [PubMed]
4. Trucharte, A.; Leon, L.; Castillo-Parra, G.; Magán, I.; Freites, D.; Redondo, M. Emotional regulation processes: Influence on pain and disability in fibromyalgia patients. *Clin. Exp. Rheumatol.* **2020**, *38*, 40–46.
5. Inanici, F.F.; Yunus, M.B. History of fibromyalgia: Past to present. *Curr. Pain Headache Rep.* **2004**, *8*, 369–378. [CrossRef]
6. Ji, R.-R.; Nackley, A.; Huh, Y.; Terrando, N.; Maixner, W. Neuroinflammation and Central Sensitization in Chronic and Widespread Pain. *Anesthesiology* **2018**, *129*, 343–366. [CrossRef]
7. Dutra, R.C.; Martins, C.P.; Paes, R.S.; Baldasso, G.M.; Ferrarini, E.G.; Scussel, R.; Zaccaron, R.P.; Machado-De-Ávila, R.A.; Silveira, P.C.L. Pramipexole, a dopamine D3/D2 receptor-preferring agonist, attenuates reserpine-induced fibromyalgia-like model in mice. *Neural Regen. Res.* **2022**, *17*, 450–458. [CrossRef]
8. Pomares, F.B.; Roy, S.; Funck, T.; Feier, N.A.; Thiel, A.; Fitzcharles, M.-A.; Schweinhardt, P. Upregulation of cortical GABAA receptor concentration in fibromyalgia. *Pain* **2019**, *161*, 74–82. [CrossRef]
9. Becker, S.; Schweinhardt, P. Dysfunctional Neurotransmitter Systems in Fibromyalgia, Their Role in Central Stress Circuitry and Pharmacological Actions on These Systems. *Pain Res. Treat.* **2011**, *2012*, 1–10. [CrossRef]
10. White, C.; Kwong, W.J.; Armstrong, H.; Behling, M.; Niemira, J.; Lang, K. Analysis of Real World Dosing Patterns for the 3 FDA-Approved Medications in the Treatment of Fibromyalgia. *Am. Health Drug Benefits* **2018**, *11*, 293–300.
11. Häuser, W.; Walitt, B.; Fitzcharles, M.-A.; Sommer, C. Review of pharmacological therapies in fibromyalgia syndrome. *Arthritis Res. Ther.* **2014**, *16*, 1–10. [CrossRef]
12. Rissardo, J.P.; Caprara, A.L.F. Pregabalin-associated movement disorders: A literature review. *Brain Circ.* **2020**, *6*, 96–106. [CrossRef]
13. Arora, M.K.; Baidya, D.K.; Agarwal, A.; Khanna, P. Pregabalin in acute and chronic pain. *J. Anaesthesiol. Clin. Pharmacol.* **2011**, *27*, 307–314. [CrossRef] [PubMed]
14. Myhre, M.; Jacobsen, H.B.; Andersson, S.; Stubhaug, A. Cognitive Effects of Perioperative Pregabalin. *Anesthesiology* **2019**, *130*, 63–71. [CrossRef]
15. Park, K.; Kim, S.; Ko, Y.-J.; Park, B.-J. Duloxetine and cardiovascular adverse events: A systematic review and meta-analysis. *J. Psychiatr. Res.* **2020**, *124*, 109–114. [CrossRef]
16. Zaccardi, F.; Khunti, K.; Marx, N.; Davies, M.J. First-line treatment for type 2 diabetes: Is it too early to abandon metformin? *Lancet* **2020**, *396*, 1705–1707. [CrossRef]
17. Baeza-Flores, G.D.C.; Guzmán-Priego, C.G.; Parra-Flores, L.I.; Murbartián, J.; Torres-López, J.E.; Granados-Soto, V. Metformin: A Prospective Alternative for the Treatment of Chronic Pain. *Front. Pharmacol.* **2020**, *11*, 558474. [CrossRef]
18. Das, V.; Kroin, J.S.; Moric, M.; McCarthy, R.J.; Buvanendran, A. Early Treatment With Metformin in a Mice Model of Complex Regional Pain Syndrome Reduces Pain and Edema. *Anesthesia Analg.* **2020**, *130*, 525–534. [CrossRef]
19. Inyang, K.E.; Szabo-Pardi, T.; Wentworth, E.; McDougal, T.A.; Dussor, G.; Burton, M.D.; Price, T.J. The antidiabetic drug metformin prevents and reverses neuropathic pain and spinal cord microglial activation in male but not female mice. *Pharmacol. Res.* **2018**, *139*, 1–16. [CrossRef]
20. Bullón, P.; Alcocer-Gómez, E.; Carrión, A.M.; Marín-Aguilar, F.; Garrido-Maraver, J.; Román-Malo, L.; Ruiz-Cabello, J.; Culic, O.; Ryffel, B.; Apetoh, L.; et al. AMPK Phosphorylation Modulates Pain by Activation of NLRP3 Inflammasome. *Antioxid. Redox Signal.* **2016**, *24*, 157–170. [CrossRef]
21. Alcocer-Gómez, E.; Garrido-Maraver, J.; Bullón, P.; Marín-Aguilar, F.; Cotán, D.; Carrión, A.M.; Alvarez-Suarez, J.M.; Giampieri, F.; Sánchez-Alcazar, J.A.; Battino, M.; et al. Metformin and caloric restriction induce an AMPK-dependent restoration of mitochondrial dysfunction in fibroblasts from Fibromyalgia patients. *Biochim. Biophys. Acta* **2015**, *1852*, 1257–1267. [CrossRef] [PubMed]
22. Shivavedi, N.; Kumar, M.; Tej, G.N.V.C.; Nayak, P.K. Metformin and ascorbic acid combination therapy ameliorates type 2 diabetes mellitus and comorbid depression in rats. *Brain Res.* **2017**, *1674*, 1–9. [CrossRef]
23. Papachristou, S.; Papanas, N. Reduction of Depression in Diabetes: A New Pleiotropic Action of Metformin? *Diabetes Ther.* **2021**, *12*, 965–968. [CrossRef]
24. Mohamed, M.A.E.; Abdel-Rahman, R.F.; Mahmoud, S.S.; Khattab, M.M.; Safar, M.M. Metformin and trimetazidine ameliorate diabetes-induced cognitive impediment in status epileptic rats. *Epilepsy Behav.* **2020**, *104*, 106893. [CrossRef]
25. Mathkhor, A.J.; Ibraheem, N.M. Prevalence and Impact of obesity on fibromyalgia syndrome and its allied symptoms. *J. Fam. Med. Prim. Care* **2023**, *12*, 123–127. [CrossRef]
26. Nagakura, Y.; Oe, T.; Aoki, T.; Matsuoka, N. Biogenic amine depletion causes chronic muscular pain and tactile allodynia accompanied by depression: A putative animal model of fibromyalgia. *Pain* **2009**, *146*, 26–33. [CrossRef] [PubMed]
27. Singh, L.; Kaur, A.; Singh, A.P.; Bhatti, R. Daphnetin, a natural coumarin averts reserpine-induced fibromyalgia in mice: Modulation of MAO-A. *Exp. Brain Res.* **2021**, *239*, 1451–1463. [CrossRef]

28. AboTaleb, H.; Hindi, E.A.; Abd El-Aziz, G.S.; Alturkistani, H.A.; Halawani, M.M.; Al-Thepyani, M.A.; Alghamdi, B.S. Evaluation of reserpine-induced fibromyalgia in mice: A comparative behavioral, neurochemical, and histological assessment of two doses. *IBRO Neurosci. Rep.* **2024**. [CrossRef]
29. Brusco, I.; Justino, A.B.; Silva, C.R.; Fischer, S.; Cunha, T.M.; Scussel, R.; Machado-De-Ávila, R.A.; Ferreira, J.; Oliveira, S.M. Kinins and their B1 and B2 receptors are involved in fibromyalgia-like pain symptoms in mice. *Biochem. Pharmacol.* **2019**, *168*, 119–132. [CrossRef]
30. Gonzalez-Cano, R.; Boivin, B.; Bullock, D.; Cornelissen, L.; Andrews, N.; Costigan, M. Up–Down Reader: An Open Source Program for Efficiently Processing 50% von Frey Thresholds. *Front. Pharmacol.* **2018**, *9*, 433. [CrossRef]
31. Christensen, S.L.; Hansen, R.B.; Storm, M.A.; Olesen, J.; Hansen, T.F.; Ossipov, M.; Izarzugaza, J.M.G.; Porreca, F.; Kristensen, D.M. Von Frey testing revisited: Provision of an online algorithm for improved accuracy of 50% thresholds. *Eur. J. Pain* **2019**, *24*, 783–790. [CrossRef] [PubMed]
32. Klein, C.P.; Sperotto, N.D.; Maciel, I.S.; Leite, C.E.; Souza, A.H.; Campos, M.M. Effects of D-series resolvins on behavioral and neurochemical changes in a fibromyalgia-like model in mice. *Neuropharmacology* **2014**, *86*, 57–66. [CrossRef]
33. Hunskaar, S.; Berge, O.-G.; Hole, K. A modified hot-plate test sensitivie to mild analgesics. *Behav. Brain Res.* **1986**, *21*, 101–108. [CrossRef]
34. Sałat, K.; Furgała-Wojas, A. Serotonergic Neurotransmission System Modulator, Vortioxetine, and Dopaminergic D_2/D_3 Receptor Agonist, Ropinirole, Attenuate Fibromyalgia-Like Symptoms in Mice. *Molecules* **2021**, *26*, 2398. [CrossRef] [PubMed]
35. Deuis, J.R.; Dvorakova, L.S.; Vetter, I. Methods Used to Evaluate Pain Behaviors in Rodents. *Front. Mol. Neurosci.* **2017**, *10*, 284. [CrossRef]
36. Taleb, H.A.A.; Alghamdi, B.S. Neuroprotective Effects of Melatonin during Demyelination and Remyelination Stages in a Mouse Model of Multiple Sclerosis. *J. Mol. Neurosci.* **2019**, *70*, 386–402. [CrossRef]
37. Altarifi, A.; Alsalem, M.; Mustafa, A. Effects of intraplantar administration of Complete Freund's Adjuvant (CFA) on rotarod performance in mice. *Scand. J. Pain* **2019**, *19*, 805–811. [CrossRef] [PubMed]
38. Dagnino, A.P.A.; da Silva, R.B.M.; Chagastelles, P.C.; Pereira, T.C.B.; Venturin, G.T.; Greggio, S.; da Costa, J.C.; Bogo, M.R.; Campos, M.M. Nociceptin/orphanin FQ receptor modulates painful and fatigue symptoms in a mouse model of fibromyalgia. *Pain* **2019**, *160*, 1383–1401. [CrossRef]
39. Nascimento, S.S.; Camargo, E.A.; DeSantana, J.M.; Araújo, A.A.S.; Menezes, P.P.; Lucca-Júnior, W.; Albuquerque-Júnior, R.L.C.; Bonjardim, L.R.; Quintans-Júnior, L. Linalool and linalool complexed in β-cyclodextrin produce anti-hyperalgesic activity and increase Fos protein expression in animal model for fibromyalgia. *Naunyn-Schmiedebergs Arch. Exp. Pathol. Pharmakol.* **2014**, *387*, 935–942. [CrossRef]
40. Serretti, A. Anhedonia and Depressive Disorders. *Clin. Psychopharmacol. Neurosci.* **2023**, *21*, 401–409. [CrossRef]
41. Yankelevitch-Yahav, R.; Franko, M.; Huly, A.; Doron, R. The Forced Swim Test as a Model of Depressive-like Behavior. *J. Vis. Exp.* **2015**, e52587. [CrossRef]
42. Sałat, K.; Siwek, A.; Starowicz, G.; Librowski, T.; Nowak, G.; Drabik, U.; Gajdosz, R.; Popik, P. Antidepressant-like effects of ketamine, norketamine and dehydronorketamine in forced swim test: Role of activity at NMDA receptor. *Neuropharmacology* **2015**, *99*, 301–307. [CrossRef]
43. Cryan, J.F.; Mombereau, C.; Vassout, A. The tail suspension test as a model for assessing antidepressant activity: Review of pharmacological and genetic studies in mice. *Neurosci. Biobehav. Rev.* **2005**, *29*, 571–625. [CrossRef]
44. Alqurashi, G.K.; Hindi, E.A.; Zayed, M.A.; El-Aziz, G.S.A.; Alturkistani, H.A.; Ibrahim, R.F.; Al-Thepyani, M.A.; Bakhlgi, R.; Alzahrani, N.A.; Ashraf, G.M.; et al. The Impact of Chronic Unpredictable Mild Stress-Induced Depression on Spatial, Recognition and Reference Memory Tasks in Mice: Behavioral and Histological Study. *Behav. Sci.* **2022**, *12*, 166. [CrossRef]
45. Vincent, A.; Benzo, R.P.; O Whipple, M.; McAllister, S.J.; Erwin, P.J.; Saligan, L.N. Beyond pain in fibromyalgia: Insights into the symptom of fatigue. *Arthritis Res. Ther.* **2013**, *15*, 221. [CrossRef]
46. Díaz-Marsá, M.; Palomares, N.; Morón, M.D.; Tajima, K.; Fuentes, M.E.; López-Ibor, J.J.; Carrasco, J.L. Psychological Factors Affecting Response to Antidepressant Drugs in Fibromyalgia. *Psychosomatics* **2011**, *52*, 237–244. [CrossRef]
47. Smith, M.T.; Moore, B.J. Pregabalin for the treatment of fibromyalgia. *Expert Opin. Pharmacother.* **2012**, *13*, 1527–1533. [CrossRef]
48. Brum, E.S.; Becker, G.; Fialho, M.F.P.; Oliveira, S.M. Animal models of fibromyalgia: What is the best choice? *Pharmacol. Ther.* **2021**, *230*, 107959. [CrossRef]
49. Li, J.; Zhang, B.; Liu, W.-X.; Lu, K.; Pan, H.; Wang, T.; Oh, C.-D.; Yi, D.; Huang, J.; Zhao, L.; et al. Metformin limits osteoarthritis development and progression through activation of AMPK signalling. *Ann. Rheum. Dis.* **2020**, *79*, 635–645. [CrossRef]
50. Augusto, P.S.; Braga, A.V.; Rodrigues, F.F.; Morais, M.I.; Dutra, M.M.; Batista, C.R.; Melo, I.S.; Costa, S.O.; Goulart, F.A.; Coelho, M.M.; et al. Metformin antinociceptive effect in models of nociceptive and neuropathic pain is partially mediated by activation of opioidergic mechanisms. *Eur. J. Pharmacol.* **2019**, *858*, 172497. [CrossRef]
51. Burton, M.D.; Tillu, D.V.; Mazhar, K.; Mejia, G.L.; Asiedu, M.N.; Inyang, K.; Hughes, T.; Lian, B.; Dussor, G.; Price, T.J. Pharmacological activation of AMPK inhibits incision-evoked mechanical hypersensitivity and the development of hyperalgesic priming in mice. *Neuroscience* **2017**, *359*, 119–129. [CrossRef]
52. Pecikoza, U.; Tomić, M.; Nastić, K.; Micov, A.; Stepanović-Petrović, R. Synergism between metformin and analgesics/vitamin B12 in a model of painful diabetic neuropathy. *Biomed. Pharmacother.* **2022**, *153*, 113441. [CrossRef]

53. Melemedjian, O.K.; Khoutorsky, A.; Sorge, R.E.; Yan, J.; Asiedu, M.N.; Valdez, A.; Ghosh, S.; Dussor, G.; Mogil, J.S.; Sonenberg, N.; et al. mTORC1 inhibition induces pain via IRS-1-dependent feedback activation of ERK. *Pain* **2013**, *154*, 1080–1091. [CrossRef]
54. Melemedjian, O.K.; Asiedu, M.N.; Tillu, D.V.; Sanoja, R.; Yan, J.; Lark, A.; Khoutorsky, A.; Johnson, J.; A Peebles, K.; Lepow, T.; et al. Targeting Adenosine Monophosphate-Activated Protein Kinase (AMPK) in Preclinical Models Reveals a Potential Mechanism for the Treatment of Neuropathic Pain. *Mol. Pain* **2011**, *7*, 70. [CrossRef]
55. Jîtcă, G.; Gáll, Z.; E Vari, C.; E Ősz, B.; Tero-Vescan, A.; Groșan, A.; Dogaru, M.T. Beneficial effects of metformin on haloperidol-induced motor deficits in rats. A behavioral assessment. *Acta Marisiensis-Ser. Medica* **2021**, *67*, 115–121. [CrossRef]
56. Ghadernezhad, N.; Khalaj, L.; Pazoki-Toroudi, H.; Mirmasoumi, M.; Ashabi, G. Metformin pretreatment enhanced learning and memory in cerebral forebrain ischaemia: The role of the AMPK/BDNF/P70SK signalling pathway. *Pharm. Biol.* **2016**, *54*, 2211–2219. [CrossRef]
57. Kan, C.; Silva, N.; Golden, S.H.; Rajala, U.; Timonen, M.; Stahl, D.; Ismail, K. A Systematic Review and Meta-analysis of the Association Between Depression and Insulin Resistance. *Diabetes Care* **2013**, *36*, 480–489. [CrossRef]
58. Troubat, R.; Barone, P.; Leman, S.; Desmidt, T.; Cressant, A.; Atanasova, B.; Brizard, B.; El Hage, W.; Surget, A.; Belzung, C.; et al. Neuroinflammation and depression: A review. *Eur. J. Neurosci.* **2020**, *53*, 151–171. [CrossRef]
59. Arnold, L.M.; Crofford, L.J.; Martin, S.A.; Young, J.P.; Sharma, U. The Effect of Anxiety and Depression on Improvements in Pain in a Randomized, Controlled Trial of Pregabalin for Treatment of Fibromyalgia. *Pain Med.* **2007**, *8*, 633–638. [CrossRef]
60. White, K.P.; Nielson, W.R.; Harth, M.; Ostbye, T.; Speechley, M. Chronic widespread musculoskeletal pain with or without fibromyalgia: Psychological distress in a representative community adult sample. *J. Rheumatol.* **2002**, *29*, 588–594.
61. Clauw, D.J.; Arnold, L.M.; McCarberg, B.H. The Science of Fibromyalgia. *Mayo Clin. Proc.* **2011**, *86*, 907–911. [CrossRef]
62. Desmeules, J.A.; Cedraschi, C.; Rapiti, E.; Baumgartner, E.; Finckh, A.; Cohen, P.; Dayer, P.; Vischer, T.L. Neurophysiologic evidence for a central sensitization in patients with fibromyalgia. *Arthritis Rheum.* **2003**, *48*, 1420–1429. [CrossRef]
63. Alnigenis, M.N.; Barland, P. Fibromyalgia syndrome and serotonin. *Clin. Exp. Rheumatol.* **2001**, *19*, 205–210.
64. Fürst, S. Transmitters involved in antinociception in the spinal cord. *Brain Res. Bull.* **1999**, *48*, 129–141. [CrossRef]
65. Pertovaara, A. Noradrenergic pain modulation. *Prog. Neurobiol.* **2006**, *80*, 53–83. [CrossRef]
66. Kendroud, S.; Fitzgerald, L.A.; Murray, I.V.; Hanna, A. *Physiology, Nociceptive Pathways*; StatPearls Publishing: Treasure Island, FL, USA, 2022. [PubMed]
67. O'mahony, L.F.; Srivastava, A.; Mehta, P.; Ciurtin, C. Is fibromyalgia associated with a unique cytokine profile? A systematic review and meta-analysis. *Rheumatology* **2021**, *60*, 2602–2614. [CrossRef]
68. Bai, B.; Chen, H. Metformin: A Novel Weapon Against Inflammation. *Front. Pharmacol.* **2021**, *12*, 622262. [CrossRef]
69. Xiang, H.-C.; Lin, L.-X.; Hu, X.-F.; Zhu, H.; Li, H.-P.; Zhang, R.-Y.; Hu, L.; Liu, W.-T.; Zhao, Y.-L.; Shu, Y.; et al. AMPK activation attenuates inflammatory pain through inhibiting NF-κB activation and IL-1β expression. *J. Neuroinflammation* **2019**, *16*, 34. [CrossRef]
70. Liu, G.; Wang, F.; Feng, Y.; Tang, H. Metformin Inhibits NLRP3 Inflammasome Expression and Regulates Inflammatory Microenvironment to Delay the Progression of Colorectal Cancer. *Recent Pat. Anticancer Drug Discov.* **2024**, *19*, 1–11. [CrossRef]
71. Smallwood, R.F.; Laird, A.R.; Ramage, A.E.; Parkinson, A.L.; Lewis, J.; Clauw, D.J.; Williams, D.A.; Schmidt-Wilcke, T.; Farrell, M.J.; Eickhoff, S.B.; et al. Structural Brain Anomalies and Chronic Pain: A Quantitative Meta-Analysis of Gray Matter Volume. *J. Pain* **2013**, *14*, 663–675. [CrossRef]
72. Abd-Ellatief, R.B.; Mohamed, H.K.; Kotb, H.I. Reactive Astrogliosis in an Experimental Model of Fibromyalgia: Effect of Dexmedetomidine. *Cells Tissues Organs* **2018**, *205*, 105–119. [CrossRef] [PubMed]
73. Emad, Y.; Ragab, Y.; Zeinhom, F.; El-Khouly, G.; Abou-Zeid, A.; Rasker, J.J. Hippocampus dysfunction may explain symptoms of fibromyalgia syndrome. A study with single-voxel magnetic resonance spectroscopy. *J. Rheumatol.* **2008**, *35*, 1371–1377. [PubMed]
74. McCrae, C.S.; O'shea, A.M.; Boissoneault, J.; Vatthauer, K.; Robinson, M.; Staud, R.; Perlstein, W.M.; Craggs, J.G.; Center, G.M.R.V.A.M. Fibromyalgia patients have reduced hippocampal volume compared with healthy controls. *J. Pain Res.* **2015**, *8*, 47–52. [CrossRef] [PubMed]
75. Muñoz-Arenas, G.; Pulido, G.; Treviño, S.; Vázquez-Roque, R.; Flores, G.; Moran, C.; Handal-Silva, A.; Guevara, J.; Venegas, B.; Díaz, A. Effects of metformin on recognition memory and hippocampal neuroplasticity in rats with metabolic syndrome. *Synapse* **2020**, *74*, e22153. [CrossRef]
76. Oliveira, W.H.; Nunes, A.K.; França, M.E.R.; Santos, L.A.; Lós, D.B.; Rocha, S.W.; Barbosa, K.P.; Rodrigues, G.B.; Peixoto, C.A. Effects of metformin on inflammation and short-term memory in streptozotocin-induced diabetic mice. *Brain Res.* **2016**, *1644*, 149–160. [CrossRef]
77. Zhang, W.; Zhao, L.; Zhang, J.; Li, P.; Lv, Z. Metformin improves cognitive impairment in diabetic mice induced by a combination of streptozotocin and isoflurane anesthesia. *Bioengineered* **2021**, *12*, 10982–10993. [CrossRef]
78. Ferrarini, E.G.; Gonçalves, E.C.D.; Menegasso, J.F.; Rabelo, B.D.; Felipetti, F.A.; Dutra, R.C. Exercise Reduces Pain and Deleterious Histological Effects in Fibromyalgia-like Model. *Neuroscience* **2021**, *465*, 46–59. [CrossRef]
79. Todd, A.J. Neuronal circuitry for pain processing in the dorsal horn. *Nat. Rev. Neurosci.* **2010**, *11*, 823–836. [CrossRef]
80. Willard, S.S.; Koochekpour, S. Glutamate, Glutamate Receptors, and Downstream Signaling Pathways. *Int. J. Biol. Sci.* **2013**, *9*, 948–959. [CrossRef]

81. Yao, X.; Li, L.; Kandhare, A.D.; Mukherjee-Kandhare, A.A.; Bodhankar, S.L. Attenuation of reserpine-induced fibromyalgia via ROS and serotonergic pathway modulation by fisetin, a plant flavonoid polyphenol. *Exp. Ther. Med.* **2020**, *19*, 1343–1355. [CrossRef] [PubMed]
82. Liang, X.; Giacomini, K.M. Transporters Involved in Metformin Pharmacokinetics and Treatment Response. *J. Pharm. Sci.* **2017**, *106*, 2245–2250. [CrossRef] [PubMed]
83. Urakami, Y.; Okuda, M.; Saito, H.; Inui, K.-I. Hormonal regulation of organic cation transporter OCT2 expression in rat kidney. *FEBS Lett.* **2000**, *473*, 173–176. [CrossRef] [PubMed]
84. Alnouti, Y.; Petrick, J.S.; Klaassen, C.D. Tissue distribution and ontogeny of organic cation transporters in mice. *Drug Metab. Dispos.* **2005**, *34*, 477–482. [CrossRef]
85. De Angelis, F.; Vacca, V.; Tofanicchio, J.; Strimpakos, G.; Giacovazzo, G.; Pavone, F.; Coccurello, R.; Marinelli, S. Sex Differences in Neuropathy: The Paradigmatic Case of MetFormin. *Int. J. Mol. Sci.* **2022**, *23*, 14503. [CrossRef]
86. Dalla, C.; Pavlidi, P.; Sakelliadou, D.-G.; Grammatikopoulou, T.; Kokras, N. Sex Differences in Blood–Brain Barrier Transport of Psychotropic Drugs. *Front. Behav. Neurosci.* **2022**, *16*, 844916. [CrossRef]
87. Mucci, V.; Demori, I.; Browne, C.J.; Deblieck, C.; Burlando, B. Fibromyalgia in Pregnancy: Neuro-Endocrine Fluctuations Provide Insight into Pathophysiology and Neuromodulation Treatment. *Biomedicines* **2023**, *11*, 615. [CrossRef]

cells

MDPI

Communication

Effects of Tryptophan and Physical Exercise on the Modulation of Mechanical Hypersensitivity in a Fibromyalgia-like Model in Female Rats

Rafael Marins Rezende [1], Roney Santos Coimbra [2], Markus Kohlhoff [2], Lukiya Silva Campos Favarato [3], Hércia Stampini Duarte Martino [4], Luciano Bernardes Leite [5,6], Leoncio Lopes Soares [5], Samuel Encarnação [7,8,9], Pedro Forte [6,9,10,11], António Miguel de Barros Monteiro [6,9], Maria do Carmo Gouveia Peluzio [4] and Antônio José Natali [5,*]

1 Department of Physiotherapy, Universidade Federal de Juiz de Fora, Governador Valadares 35020-360, MG, Brazil; faelfisio@yahoo.com.br
2 Instituto Rene Rachou–Fiocurz Minas, Belo Horizonte 30190-009, MG, Brazil; roney.coimbra@fiocruz.br (R.S.C.); markus.kohlhoff@fiocruz.br (M.K.)
3 Department of Veterinary Medicine, Universidade Federal de Viçosa, Viçosa 36570-900, MG, Brazil; lscampos@ufv.br
4 Department of Nutrition and Health, Universidade Federal de Viçosa, Viçosa 36570-900, MG, Brazil; hercia@ufv.br (H.S.D.M.); mpeluzio@ufv.br (M.d.C.G.P.)
5 Exercise Biology Laboratory, Department of Physical Education, Universidade Federal de Viçosa, Viçosa 36570-900, MG, Brazil; luciano.leite@ufv.br (L.B.L.); leoncio.lopes@ufv.br (L.L.S.)
6 Department of Sports Sciences, Instituto Politécnico de Bragança, 5300-253 Bragança, Portugal; pedromiguelforte@gmail.com (P.F.); mmonteiro@ipb.pt (A.M.d.B.M.)
7 Instituto Politécnico de Bragança, Campus de Santa Apolónia, 5300-253 Bragança, Portugal; samuel01.encarnacao@gmail.com
8 Department of Physical Education, Sport and Human Movement, Universidad Autónoma de Madrid, 28049 Madrid, Spain
9 Research Centre for Active Living and Wellbeing (Livewell), Instituto Politécnico de Bragança, 5300-253 Bragança, Portugal
10 CI-ISCE, Instituto Superior de Ciências Educativas do Douro (ISCE Douro), 4560-547 Penafiel, Portugal
11 Research Centre in Sports Sciences, Health Sciences and Human Development, 5001-801 Vila Real, Portugal
* Correspondence: anatali@ufv.br

Abstract: Though the mechanisms are not fully understood, tryptophan (Trp) and physical exercise seem to regulate mechanical hypersensitivity in fibromyalgia. Here, we tested the impact of Trp supplementation and continuous low-intensity aerobic exercise on the modulation of mechanical hypersensitivity in a fibromyalgia-like model induced by acid saline in female rats. Twelve-month-old female Wistar rats were randomly divided into groups: [control (n = 6); acid saline (n = 6); acid saline + exercise (n = 6); acid saline + Trp (n = 6); and acid saline + exercise + Trp (n = 6)]. Hypersensitivity was caused using two intramuscular jabs of acid saline (20 μL; pH 4.0; right gastrocnemius), 3 days apart. The tryptophan-supplemented diet contained 7.6 g/hg of Trp. The three-week exercise consisted of progressive (30–45 min) treadmill running at 50 to 60% intensity, five times (Monday to Friday) per week. We found that acid saline induced contralateral mechanical hypersensitivity without changing the levels of Trp, serotonin (5-HT), and kynurenine (KYN) in the brain. Hypersensitivity was reduced by exercise (~150%), Trp (~67%), and its combination (~160%). The Trp supplementation increased the levels of Trp and KYN in the brain, and the activity of indoleamine 2,3-dioxygenase (IDO), and decreased the ratio 5-HT:KYN. Exercise did not impact the assessed metabolites. Combining the treatments reduced neither hypersensitivity nor the levels of serotonin and Trp in the brain. In conclusion, mechanical hypersensitivity induced by acid saline in a fibromyalgia-like model in female rats is modulated by Trp supplementation, which increases IDO activity and leads to improved Trp metabolism via the KYN pathway. In contrast, physical exercise does not affect mechanical hypersensitivity through brain Trp metabolism via either the KYN or serotonin pathways. Because this is a short study, generalizing its findings warrants caution.

Keywords: fibromyalgia; treadmill running; brain; hypersensitivity; tryptophan; serotonin; kynurenine; indolamine

1. Introduction

A syndrome of persistent and diffuse pain defines fibromyalgia. Two to five percent of the population experience such discomfort, especially middle-aged and old women [1,2]. Fibromyalgia is connected to deficits in the endogenous systems that modulate pain [3–5]. For instance, because efferent serotoninergic neurons negatively modulate the levels of substance P, low and high levels of serotonin and substance P, respectively, in the fluid in the brain ventricles and surrounding the spinal cord of individuals with fibromyalgia indicate modifications in the central neurotransmissions, thus resulting in alterations in the central response to pain [6–9].

Since tryptophan (Trp) is an originator of serotonin, disruptions in Trp metabolism appear to play an important part in the etiology of fibromyalgia. Trp metabolism has two main pathways that result in two different metabolites: serotonin (5-HT) and kynurenine (KYN). While the breakdown of 5-HT produces 5-hidroxiindolacetic acid (5-HIAA), KYN is further converted into neuroactive metabolites (i.e., quinolinic acid) [10]. An inverse relationship between the brain levels of Trp, 5-HT, 5-HIAA, and clinical pain measures and a positive interrelationship between the levels of KYN and the perception of pain are suggested [11–13].

An animal model of acid saline-induced fibromyalgia has been used to evaluate prophylactic and curative therapies and to investigate its underlying mechanisms. For example, diet and physical exercise have been tested as adjuvant non-pharmacological strategies to care for fibromyalgia. Concerning diet, a diet supplemented with Trp has been tested in acid saline-induced fibromyalgia to augment its availability in the brain, which would boost the modulatory actions of the serotoninergic system in pain control [14–18]. Supplementation with Trp thus appears to improve serotonin function in the brain, which reduces cortisol secretion during situations of high stress [19]. Regarding physical exercise, previous studies demonstrated that animals with acid saline-induced fibromyalgia had their mechanical hyperalgesia attenuated after performing aerobic exercise regimes [20–22]. It appears that aerobic exercise activates endogenous inhibitory mechanisms, including opioids and serotonin, and reduces the systemic levels of biomarkers for anxiety (e.g., cortisol, norepinephrine) and infection (e.g., cytokines).

Despite such evidence, the efficiency of supplementation with Trp and aerobic exercise on the modulation of mechanical hypersensitivity is not well understood and requires further investigation. Thus, we evaluate the impact of a Trp-supplemented diet and continuous low-intensity aerobic exercise on modulating mechanical hypersensitivity in a fibromyalgia-like model in female rats.

2. Materials and Methods

2.1. Animals and Study Design

Because fibromyalgia is experienced mainly by middle-aged and old women [1,2], female Wistar rats at 12 months of age were assigned to one of the following groups: C (control, n = 6); AS (acid saline, n = 6); ASE (acid saline + exercise, n = 6); ASTrp (acid saline + Trp, n = 6); and ASETrp (acid saline + exercise + Trp, n = 6). Throughout experiments (21 days), rats were housed individually in cages in the animal house (22 ± 2 °C), with a 12/12 h light/dark cycle and water and food ad libitum (see below). One day after the acid saline injection, exercised rats underwent a continuous aerobic exercise program (see below). All procedures in the experiments followed the National Guidance for the Care and Use of Laboratory Animals and had the consent of the Institutional Ethics Committee (CEUA-UFV 21/2015).

2.2. Induction of Hypersensitivity

Hypersensitivity was induced as described by Sluka et al. [23]. The rats were kept in sedation using isoflurane, 100% FiO2, with instinctive ventilation. Rats in the AS groups were injected two doses of acid saline (20 µL; pH 4.0) intramuscularly in the right hind limb (gastrocnemius) 3 days apart. The rats used as controls were injected with the same dose of saline (pH 7.4).

2.3. Dietary Supplementation

Rats were given water and food ad libitum over the experiment (21 days). Purified diets (i.e., control and supplemented) were pelletized (RHOSTER, São Paulo, Brazil). Food ingestion and efficiency and weight gain were determined. Diet composition followed the AIN-93M (American Institute of Nutrition, Rockville, MD, USA) for maintaining adult rats [24]. The increase in the levels of Trp in the casein amino acid profile was adapted (RHOSTER, São Paulo, Brazil) based on studies previously reported [15–17]. Rats receiving a diet supplemented with Trp had 7.6 g of Trp/kg, whereas those receiving a control diet had 2.5 g of Trp/kg.

2.4. Exercise Training Protocol

The maximum running speed (MRS) was determined using a test on a treadmill (AVS, São Paulo-SP, Brazil), as described elsewhere [25]. The MRS reached by the rats was considered 100% to calculate the running intensity in training. The low-intensity continuous aerobic exercise protocol (Modified, Sharma et al. [26]) was performed five times/wk for 21 days. In week one, the rats exercised at the speed of 10 m/min (50% MRS) for 30 min. During week 2, the rats exercised at 11 m/min (55% MRS) for min. During week 3, the rats exercise at 12 m/min (60% MRS), for 45 min.

2.5. Mechanical Hypersensitivity Test

Mechanical hypersensitivity was determined using a von Frey filament (Insight, Ribeirao Preto—SP, Brazil) as described by others [20,23]. In summary, after 30 min of adaptation to the equipment, hypersensitivity was assessed in the contralateral (i.e., left) hind paw. The stimulus was imposed 3 times with an interval of 30 s. The average value of 3 measurements was considered the threshold of the paw's withdrawal (g) and adopted as our mechanical hypersensitivity index. Hypersensitivity was assessed before (baseline) and 1, 7 and 21 days after the injections. These evaluations were carried out in an appropriate room (i.e., silent and controlled temperature), on the same weekday (i.e., Wednesdays), and on time (i.e., 8 to 10 a.m.) by one unique evaluator blind to the treatments.

2.6. Sample Collection

Two days after the final exercise training session, the rats were killed in an appropriate room (i.e., clean and quiet). Each group was euthanized on different days of the week, between 8 and 10 a.m. After decapitation, a sample of blood was harvested and centrifugalized at 704 g for 10 min. The serum was separated for further analyses of substance P. To analyze the levels of Trp, 5-HT and KYN, the brain was immediately removed and saved at −80 °C. One brain sample was missed in groups C and ASETrp during the harvesting.

2.7. Determination of Brain Tryptophan, Serotonin and Kynurenine Levels

Whole brains were homogenized in three volumes of cold methanol containing phenacetin (100 ng/mL) as a control for the extraction procedure (weight per volume), vortexed and centrifugalized at 13,000× g for 10 min (4 °C). Activated resin C18 was added to the supernatant (1/5 of the sample weight—10 mg) to remove hydrophobic compounds (fatty acids, remaining proteins). This mixture was then vortexed and centrifugalized again. The buoyant was passed on to another tube, and the solvent was separated using a SpeedVac (Thermo Fisher Scientific, Waltham, MA, USA). The specimens were reconstructed in the same amount of ultrapure water, passed on to polypropylene flasks (100 µL) and deposited in a chilled autosampler (5 °C). The standards of Trp, 5-HT and KYN (96% purity), as well as the other chemicals, were acquired in a Sigma Aldrich (St. Louis, MO, USA) store. We used a Nexera ultra high-performance liquid chromatography (UHPLC) machine (Shimadzu, Kyoto, Japan) coupled with a Shimadzu Shim-Pack XR-ODSIII (C18, 2.2 μm, 80 Å, 2.0 × 150 mm) column, hyphenated to a maXis ETD high-resolution Electrospray ionization quadrupole time-of-flight (ESI-QTOF) mass spectrometer (Bruker, Billerica, MA, USA), and regulated by the Compass 1.7 software package (Bruker) as reported by Danielski et al. [27] for the liquid chromatography coupled with mass spectrometry tests. The KYN/Trp, 5-HT/KYN and 5-HT/Trp ratios were calculated to assess the participation of the kynurenine and serotonin pathways in the brain Trp metabolism [28–30].

2.8. Substance P Determination

The serum levels of substance P were measured via commercially available ELISA kits (CEA393RA Substance P, Uscn®, Wuhan, China), following the manufacturer's instructions. After the preparatory procedures, the yellow color generated was read at 405 nm (Stat Fax 2100; Awareness Technology, Palm City, FL, USA), and the levels of substance P were expressed as pg/mL.

2.9. Statistical Analyses

Data were checked for normality using the Kolmogorov–Smirnov test. The mechanical hypersensitivity data of the experimental groups were compared utilizing two-way repeated measures ANOVA and the post hoc test of Bonferroni. Data for the levels of Trp, 5-HT, KYN and substance P of the experimental groups were compared utilizing one-way ANOVA and Tukey's post hoc test. All analyses used the SPSS software (IBM SPSS Statistics for Windows, Version 19.0. Armonk, NY: IBM Corp) and p values ≤ 0.05 denote statistical significance.

3. Results

The contralateral mechanical hypersensitivity was lower in rats injected with acid saline than in control rats at one day post-induction (Table 1). Aerobic exercise reduced the levels of hypersensitivity of the control rats on the 7th and 21st days post-induction. At these time points, supplementation with Trp diminished hypersensitivity, although to a minor degree. Combining the two interventions did not add to the benefit observed with exercise alone.

Table 1. Mechanical hypersensitivity in the contralateral paw measured by von Frey filaments.

	Baseline	1 Day Post-Injection	7 Days Post-Injection	21 Days Post-Injection
C (g)	36.69 ± 0.76	36.44 ± 2.45	35.24 ± 0.33	36.11 ± 1.38
AS (g)	35.95 ± 0.84	17.14 ± 1.36 #	14.36 ± 1.39 #	14.92 ± 2.09 #
ASE (g)	37.09 ± 0.45	13.42 ± 1.49 #	38.01 ± 0.79 *	36.54 ± 0.99 *
ASTrp (g)	36.98 ± 0.35	12.81 ± 1.72 #	18.49 ± 3.99 *#	21.89 ± 3.69 *#
ASETrp (g)	36.32 ± 0.46	11.64 ± 1.27 #	36.96 ± 0.41 *	38.15 ± 1.82 *

Data are means ± SD; C, control; AS, acid saline; ASE, acid saline + exercise; ASTrp, acid saline + Trp supplementation; ASETrp, acid saline + exercise and Trp supplementation. # $p < 0.05$ vs. C group. * $p < 0.05$ vs. AS group.

Table 2 presents the monoisotopic masses of molecular ions and in-source fragments of Trp metabolites.

After 21 days of treatments, the brain levels of Trp, 5-HT and KYN did not alter when comparing control and AS groups (Figure 1). Concerning the impact of interventions, rats from the ASTrp group displayed higher levels of Trp than the ASE and ASETrp groups (Figure 1A). The levels of 5-HT, however, did not alter between groups (Figure 1B). Rats in the ASTrp and ASETrp groups presented higher levels of KYN (Figure 1C) than those in the ASE group.

Table 2. Monoisotopic masses of molecular ions and in-source fragments of Trp metabolites.

Metabolite	*m*/*z* [M + H]$^+$	*m*/*z* In-Source Fragments
Tryptophan	205.0972	188.071 (100)[1]
Kynurenine	209.0920	192.066 (37). 136.076 (15). 94.065 (8)
Serotonin	177.1022	160.076 (280)

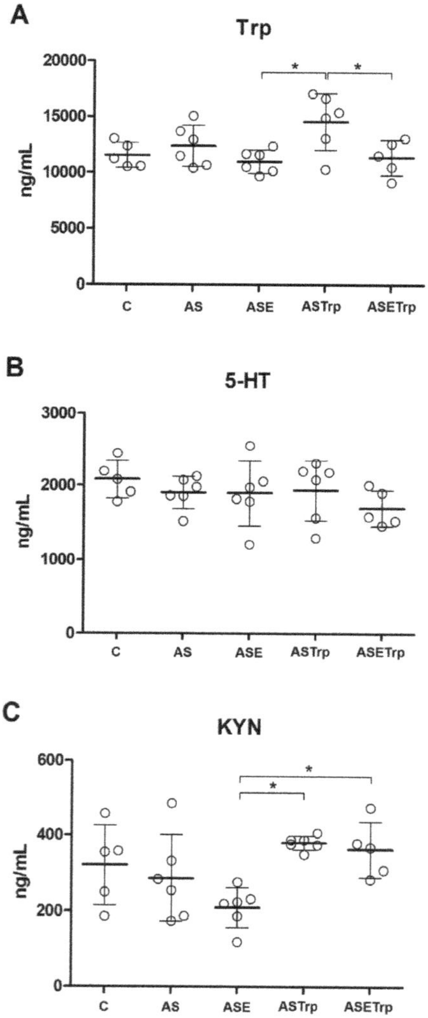

Figure 1. The brain concentrations of tryptophan (**A**) are higher in rats supplemented with tryptophan than in the ASE and ASETrp groups. The brain concentrations of serotonin (**B**) do not differ between groups. The brain concentrations of kynurenine (**C**) are lower in the ASE group than in the ASTrp and ASETrp groups. Trp, tryptophan. 5-HT, serotonin. KYN, kynurenine. C, control; AS, acid saline; ASE, acid saline + exercise; ASETrp, acid saline + exercise + tryptophan; ASTrp, acid saline + tryptophan. Data are means ± SD. * $p < 0.032$.

Despite the differences in the brain levels of Trp, curiously, the activity of IDO (i.e., KYN:Trp ratio) (Figure 2A) and the 5-HT:KYN ratio (Figure 2B) did not differ between the

ASTrp and ASETrp groups. However, these groups had higher brain KYN concentrations (see Figure 1C), higher inferred IDO activity (Figure 2A) and a lower 5-HT:KYN ratio (Figure 2B) than the ASE group. The 5-HT:Trp ratio was not statistically distinct between groups (Figure 2C).

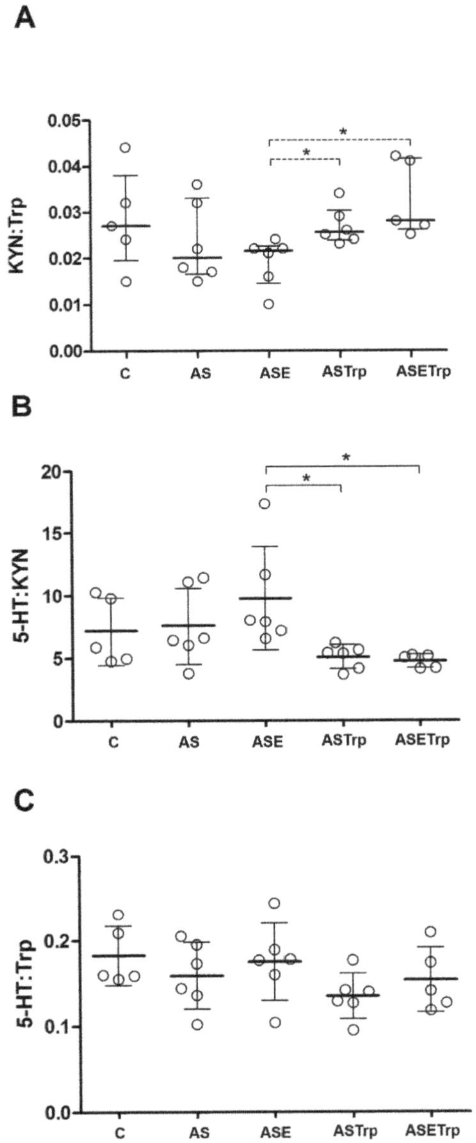

Figure 2. The indolamine 2,3-dioxygenase activity in the brain inferred by the KYN:Trp ratio (**A**) is higher in the ASTrp and ASETrp group than in the ASE group. The brain 5-HT:KYN ratio (**B**) is lower in the ASTrp and ASETrp than in the ASE group. The brain 5-HT:Trp ratio (**C**) is not different between groups. Trp, tryptophan. 5-HT, serotonin. KYN, kynurenine. C, control; AS, acid saline; ASE, acid saline + exercise; ASETrp, acid saline + exercise + tryptophan; ASTrp, acid saline + tryptophan. Data are means ± SD. * $p < 0.048$.

After the period of interferences, the serum levels of substance P were not statistically different amongst the groups in this experiment (Figure 3).

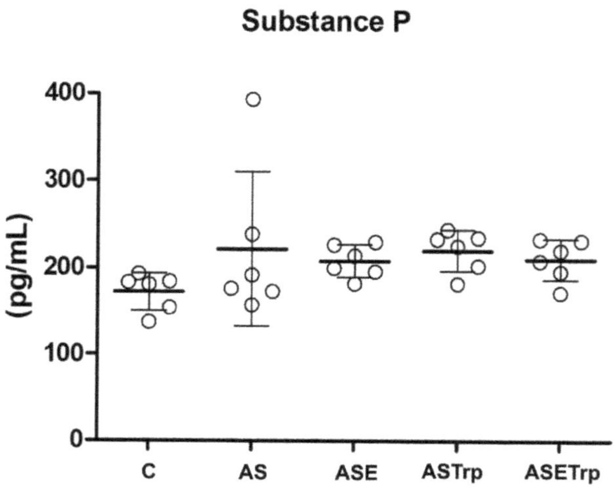

Figure 3. The serum concentrations of substance P are not different between groups. C, control; AS, acid saline; ASE, acid saline + exercise; ASETrp, acid saline + exercise + tryptophan; ASTrp, acid saline + tryptophan. Data are means ± SD.

4. Discussion

We submitted female rats injected with acid saline to supplementation with Trp and a low-intensity continuous aerobic exercise, either solely or in conjunction, for 21 days to test its effects on the modulation of mechanical hypersensitivity.

The acid saline model used here caused contralateral mechanical hypersensitivity for around 21 days. Nevertheless, our results demonstrate that the brain levels of Trp, 5-HT, and KYN were not influenced (C = AS). Such findings indicate that mechanical hypersensitivity may not be related to disruptions in the brain metabolism of Trp and hence in its metabolites' homeostasis in the studied model. Nonetheless, our results are in alignment with the hypothesis that the model for chronic pain induced by acid saline reduces the central concentration of serotonin and augments that of KYN, as reported elsewhere [7,21]. In a previous study [20], however, increased levels of muscle interleukin 6 (IL-6) and serum cortisol in the acid saline model were reported, which suggests that the HPA (hypothalamic–pituitary–adrenal) axis is predominantly affected. Indeed, there is evidence that high serum concentrations of cortisol associated with stress might worsen ache in individuals with fibromyalgia [31] and potentialize the pro-nociceptive influence of inflammatory markers like IL-6 and tumor necrosis factor-alpha (TNF-α) [32,33]. Moreover, here, we noted that the serum levels of substance P were similar in both the control and AS groups. Such results indicate that the serotoninergic modulation system appears not to be affected by acid saline-induced hypersensitivity since substance P concentrations are negatively modulated by efferent serotoninergic neurons [9,34].

Regarding the Trp supplementation effects, it significantly reduced (~67%) the hypersensitivity in the ASTrp group (ASTrp < AS), and in the brain it augmented the levels of Trp in animals that had not exercised (ASTrp > ASE and ASETrp). In this experiment, the ratio of Trp to large neutral amino acid (LNAA) in the diet supplemented with Trp (Trp/LNAA: 15.32) was higher in comparison with that in the control aliment (Trp/LNAA: 5.4). It seems, therefore, that Trp addition favored its influx to the brain.

Despite the increased brain Trp concentrations, the levels of 5-HT in the ASTrp group did not augment, which diverges from other studies [35,36]. This result must be interpreted with caution. The elevation in the brain 5-HT concentration seems to occur in specific regions of pain modulation by the serotoninergic system (i.e., hypothalamus, ventromedial rostral bulb, brain stem magnum raphe nucleus) [37,38]. Thus, the dilution of these brain structures rich in this neurotransmitter by the total brain mass to produce homogenates has masked some differences between groups in this work.

Although the brain Trp concentrations were different among experimental groups (ASTrp > ASE and ASETrp), curiously, the levels of KYN, the activity of IDO and the 5-HT/Trp ratio were not divergent between the ASTrp and ASETrp groups. These groups had higher brain KYN and IDO activity levels and a lower 5-HT/KYN ratio than the ASE group. It is conceivable that mediators of inflammation, like interleukin 1β, IL-6 and TNF-α, that are elevated in the brain of animals with acid saline-induced fibromyalgia [39] may activate IDO [40], which, in the presence of an excess of Trp, would lead to increased KYN levels in the brain. This hypothesis is supported by the literature since around 95% of Trp in the diet is metabolized via the KYN pathway. In comparison, only 1–2% are processed via the serotonin pathway [41].

Concerning the aerobic exercise employed here, our results show its efficacy in restoring the hypersensitivity to the degrees of control animals. Regular aerobic exercise is suggested to augment the plasma levels of free Trp (Trp-F). In contrast, the levels of LNAA are diminished because of its elevated muscular absorption and metabolism during exercise [42]. Therefore, the ratio Trp-F/LNAA increases, which results in elevated Trp-F influx towards the brain, leading to an increase in the levels of serotonin in the brain [43,44]. Endogenous opioids and serotonin are suggested to play important roles in analgesia produced by exercise in humans and animals [45,46]. However, we found no increase in the brain levels of Trp and 5-HT in animals with hypersensitivity who underwent the exercise program. It is conceivable that the restored hypersensitivity demonstrated here is linked to the opposed influence of the applied exercise program to inflammation (i.e., reduced muscle IL-6) and to stress (i.e., reduced cortisol) [14,20], which indicates normalization of the HPA axis activity.

The association of interventions, nevertheless, did not diminish hypersensitivity notably (~11%), in comparison with exercise alone (ASE = ASETrp). In addition, the brain levels of Trp and 5-HT were not influenced by the combined interventions. Animals from the ASETrp group had probably used the excess of Trp given by supplementation as an energy source to exercise by converting Trp into acetyl-CoA via Trp metabolism (i.e., Kegg:map00380). Acetyl-CoA is fuel to the Krebs cycle and, hence, oxidative phosphorylation. Thus, less Trp would reach these animals' brains.

Finally, this study has limitations. First, we did not determine the levels of 3-hydroxykykynurenine, which could indicate the possible neuroprotective effect of the treatments in the model using the kynurenic acid to 3-hydroxykykynurenine ratio. Moreover, the acid quinolinic acid levels, which are important in the KYN pathway, were not measured. Such analyses would help to have a clear window into the fibromyalgia-like model in female rats. And second, we did not evaluate spontaneous behavior like locomotor activity in the animals. Such measurement would add important information about the effects of the treatments on the used model. Therefore, we encourage the inclusion of assessments of these parameters in future studies.

5. Conclusions

In conclusion, mechanical hypersensitivity in the studied fibromyalgia-like model in female rats appears to be modulated by Trp supplementation because of increased brain IDO activity that leads to improved Trp metabolism via the KYN pathway. However, mechanical hypersensitivity is not modulated by low-intensity continuous aerobic exercise through brain Trp metabolism via the KYN or serotonin pathways. Finally, the conjunction of interventions does not have positive synergic actions in modulating mechanical hypersensitivity in female rats. Because this is a short study, generalizing its findings warrants caution.

Author Contributions: Conceptualization and essentially intellectual contribution, R.M.R., A.J.N. and R.S.C.; Research design and methodology, R.M.R., A.J.N., M.d.C.G.P. and H.S.D.M.; Data curation, formal data analysis and investigation, L.S.C.F., R.M.R., R.S.C. and M.K.; Project administration, R.M.R. and A.J.N.; Validation, R.M.R., R.S.C., A.J.N. and M.K.; Funding acquisition, A.J.N.; Writing—original draft, R.M.R. and A.J.N.; Writing—review and editing, A.J.N., R.S.C., L.B.L., L.L.S., S.E., P.F. and A.M.d.B.M.; Supervision, R.M.R. and A.J.N. All authors have read and agreed to the published version of the manuscript.

Funding: This study was financed partially by the Higher Education Personnel Improvement Coordination (CAPES-Brazil; Finance Code: PROEX/683/2018) and the Research Support Foundation of the State of Minas Gerais (FAPEMIG-Brazil; Grant number: CDS-APQ-01635-15). AJ Natali, MCG Peluzio and HSD Martino are thankful to the National Council for Scientific and Technological Development (CNPq-Brasil) for the fellowship.

Institutional Review Board Statement: The animal study protocol was approved by the Animal Use Ethics Committee of the Federal University of Viçosa (CEUA-UFV 21/2015).

Informed Consent Statement: Not applicable.

Data Availability Statement: The data will be shared on reasonable request to the corresponding author.

Conflicts of Interest: The authors declare no conflicts of interest.

References

1. Wolfe, F.; Brähler, E.; Hinz, A.; Häuser, W. Fibromyalgia prevalence, somatic symptom reporting, and the dimensionality of polysymptomatic distress: Results from a survey of the general population. *Arthritis Care Res.* **2013**, *65*, 777–785. [CrossRef] [PubMed]
2. Plesner, K.B.; Vaeger, H.B. Symptoms of fibromyalgia according to the 2016 revised fibromyalgia criteria in chronic pain patients referred to multidisciplinary pain rehabilitation: Influence on clinical and experimental pain sensitivity. *J. Pain* **2018**, *19*, 777–786. [CrossRef] [PubMed]
3. Harbeck, B.; Süfke, S.; Harten, P.; Haas, C.; Lehnert, H.; Mönig, H. High prevalence of fibromyalgia-associated symptoms in patients with hypothalamic-pituitary disorders. *Clin. Exp. Rheumatol.* **2012**, *31*, S16–S21.
4. Tak, L.M.; Cleare, A.J.; Ormel, J.; Manoharan, A.; Kok, I.C.; Wessely, S.; Rosmalen, J.G. Meta-analysis and meta-regression of hypothalamic-pituitary-adrenal axis activity in functional somatic disorders. *Biol. Psychol.* **2011**, *87*, 183–194. [CrossRef] [PubMed]
5. Dadabhoy, D.; Crofford, L.J.; Spaeth, M.; Russell, I.J.; Clauw, D.J. Biology and therapy of fibromyalgia: Evidence-based biomarkers for fibromyalgia syndrome. *Arthritis Res. Ther.* **2008**, *10*, 211. [CrossRef] [PubMed]
6. Karlsson, B.; Burell, G.; Kristiansson, P.; Björkegren, K.; Nyberg, F.; Svärdsudd, K. Decline of substance P levels after stress management with cognitive behaviour therapy in women with the fibromyalgia syndrome. *Scand. J. Pain.* **2019**, *19*, 473–482. [CrossRef]
7. Russell, I.J. Neurochemical pathogenesis of fibromyalgia syndrome. *Z. Rheumatol.* **1998**, *57*, S63–S66. [CrossRef]
8. Russell, I.J.; Orr, M.D.; Littman, B.; Vipraio, G.A.; Alboukrek, D.; Michalek, J.E.; Lopez, Y.; Mackillip, F. Elevated cerebrospinal fluid levels of substance P in patients with the fibromyalgia syndrome. *Arthritis Rheum.* **1994**, *37*, 1593–1601. [CrossRef]
9. Larson, A.A.; Igwe, O.J.; Seybold, V.S. Effects of lysergic acid diethylamide (LSD) and adjuvant-induced inflammation on desensitization to and metabolism of substance P in the mouse spinal cord. *Pain* **1989**, *37*, 365–373. [CrossRef]
10. Schwarz, M.J.; Offenbaecher, M.; Neumeister, A.; Ewert, T.; Willeit, M.; Praschak-Rieder, N.; Zach, J.; Zacherl, M.; Lossau, K.; Weisser, R.; et al. Evidence for an altered Trp metabolism in fibromyalgia. *Neurobiol. Dis.* **2002**, *11*, 434–442. [CrossRef]
11. Schwarz, M.J.; Spaeth, M.; Bardorff, M.H.; Pongratz, D.; Bondy, B.; Ackenheil, M. Relationship of substance P, 5-hydroxyindoleacetic acid and tryptophan in serum of fibromyalgia patients. *Neurosci. Lett.* **1999**, *259*, 196–198. [CrossRef] [PubMed]
12. Russell, I.J.; Michalek, J.E.; Vipraio, G.A.; Fletcher, E.M.; Javors, M.A.; Bowden, C.A. Platelet 3H-imipramine uptake receptor density and serum serotonin levels in patients with fibromyalgia/fibrositis syndrome. *J. Rheumatol.* **1992**, *19*, 104–109. [PubMed]
13. Moldofsky, H. Rheumatic pain modulation syndrome: The interrelationships between sleep, central nervous system serotonin and pain. *Adv. Neurol.* **1982**, *33*, 51–57. [PubMed]
14. Gleeson, M.; Bishop, N.C.; Stensel, D.J.; Lindley, M.R.; Mastana, S.S.; Nimmo, M.A. The anti-inflammatory effect of exercise: Mechanisms and implications for the prevention and treatment of diseases. *Nat. Rev. Immunol.* **2011**, *11*, 607–615. [CrossRef]
15. Yunus, M.B.; Dailey, J.W.; Aldag, J.C.; Masi, A.T.; Jobe, P.C. Plasma tryptophan and other amino acids in primary fibromyalgia: A controlled study. *J. Rheumatol.* **1992**, *9*, 90–94.

16. Markus, C.R.; Olivier, B.; Panhuysen, G.E.; Van Der Gugten, J.; Alles, M.S.; Tuiten, A.; Westenberg, H.G.; Fekkes, D.; Koppeschaar, H.F.; de Haan, E.E. The bovine protein alpha-lactalbumin increases the plasma ratio of tryptophan to the other large neutral amino acids, and in vulnerable subjects raises brain serotonin activity, reduces cortisol concentration, and improves mood under stress. *Am. J. Clin. Nutr.* **2000**, *71*, 1536–1544. [CrossRef]
17. Markus, C.R.; Olivier, B.; de Haan, E.H. Whey protein rich in alpha-lactalbumin increases the ratio of plasma tryptophan to the sum of the other large neutral amino acids and improves cognitive performance in stress-vulnerable subjects. *Am. J. Clin. Nutr.* **2002**, *75*, 1051–1056. [CrossRef]
18. Schmitt, J.A.; Jorissen, B.L.; Dye, L.; Markus, C.R.; Deutz, N.E.; Riedel, W.J. Memory function in women with premenstrual complaints and the effect of serotonergic stimulation by acute administration of an alpha-lactalbumin protein. *J. Psychopharmacol.* **2005**, *19*, 375–384. [CrossRef]
19. Firk, C.; Markus, C.R. Mood and cortisol responses following tryptophan-rich hydrolyzed protein and acute stress in healthy subjects with high and low cognitive reactivity to depression. *Clin. Nutr.* **2009**, *28*, 266–271. [CrossRef]
20. Rezende, R.M.; Gouveia Pelúzio, M.D.; de Jesus Silva, F.; Della Lucia, E.M.; Silva Campos Favarato, L.; Stampini Duarte Martino, H.; Natali, A.J. Does aerobic exercise associated with tryptophan supplementation attenuates hyperalgesia and inflammation in female rats with experimental fibromyalgia? *PLoS ONE* **2019**, *14*, e0211824. [CrossRef]
21. Brito, R.G.; Rasmussen, L.A.; Sluka, K.A. Regular physical activity prevents development of chronic muscle pain through modulation of supraspinal opioid and serotonergic mechanisms. *Pain Rep.* **2017**, *2*, 618–630. [CrossRef] [PubMed]
22. Bote, M.E.; García, J.J.; Hinchado, M.D.; Ortega, E. Fibromyalgia. anti-Inflammatory and stress responses after acute moderate exercise. *PLoS ONE* **2013**, *8*, e74524. [CrossRef] [PubMed]
23. Sluka, K.A.; Kalra, A.; Moore, S.A. Unilateral intramuscular injections of acidic saline produce a bilateral, long-lasting hyperalgesia. *Muscle Nerve* **2001**, *24*, 37–46. [CrossRef] [PubMed]
24. Reeves, P.G.; Nielsen, F.H.; Fahey, G.C. AIN-93 purified diets for laboratory rodents: Final report of the American Institute of Nutrition ad hoc writing committee on the reformulation of the AIN-76A rodent diet. *J. Nutr.* **1993**, *123*, 1939–1951. [CrossRef] [PubMed]
25. Lacerda, A.C.; Marubayashi, U.; Balthazar, C.H.; Leite, L.H.; Coimbra, C.C. Central nitric oxide inhibition modifies metabolic adjustments induced by exercise in rats. *Neurosci. Lett.* **2005**, *410*, 152–156. [CrossRef]
26. Sharma, N.K.; Ryals, J.M.; Gajewski, B.J.; Wright, D.E. Aerobic exercise alters analgesia and neurotrophin-3 synthesis in an animal model of chronic widespread pain. *Phys. Ther.* **2010**, *90*, 714–725. [CrossRef]
27. Danielski, L.G.; Giustina, A.D.; Goldim, M.P.; Florentino, D.; Mathias, K.; Garbossa, L.; de Bona Schraiber, R.; Laurentino, A.O.; Goulart, M.; Michels, M.; et al. Vitamin B6 reduces neurochemical and long-term cognitive alterations after polymicrobial sepsis: Involvement of the kynurenine pathway modulation. *Mol. Neurobiol.* **2017**, *55*, 5255–5268. [CrossRef]
28. Myint, A.M.; Kim, Y.K. Network beyond IDO in psychiatric disorders: Revisiting neurodegeneration hypothesis. *Prog. Neuropsychopharmacol. Biol. Psychiatry* **2014**, *48*, 304–313. [CrossRef]
29. Miura, H.; Shirokawa, T.; Isobe, K.; Ozaki, N. Shifting the balance of brain tryptophan metabolism elicited by isolation housing and systemic administration of lipopolysaccharide in mice. *Stress* **2009**, *12*, 206–214. [CrossRef]
30. Braun, D.; Longman, R.S.; Albert, M.L. A two-step induction of indoleamine 2,3 dioxygenase (IDO) activity during dendritic-cell maturation. *Blood* **2005**, *106*, 2375–2381. [CrossRef]
31. Fischer, S.; Doerr, J.M.; Strahler, J.; Mewes, R.; Thieme, K.; Nater, U.M. Stress exacerbates pain in the everyday lives of women with fibromyalgia syndrome-the role of cortisol and alpha-amylase. *Psychoneuroendocrinology* **2015**, *63*, 68–77. [CrossRef] [PubMed]
32. Ross, R.L.; Jones, K.D.; Bennett, R.M.; Ward, R.L.; Druker, B.J.; Wood, L.J. Preliminary evidence of increased pain and elevated cytokines in fibromyalgia patients with defective growth hormone response to exercise. *Open Immunol. J.* **2010**, *3*, 9–18. [CrossRef] [PubMed]
33. Dina, O.A.; Levine, J.D.; Green, P.G. Enhanced cytokine-induced mechanical hyperalgesia in skeletal muscle produced by a novel mechanism in rats exposed to unpredictable sound stress. *Eur. J. Pain.* **2011**, *15*, 796–800. [CrossRef] [PubMed]
34. Naranjo, J.R.; Arnedo, A.; Molinero, M.T.; Del, R.J. Involvement of spinal monoaminergic pathways in antinociception produced by substance P and neurotensin in rodents. *Neuropharmacology* **1989**, *28*, 291–298. [CrossRef] [PubMed]
35. Markus, C.R. Dietary amino acids and brain serotonin function; implications for stress-related affective changes. *Neuromol Med.* **2008**, *10*, 247–258. [CrossRef]
36. Orosco, M.; Rouch, C.; Beslot, F.; Feurte, S.; Regnault, A.; Dauge, V. Alpha-lactalbuminenriched diets enhance serotonin release and induce anxiolytic and rewarding effects in the rat. *Behav. Brain Res.* **2004**, *148*, 1–10. [CrossRef]
37. Tillu, D.V.; Gebhart, G.F.; Sluka, K.A. Descending facilitatory pathways from the RVM initiate and maintain bilateral hyperalgesia after muscle insult. *Pain* **2008**, *136*, 331–339. [CrossRef]
38. Inase, M.; Nakahama, H.; Otsuki, T.; Fang, J.Z. Analgesic effects of serotonin microinjection into nucleus raphe magnus and nucleus raphe dorsalis evaluated by the monosodium urate (MSU) tonic pain model in the rat. *Brain Res.* **1987**, *426*, 205–211. [CrossRef]
39. Abd-Ellatief, R.B.; Mohamed, H.K.; Kotb, H.I. Reactive Astrogliosis in an Experimental Model of Fibromyalgia: Effect of Dexmedetomidine. *Cells Tissues Organs* **2018**, *205*, 105–119. [CrossRef]

40. Fujigaki, S.; Saito, K.; Sekikawa, K.; Tone, S.; Takikawa, O.; Fujii, H.; Wada, H.; Noma, A.; Seishima, M. Lipopolysaccharide induction of indoleamine 2,3-dioxygenase is mediated dominantly by an IFN-gamma-independent mechanism. *Eur. J. Immunol.* **2001**, *31*, 2313–2318. [CrossRef]
41. Sadok, I.; Gamian, A.; Staniszewska, M.M. Chromatographic analysis of tryptophan metabolites. *J. Sep. Sci.* **2017**, *40*, 3020–3045. [CrossRef] [PubMed]
42. Curzon, G.; Knott, P.J. Effects on plasma and brain tryptophan in the rat of drugs and hormones that influence the concentration of unesterified fatty acid in the plasma. *Br. J. Pharmacol.* **1974**, *50*, 197–204. [CrossRef] [PubMed]
43. Davis, J.M. Carbohydrates, branched-chain amino acids, and endurance: The central fatigue hypothesis. *Int. J. Sports Nutr.* **1995**, *5*, S29–S38. [CrossRef] [PubMed]
44. Costil, D.L.; Bowers, R.; Braunam, G. Muscle glycogen utilization during prolonged exercise on successive days. *J. Appl. Physiol.* **1971**, *31*, 834–838. [CrossRef] [PubMed]
45. Bobinski, F.; Ferreira, T.A.A.; Córdova, M.M.; Dombrowski, P.A.; da Cunha, C.; Santo, C.C.D.E.; Poli, A.; Pires, R.G.W.; Martins-Silva, C.; Sluka, K.A.; et al. Role of brainstem serotonin in analgesia produced by low-intensity exercise on neuropathic pain after sciatic nerve injury in mice. *Pain.* **2015**, *156*, 2595–2606. [CrossRef]
46. Millan, M.J. Descending control of pain. *Prog. Neurobiol.* **2002**, *66*, 355–474.

cells

Article

Intranasal Treatment with Cannabinoid 2 Receptor Agonist HU-308 Ameliorates Cold Sensitivity in Mice with Traumatic Trigeminal Neuropathic Pain

Simeng Ma †, Yoki Nakamura *,†, Suzuna Uemoto, Kenta Yamamoto, Kazue Hisaoka-Nakashima and Norimitsu Morioka *

Department of Pharmacology, Graduate School of Biomedical and Health Sciences, Hiroshima University, 1-2-3 Kasumi, Minami-ku, Hiroshima 734-8553, Japan; d225717@hiroshima-u.ac.jp (S.M.); b180856@hiroshima-u.ac.jp (S.U.); m234539@hiroshima-u.ac.jp (K.Y.); hisaokak@hiroshima-u.ac.jp (K.H.-N.)

* Correspondence: nakayoki@hiroshima-u.ac.jp (Y.N.); mnori@hiroshima-u.ac.jp (N.M.); Tel.: +81-082-257-5312 (Y.N.); +81-082-257-5310 (N.M.)

† These authors contributed equally to this work.

Abstract: Post-traumatic trigeminal neuropathy (PTTN) is a sensory abnormality caused by injury to the trigeminal nerve during orofacial surgery. However, existing analgesics are ineffective against PTTN. Abnormal microglial activation in the caudal part of the spinal trigeminal nucleus caudal part (Sp5C), where the central trigeminal nerve terminals reside, plays an important role in PTTN pathogenesis. Therefore, regulating microglial activity in Sp5C appears to be an important approach to controlling pain in PTTN. Cannabinoid receptor 2 (CB_2) is expressed in immune cells including microglia, and its activation has anti-inflammatory effects. The current study demonstrates that the repeated intranasal administration of CB_2 agonist HU-308 ameliorates the infraorbital nerve cut (IONC)-induced hyperresponsiveness to acetone (cutaneous cooling). The therapeutic efficacy of oral HU-308 was found to be less pronounced in alleviating cold hypersensitivity in IONC mice compared to intranasal administration, indicating the potential advantages of the intranasal route. Furthermore, repeated intranasal administration of HU-308 suppressed the activation of Sp5C microglia in IONC mice. Additionally, pretreatment with the CB_2 antagonist, SR 144528, significantly blocked the anti-nociceptive effect of repeated intranasal administration of HU-308 on cold hypersensitization in IONC mice. These data suggest that the continuous stimulation of CB_2 ameliorates PTTN-induced pain via the inhibition of microglial activation. Thus, CB_2 agonists are potential candidates for novel therapeutic agents against PTTN.

Keywords: cannabinoid receptor 2; cold sensitivity; intranasal treatment; post-traumatic trigeminal neuropathy; microglia

1. Introduction

Various studies have shown that more than 30% of the world's population suffers from chronic pain [1]. Chronic pain imposes a significant psychological and socioeconomic burden on patients and reduces quality of life. However, it is resistant to existing analgesics, and thus, it is difficult to treat [2,3]. Therefore, there is an urgent need to search for new therapeutic approaches. The trigeminal nerve is the fifth cranial nerve, which is divided into three branches, ophthalmic, maxillary, and mandibular nerves, and it is involved in the sensation of the face, oral cavity, and nasal cavity [4]. Post-traumatic trigeminal neuropathy (PTTN) is a chronic pain condition caused by trigeminal nerve injury, which is associated with chronic hypoesthesia, allodynia, or both in the oral–facial region. PTTN develops in 3–7% of patients following dental or oral–facial surgery owing to trigeminal nerve injury [5]. However, non-steroidal anti-inflammatory drugs and opioid analgesics are not effective against the pain associated with PTTN [6,7]. Thus, other treatment modalities need to be developed.

Cannabinoid receptors are characterized as Gi protein-coupled receptors that are specifically activated upon the binding of tetrahydrocannabinol (THC), the primary psychoactive component of cannabis [8,9]. There are two subtypes of cannabinoid receptors, cannabinoid receptor 1 (CB_1) and cannabinoid receptor 2 (CB_2). CB_1 is mainly expressed on neurons and has been reported to be involved in the analgesic and psychoactive effects of Δ9-THC, while CB_2 is expressed on immune cells such as macrophages and microglia, and its activation is associated with anti-inflammatory, but not psychoactive, effects [9–11]. Thus, given that the stimulation of CB_2 does not result in drug abuse, it is considered a promising target for the development of analgesic drugs. The interaction between the immune and nervous systems is important in neuropathic pain [12]. Microglia, the primary immune cells of the central nervous system, participate in the pathogenesis of chronic pain by releasing cytokines, chemokines, and other inflammatory mediators [13–17]. In our previous study, we found that abnormal microglial activation in the spinal trigeminal nucleus caudal part (Sp5C), a part of the medulla oblongata where the central trigeminal nerve terminals are located, plays an important role in the pathogenesis and exacerbation of PTTN [17,18]. However, whether CB_2 is involved in PTTN-induced pain has not been investigated, to our knowledge.

To assess Sp5C microglial function in the brain using drugs, it is necessary to efficiently deliver them to that site. In previous studies, intranasal administration has been shown to efficiently deliver drugs to the olfactory bulb and the medulla oblongata via the olfactory and trigeminal nerves, respectively, offering advantages such as avoiding the blood–brain barrier, reducing dosage, and minimizing systemic side effects [19–21]. Therefore, intranasal administration is expected to deliver drugs to Sp5C microglia, which are abnormally activated in PTTN, and thus, this administration method may be a new therapeutic strategy against PTTN.

The present study focused on CB_2 as a new therapeutic target for PTTN. We evaluated the effects of the intranasal administration of a CB_2 agonist on pain-like behaviors and the hyperactivated state of Sp5C microglia in an infraorbital nerve cut (IONC) mouse model.

2. Materials and Methods

2.1. Animals

Male ddY mice (RRID: MGI:5558113; 8 weeks old) were obtained from Japan SLC, Inc. (Hamamatsu, Shizuoka, Japan), and housed under standard laboratory conditions (temperature: 23.5 ± 2 °C; humidity: 50 ± 10%; 12 h light/dark cycle). The animals were provided with standard rodent chow and water ad libitum. All experiments were conducted in accordance with the Guidelines for the Care and Use of Laboratory Animals established by the Japanese Pharmacological Society and approved by the Committee of Research Facilities for Laboratory Animal Science of Hiroshima University (approval numbers A19-65 and A20-163). All experimental procedures were performed through blinded experimenters. Experiments were conducted using groups of 5 mice to examine the effects of intranasal HU-308 administration (Figure 1A), oral HU-308 administration (Figure 1B), and pretreatment with SR 144528 prior to HU-308 administration (Figure 1C).

2.2. Infraorbital Nerve Cut Model as Post-Traumatic Trigeminal Neuropathy in Mice

In this study, the IONC model was used as the PTTN model [22]. Sodium pentobarbital (50 mg/kg, i.p., Cat. #: 26427-14; Nacalai Tesque, Kyoto, Japan) and 2% isoflurane (induction, 5%; maintenance, 2–3%; Cat. #: 099-06571; FUJIFILM Wako Pure Chemical Corporation, Osaka, Japan) were used to anesthetize ddY mice. A 3–4 mm incision was made in the skin overlying the left infraorbital nerve, which was then exposed and transected at two points 2 mm apart to induce neuropathic pain. For sham-operated controls, an identical incision was made but the nerve was left intact. Postoperatively, mice were closely monitored for recovery from anesthesia.

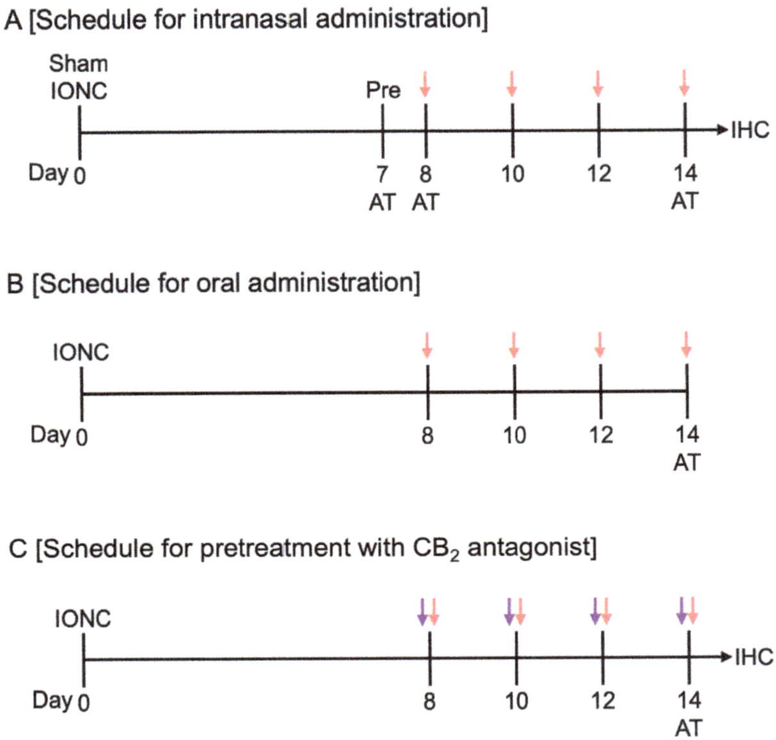

Figure 1. Drug administration and behavioral experiment schedule. (**A**) Intranasal administration of HU-308 (30 nmole, 10 μL) was performed 8, 10, 12, and 14 days after nerve injury. The red arrows indicate when HU-308 was administered. An acetone test (AT) was conducted one day before the initial drug administration (day 7), immediately after the initial administration (day 8), and after repeated administrations (day 14). Following the behavioral tests (day 15), brain tissue samples were collected from the mice for immunohistochemical analysis (IHC). (**B**) Oral drug administration was performed 8, 10, 12, and 14 days after nerve injury. The red arrows indicate when the drug was added. An acetone test (AT) was conducted after repeated administrations (day 14). (**C**) Pretreatment with SR 144528 was intranasally administered 15 min before each HU-308 administration. The red and purple arrows indicate when the HU-308 and SR 144528 were administered, respectively. An acetone test (AT) was conducted after repeated administrations (day 14).

2.3. Acetone Test

The acetone test was used to measure cold hypersensitivity in mice [17,23]. Acetone sensitivity was evaluated at 7, 8, and 14 days post nerve injury (Figure 1). Prior to testing, the mice were habituated to a Plexiglas chamber (12 × 12 × 20 cm) for 15 min. Cold hypersensitivity was assessed 1, 3, 6, and 24 h after drug treatments. Fifteen microliters of acetone was applied to the left whisker pad, and the reaction time of the mouse was measured within 1 min. The response to the cold stimulation was defined as the duration that the mouse touched the left side of its face and rubbed its left cheek against the wall.

2.4. Drug Treatment

Mice were treated with intranasal administration (i.n.) or oral administration (p.o.) 8, 10, 12, and 14 days after the nerve injury. The temporal profile of analgesic effects was evaluated on day 8 after the initial drug administration for the single-administration group. For the repeated administration group, analgesic efficacy was assessed 14 days after the

4th drug administration (Figure 1). A pipette was used to administer 10 μL of the indicated solution(s) via intranasal administration. Considering the potential for drug reflux, the administration volume was decreased to 10 μL, which is lower than the typical volume of 20–30 μL [24]. Following intranasal administration, the animals were carefully monitored to ensure that no drug reflux occurred prior to behavioral testing. A mouse gavage needle was used to administer 100 μL of the indicated solution(s) via oral administration. Regardless of the administration methods, 30 nmol of CB_2 agonist HU-308 (Cat. #: 90086; Cayman Chemical, Ann Arbor, MI, USA) and 100 nmol of CB_2 antagonist SR 144528 (Cat. #: 9000491; Cayman Chemical, Ann Arbor, MI, USA) were used in this study. SR 144528, which has been reported to abolish the effects of HU-308 [25], was intranasally administered 15 min before each HU-308 administration.

2.5. Immunohistochemistry

Following behavioral testing, mice were euthanized with sodium pentobarbital (50 mg/kg) and 2% isoflurane and perfused transcardially with saline (50 mL), followed by 4% (w/v) paraformaldehyde (8% [w/v] paraformaldehyde dissolved 1:1 in 0.2 M phosphate buffer [pH 7.4], 20 mL). Brains were removed, post-fixed in 4% paraformaldehyde for 24 h, and cryoprotected in 30% (w/v) sucrose solution (60% [w/v] sucrose in 1:1 in 0.2 M phosphate buffer [pH 7.4]) at 4 °C for 3 days. Coronal brain sections (30 μm) were cut on a cryostat (CM3050S II, Leica Microsystems, Wetzlar, Germany). After air-drying, the sections were washed three times with 10 mM glycine/phosphate-buffered saline (PBS) at 5 min intervals and incubated in blocking solution (3% bovine serum albumin [BSA], 10% goat serum, 0.1% Triton X-100, and 0.05% Tween20) for 2 h at room temperature. The primary antibody used was rabbit polyclonal anti-ionized calcium-binding adapter molecule 1 (Iba-1; a microglia marker) antibody (1:500, Catalog # 019-19741, FUJIFILM Wako Pure Chemical Corporation, Osaka, Japan), diluted in 3% BSA in PBS. After incubation at 4 °C for three days, the primary antibody was removed, and the sections were washed six times with 0.1% BSA-PBS for 5 min per wash. Goat anti-rabbit IgG secondary antibody with Alexa FluorTM 555 (1: 500, Catalog # A21429, Thermo Fisher Scientific, Waltham, MA, USA) in 3% BSA-PBS was applied onto tissue sections on glass slides and incubated at 4 °C for 2 h. The antibody solution was then discarded, and the sections were washed six times with 0.1% BSA for 5 min per wash. Finally, the slides were sealed with a coverslip and observed using a BZ-900 Biorevo all-in-one fluorescence microscope (Keyence Corporation, Osaka, Japan). Individual Iba1-positive cells in the images were identified using CellProfiler™ [26,27]. Briefly, for each cell, the cell number, the integrated fluorescence intensity of Iba1, and the average cell area were calculated using the MeasureObjectIntensity and MeasureObjectSizeShape module of Cell Profiler™.

2.6. Statistical Analysis

Statistical analysis was performed using GraphPad Prism 7 (GraphPad Software, Boston, MA, USA). The results are presented as mean ± standard error of the mean (SEM). An unpaired t-test was used to compare two groups. The time course of the drug's effects was analyzed using a two-way repeated measures (RM) analysis of variance (ANOVA) to confirm significance, followed by multiple comparisons using Tukey's or Sidak's method. Two-way ANOVA was used to compare the effects of different administration methods. Other data were analyzed by one-way ANOVA to confirm significance, followed by multiple comparisons using Tukey's or Dunnett's method. $p < 0.05$ was considered a significant difference. All the statistical details can be found in the Supplementary File.

3. Results

3.1. Intranasal Administration of CB_2 Agonist Alleviates Cold Hypersensitization in IONC Mice

Cold hypersensitivity is a common clinical manifestation in patients with trigeminal nerve injury [28]. Consistent with clinical findings, our experimental data demonstrate that IONC mice exhibited a significant increase in responsiveness to acetone, a cold stimulus, at

7–14 days post nerve injury (Figure 2A). Moreover, to assess the efficacy of CB_2 agonist HU-308 in ameliorating this condition, we administered HU-308 intranasally to IONC mice in either a single or repeated dose regimen. While a single intranasal administration of HU-308 (30 nmol) failed to reverse cold hypersensitization, repeated intranasal administration significantly alleviated cold hypersensitization (Figure 2A). Additionally, the area under the curve analysis demonstrated that repeated administration of HU-308 produced a significantly greater reduction in cold hypersensitization in IONC mice compared with a single administration (Figure 2B,C).

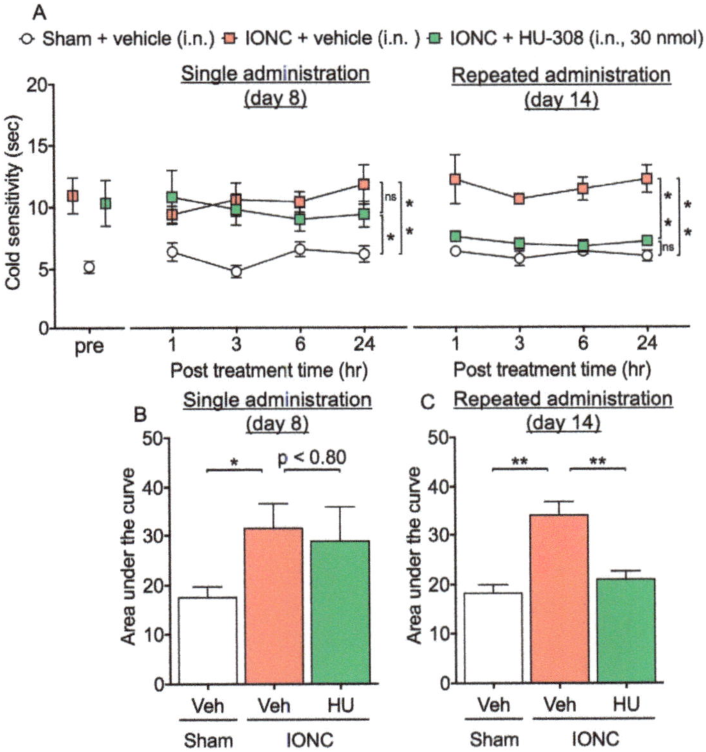

Figure 2. Effect of intranasal administration of HU-308 on cold hypersensitization in IONC mice. (**A**) The responsiveness to acetone was assessed at 1, 3, 6, and 24 h following a single and repeated intranasal administration of HU-308 (30 nmole, 10 μL, i.n.). The area under the curve was determined from the acetone test data following either single (**B**) or repeated (**C**) intranasal administration and compared. Individual data and mean ± SEM are shown. N = 5 (A); * $p < 0.05$, ** $p < 0.01$; n.s., not significant (two-way RM ANOVA followed by Tukey's multiple comparisons test). (**B**,**C**) * $p < 0.05$, ** $p < 0.01$ (one-way ANOVA followed by Tukey's multiple comparisons test).

3.2. *Oral Administration of CB_2 Agonist Does Not Affect Cold Hypersensitization in IONC Mice*

To evaluate the therapeutic potential and use of intranasal administration for PTTN, we investigated whether a repeated oral administration with an equivalent dose of HU-308 can alleviate cold hypersensitization. Although no significant changes were observed in the time course of drug administration (Figure 3A), a comparison of the area under the curve values revealed that repeated oral administration of HU-308 resulted in a slight but significant improvement in cold hypersensitivity in IONC mice, similar to the effects observed with the intranasal administration (Figure 3B). The area under the curve analysis indicated that repeated intranasal administration of HU-308 produced a significantly greater amelioration of cold hypersensitization compared with repeated oral administration, suggesting a superior therapeutic effect of the intranasal route (Figure 4).

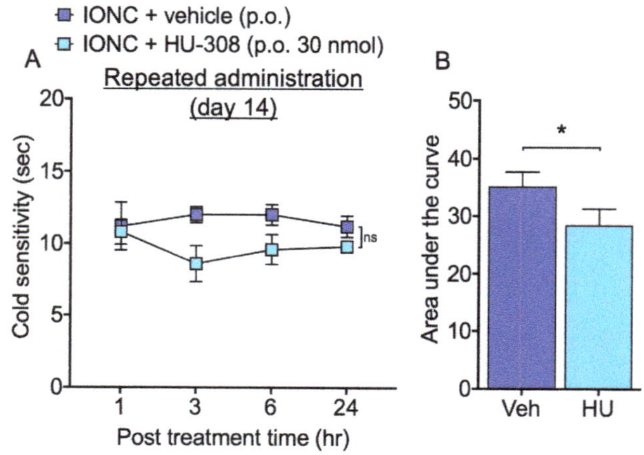

Figure 3. Effect of oral administration of HU-308 on cold hypersensitization in IONC mice. (**A**) The responsiveness to acetone was assessed at 1, 3, 6, and 24 h following repeated oral administration of HU-308 (HU, 30 nmole, 100 µL, p.o.). (**B**) The area under the curve was determined from the acetone test data following repeated oral administration and compared. Individual data and mean ± SEM are shown. N = 5 (**A**); n.s., not significant (two-way RM ANOVA followed by Sidak's multiple comparisons test). (**B**); * $p < 0.05$ (Unpaired t-test).

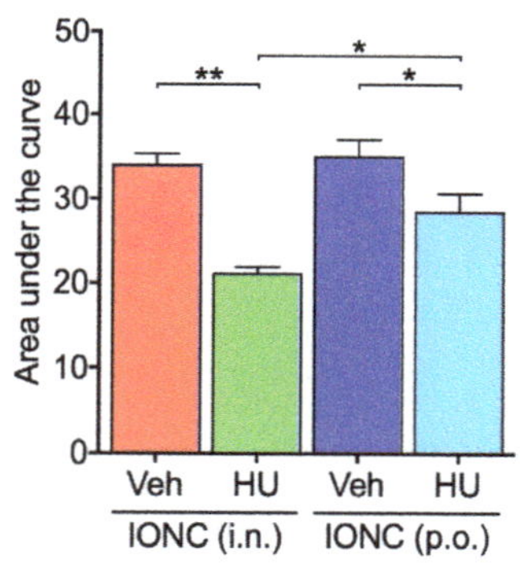

Figure 4. Comparative analysis of the effects of intranasal and oral administration of CB_2 agonist on cold hypersensitivity in IONC mice. The area under the curve was determined from the acetone test data following either intranasal (i.n.) or oral (p.o.) repeated administration of HU-308 (HU, 30 nmole, 10 µL) and compared. Individual data and mean ± SEM are shown. N = 5; * $p < 0.05$, ** $p < 0.01$ (two-way ANOVA followed by Tukey's multiple comparisons test).

3.3. Intranasal Administration of CB_2 Agonist Alleviates Sp5C Microglial Activation in IONC Mice

Given the established role of microglial activation, as evidenced by increased Iba-1 expression, in the pathogenesis of IONC-induced hypersensitivity [22], we examined the effects of HU-308 treatment on microglial activation in Sp5C. In the Sp5C region, IONC mice exhibited a significant elevation in the number of Iba1-positive cells, Iba1 fluorescence intensity, and cellular area compared to sham mice. The intranasal administration of HU-308 demonstrated a trend toward reducing these parameters (Figure 5).

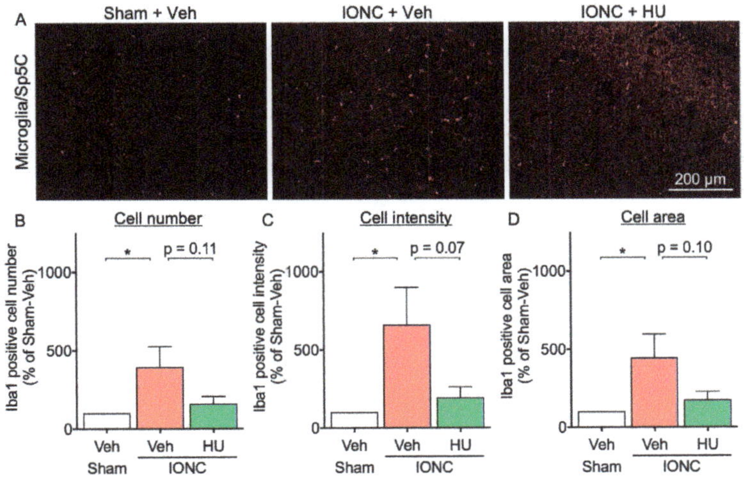

Figure 5. Effect of intranasal administration of HU-308 on Sp5C microglial activation in IONC mice. (**A**) Fluorescent photomicrographs of Iba-1 in Sp5C from sham and IONC mice following repeated intranasal administration of HU-308 (HU, 30 nmole, 10 µL). Scale bar = 200 µm. The number of Iba1-positive cells (**B**), the Iba1 intensity (**C**), and the cell area (**D**) were calculated from N = 5 mice. Individual data and mean ± SEM are shown. * $p < 0.05$ (one-way ANOVA followed by Dunnett's multiple comparisons test).

3.4. Pretreatment with CB_2 Antagonist Inhibits the CB_2 Agonist-Induced Anti-Nociceptive Effect in IONC Mice

It is necessary to ascertain whether the analgesic effect of HU-308 is a CB_2-mediated response. It has previously been reported that the analgesic efficacy of HU-308 for formalin-induced peripheral pain was attenuated by pretreatment with CB_2 antagonist SR 144528 [25]. Similarly to previous findings, pretreatment with SR 144528 15 min prior to each HU-308 intranasal administration reduced the effects of HU-308 on cold hypersensitivity and the microglial activation in Sp5C (Figures 6 and 7), suggesting that the analgesic effects of HU-308 are mediated by CB_2 activation in IONC mice.

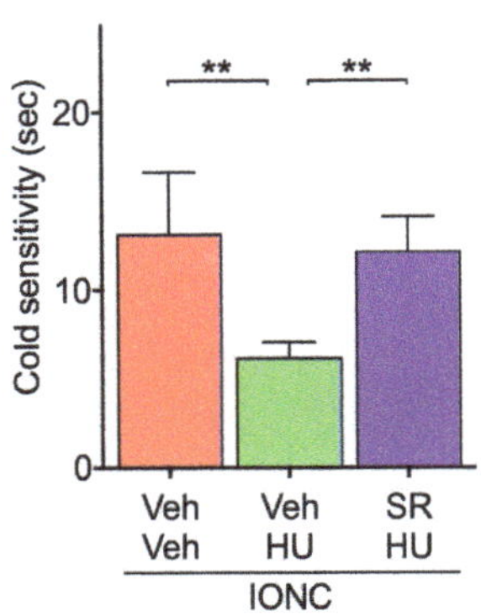

Figure 6. Effect of pretreatment with SR144528 on HU-308-inhibited cold sensitization in IONC mice. The responsiveness to acetone was assessed 3 h following repeated intranasal administration of HU-308 (HU, 30 nmole, 10 µL). SR 144528 (SR, 100 nmole, 10 µL) was administered 15 min before HU-308. Individual data and mean ± SEM are shown. N = 5, ** $p < 0.01$ (one-way ANOVA followed by Dunnett's multiple comparisons test).

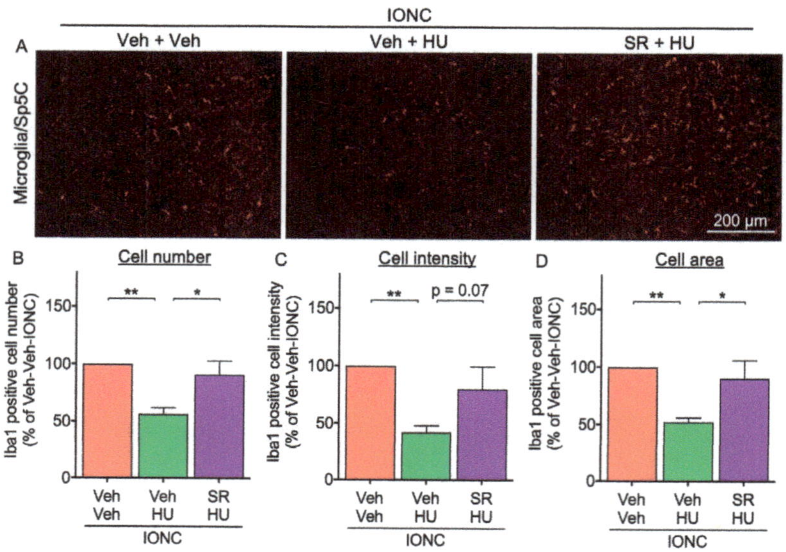

Figure 7. Effect of pretreatment with SR 144528 on HU-308-inhibited microglial activation in IONC mice. (**A**) Fluorescent photomicrographs of Iba-1 in Sp5C from IONC mice were assessed 3 h following repeated intranasal administration of HU-308 (HU, 30 nmole, 10 μL). SR 144528 (SR, 100 nmole, 10 μL) was administered 15 min before HU-308. Scale bar = 200 μm. The number of Iba1-positive cells (**B**), the Iba1 intensity (**C**), and the cell area (**D**) were calculated from N = 5 mice. Individual data and mean ± SEM are shown. * $p < 0.05$, ** $p < 0.01$ (one-way ANOVA followed by Dunnett's multiple comparisons test).

4. Discussion

The present study explored the therapeutic efficacy of CB_2 receptor activation in a preclinical model of PTTN. Repeated intranasal administration of HU-308, a selective CB_2 agonist, significantly alleviated acetone-induced cold hypersensitivity and attenuated microglial activation in IONC mice. These findings provide compelling evidence supporting the therapeutic potential of CB_2 agonists for the treatment of PTTN.

Numerous studies have demonstrated that the aberrant activation of microglia in the spinal dorsal horn or the Sp5C region is pivotal in the pathogenesis of neuropathic pain [13–17]. Microglia contribute to chronic pain conditions by inducing changes in neuronal plasticity, including long-term potentiation, through the release of inflammatory cytokines [15]. Given these findings, targeting microglia with anti-inflammatory drugs presents a compelling strategy for the development of novel therapeutics for neuropathic pain. CB_2 proteins, expressed on immune cells including microglia, have been shown to mediate anti-inflammatory effects [29]. Previous studies have indicated that CB_2 activation can significantly suppress inflammatory responses induced by various stimuli in microglia [30–34]. For instance, CB_2 activation has been shown to suppress the Janus kinase (JAK)/signal transducer and activator of transcription 1 (STAT1) pathway induced by interferon-gamma and the mitogen-activated protein (MAP) kinase pathway triggered by Toll-like receptor 4 activation, leading to the decreased production of inflammatory cytokines and exertion of anti-inflammatory effects [30,31]. In the present study, while a single administration of HU-308 failed to alleviate cold hypersensitivity in IONC mice, repeated administration produced a significant improvement. Additionally, the significant attenuation of the therapeutic effects by pretreatment with SR 144528, a CB_2 antagonist, strongly supports the notion that HU-308 exerts its therapeutic effects via CB_2. These findings suggest that the transient stimulation of CB_2 may be inadequate to counteract the multifaceted effects of inflammation, and that sustained CB_2 activation and its associated anti-inflammatory effects are essential for ameliorating pain conditions. The observation that only repeated administration of HU-308 demonstrated therapeutic effects can be attributed to the fact that inflammatory responses form a feed-forward loop, perpetuating the pathological state. A previous study has also demonstrated that minocycline, a microglial inhibitor, exhibited greater efficacy with repeated administration compared with a single dose in the treatment of oxaliplatin-induced mechanical allodynia [35]. Thus, the transient inhibition of microglial activity may be insufficient to achieve adequate therapeutic outcomes, because microglia were continuously activated before the treatment and sustained inflammation; hence, the continuous suppression of microglia may be essential. Another possibility is that repeated administration of HU-308 enabled its accumulation in brain parenchyma, reaching effective concentrations. Although the details of HU-308 accumulation in brain tissue are unknown, the calculated logarithm of the octanol–water partition coefficient (cLogP) of 8.97 indicates that it is highly lipophilic and may accumulate in the lipid-rich central nervous system [36].

Intranasal administration offers a non-invasive route for drug delivery to the central nervous system, specifically targeting the olfactory bulb and the pons and the medulla oblongata via the olfactory and trigeminal nerve pathways, respectively [19–21]. This is supported by studies demonstrating the accumulation of radiolabeled antibodies within these brain regions after intranasal administration [20]. The current study demonstrated in IONC mice that repeated intranasal administration of HU-308 attenuated microglial activation in Sp5C, a brain region located around the pons and the medulla oblongata. Given the analogous roles of Sp5C and the spinal dorsal horn in pain processing and the established role of microglial activation in neuropathic pain [14–17], we propose that HU-308 exerts its analgesic effects by targeting CB_2 on Sp5C microglia. However, considering that intranasal administration can deliver drugs directly to the brain via the trigeminal nerve pathway, it is conceivable that HU-308 exerts its therapeutic effects by interacting with the damaged nerve or the trigeminal ganglion during its transit [19]. Previous studies have shown that macrophage infiltration and accumulation in the trigeminal ganglion following nerve injury play a crucial role in the pathogenesis of PTTN [37]. Additionally, CB_2 activation has been reported to inhibit tumor necrosis factor-α-induced inflammatory responses and monocyte migration [38]. Considering these findings, it is plausible that the activation of CB_2 on macrophages accumulated in the trigeminal ganglion alleviates chronic pain in PTTN by suppressing inflammation and reducing cellular infiltration. Further studies are required to elucidate the mechanisms underlying the analgesic effects of CB_2 activation in PTTN.

In this study, intranasal administration was used to deliver drugs to the medulla oblongata, the injured trigeminal nerve, and the trigeminal ganglion, which are target tissues for the treatment of PTTN. The intranasal administration of HU-308 ameliorated cold hypersensitization and suppressed the abnormal activation of Sp5C microglia in IONC mice. In contrast, repeated oral administration of HU-308 resulted in only a slight improvement in cold hypersensitivity in IONC mice. While oral administration is a convenient method that enables widespread distribution of drugs throughout the body, it has several drawbacks, including the requirement for larger drug doses compared with local administration and the potential for unexpected systemic side effects. Additionally, CB_2 is expressed not only on immune cells but also in some parts of the gastrointestinal tract [39]. Therefore, achieving sufficient analgesic effects with oral administration may require higher doses of HU-308, increasing the risk of inducing unexpected systemic side effects. These results suggest that the intranasal administration of CB_2 agonists may be a promising therapeutic approach for the treatment of PTTN, not only in terms of therapeutic efficacy but also in terms of avoiding side effects and cost-effectiveness.

5. Conclusions

Neuropathic pain, characterized by its resistance to conventional analgesics, remains a significant clinical challenge. In this study, we used IONC mice, a well-established animal model of PTTN, to investigate the therapeutic potential of a selective CB_2 agonist. Remarkably, repeated intranasal administration of HU-308 significantly attenuated pain-related behavior and pathological changes. These findings strongly support the hypothesis that CB_2 may be a novel therapeutic target for the treatment of trigeminal neuropathic pain, and intranasal delivery may offer a promising approach for targeting the Sp5C microglia.

Supplementary Materials: The following supporting information can be downloaded at: https://www.mdpi.com/article/10.3390/cells13231943/s1, Table S1: Summary of statistical analysis.

Author Contributions: Conceptualization, Y.N. and N.M.; methodology, Y.N.; validation, Y.N., K.H.-N. and N.M.; investigation, S.M., S.U. and K.Y.; data curation, Y.N. and N.M.; writing—original draft preparation, Y.N. and S.M.; writing—review and editing, Y.N., K.H.-N. and N.M.; visualization, Y.N.; supervision, Y.N. and N.M.; project administration, Y.N. and N.M.; funding acquisition, Y.N. All authors have read and agreed to the published version of the manuscript.

Funding: This work was supported by Grant-in-Aid for Research Activity start-up (19K21225) and Grant-in-Aid for Early-Career Scientists (20K18476) from Japan Society for the Promotion of Science.

Institutional Review Board Statement: All experiments using animals were conducted in accordance with the Guidelines for the Care and Use of Laboratory Animals established by the Japanese Pharmacological Society. The animal study protocol was approved by the Committee of Research Facilities for Laboratory Animal Science of Hiroshima University (A19-65, A20-163).

Informed Consent Statement: Not applicable.

Data Availability Statement: The data supporting the findings of this study are available from the corresponding author upon reasonable request.

Acknowledgments: This work was supported in part by the Natural Science Center for Basic Research and Development (NBARD-00069). We thank the Analysis Center of Life Science and the Natural Science Center for Basic Research and Development, Hiroshima University, for the use of equipment for the experiments, and Michal Bell, from Edanz (https://jp.edanz.com/ac, accessed on 3 September 2024) for editing a draft of this manuscript. The graphical abstract was created with BioRender.com (Agreement No. OP27KUQ5HO).

Conflicts of Interest: The authors declare no conflicts of interest.

References

1. Cohen, S.P.; Vase, L.; Hooten, W.M. Chronic pain: An update on burden, best practices, and new advances. *Lancet* **2021**, *397*, 2082–2097. [CrossRef] [PubMed]
2. McNicol, E.D.; Midbari, A.; Eisenberg, E. Opioids for neuropathic pain. *Cochrane Database Syst. Rev.* **2013**, *8*, CD006146. [CrossRef] [PubMed]
3. Finnerup, N.B.; Attal, N.; Haroutounian, S.; McNicol, E.; Baron, R.; Dworkin, R.H.; Gilron, I.; Haanpää, M.; Hansson, P.; Jensen, T.S.; et al. Pharmacotherapy for neuropathic pain in adults: A systematic review and meta-analysis. *Lancet Neurol.* **2015**, *14*, 162–173. [CrossRef] [PubMed]
4. Joo, W.; Yoshioka, F.; Funaki, T.; Mizokami, K.; Rhoton, A.L. Microsurgical anatomy of the trigeminal nerve. *Clin. Anat.* **2014**, *27*, 61–88. [CrossRef]
5. Nasri-Heir, C.; Khan, J.; Benoliel, R.; Feng, C.; Yarnitsky, D.; Kuo, F.; Hirschberg, C.; Hartwell, G.; Huang, C.Y.; Heir, G.; et al. Altered pain modulation in patients with persistent postendodontic pain. *Pain* **2015**, *156*, 2032–2041. [CrossRef]
6. Moulin, D.E.; Clark, A.J.; Gordon, A.; Lynch, M.; Morley-Forster, P.K.; Nathan, H.; Smyth, C.; Toth, C.; VanDenKerkhof, E.; Gilani, A.; et al. Long-Term Outcome of the Management of Chronic Neuropathic Pain: A Prospective Observational Study. *J. Pain* **2015**, *16*, 852–861. [CrossRef]
7. Bates, D.; Schultheis, B.C.; Hanes, M.C.; Jolly, S.M.; Chakravarthy, K.V.; Deer, T.R.; Levy, R.M.; Hunter, C.W. A Comprehensive Algorithm for Management of Neuropathic Pain. *Pain Med.* **2019**, *20*, S2–S12. [CrossRef]
8. Munro, S.; Thomas, K.L.; Abu-Shaar, M. Molecular characterization of a peripheral receptor for cannabinoids. *Nature* **1993**, *365*, 61–65. [CrossRef]
9. Shahbazi, F.; Grandi, V.; Banerjee, A.; Trant, J.F. Cannabinoids and Cannabinoid Receptors: The Story so Far. *iScience* **2020**, *23*, 101301. [CrossRef]

10. Li, X.; Shen, L.; Hua, T.; Liu, Z.J. Structural and Functional Insights into Cannabinoid Receptors. *Trends Pharmacol. Sci.* **2020**, *41*, 665–677. [CrossRef]
11. Li, X.; Chang, H.; Bouma, J.; de Paus, L.V.; Mukhopadhyay, P.; Paloczi, J.; Mustafa, M.; van der Horst, C.; Kumar, S.S.; Wu, L.; et al. Structural basis of selective cannabinoid CB. *Nat. Commun.* **2023**, *14*, 1447. [CrossRef] [PubMed]
12. Fiore, N.T.; Debs, S.R.; Hayes, J.P.; Duffy, S.S.; Moalem-Taylor, G. Pain-resolving immune mechanisms in neuropathic pain. *Nat. Rev. Neurol.* **2023**, *19*, 199–220. [CrossRef] [PubMed]
13. Nakamura, Y.; Morioka, N.; Abe, H.; Zhang, F.F.; Hisaoka-Nakashima, K.; Liu, K.; Nishibori, M.; Nakata, Y. Neuropathic pain in rats with a partial sciatic nerve ligation is alleviated by intravenous injection of monoclonal antibody to high mobility group box-1. *PLoS ONE* **2013**, *8*, e73640. [CrossRef] [PubMed]
14. Inoue, K.; Tsuda, M. Microglia in neuropathic pain: Cellular and molecular mechanisms and therapeutic potential. *Nat. Rev. Neurosci.* **2018**, *19*, 138–152. [CrossRef] [PubMed]
15. Chen, G.; Zhang, Y.Q.; Qadri, Y.J.; Serhan, C.N.; Ji, R.R. Microglia in Pain: Detrimental and Protective Roles in Pathogenesis and Resolution of Pain. *Neuron* **2018**, *100*, 1292–1311. [CrossRef]
16. Nakamura, Y.; Fukuta, A.; Miyashita, K.; Zhang, F.F.; Wang, D.; Liu, K.; Wake, H.; Hisaoka-Nakashima, K.; Nishibori, M.; Morioka, N. Perineural high-mobility group box 1 induces mechanical hypersensitivity through activation of spinal microglia: Involvement of glutamate-NMDA receptor dependent mechanism in spinal dorsal horn. *Biochem. Pharmacol.* **2021**, *186*, 114496. [CrossRef]
17. Kochi, T.; Nakamura, Y.; Ma, S.; Hisaoka-Nakashima, K.; Wang, D.; Liu, K.; Wake, H.; Nishibori, M.; Irifune, M.; Morioka, N. Pretreatment with High Mobility Group Box-1 Monoclonal Antibody Prevents the Onset of Trigeminal Neuropathy in Mice with a Distal Infraorbital Nerve Chronic Constriction Injury. *Molecules* **2021**, *26*, 2035. [CrossRef]
18. Ma, S.; Nakamura, Y.; Kochi, T.; Uemoto, S.; Hisaoka-Nakashima, K.; Wang, D.; Liu, K.; Wake, H.; Nishibori, M.; Morioka, N. Perineural Treatment with High Mobility Group Box-1 Monoclonal Antibody Prevents Initiation of Pain-Like Behaviors in Female Mice with Trigeminal Neuropathy. *Biol. Pharm. Bull.* **2024**, *47*, 221–226. [CrossRef]
19. Erdő, F.; Bors, L.A.; Farkas, D.; Bajza, Á.; Gizurarson, S. Evaluation of intranasal delivery route of drug administration for brain targeting. *Brain Res. Bull.* **2018**, *143*, 155–170. [CrossRef]
20. Kumar, N.N.; Lochhead, J.J.; Pizzo, M.E.; Nehra, G.; Boroumand, S.; Greene, G.; Thorne, R.G. Delivery of immunoglobulin G antibodies to the rat nervous system following intranasal administration: Distribution, dose-response, and mechanisms of delivery. *J. Control Release* **2018**, *286*, 467–484. [CrossRef]
21. Sato, F.; Nakamura, Y.; Ma, S.; Kochi, T.; Hisaoka-Nakashima, K.; Wang, D.; Liu, K.; Wake, H.; Nishibori, M.; Morioka, N. Central high mobility group box-1 induces mechanical hypersensitivity with spinal microglial activation in a mouse model of hemi-Parkinson's disease. *Biomed. Pharmacother.* **2022**, *145*, 112479. [CrossRef] [PubMed]
22. Ueta, Y.; Miyata, M. Brainstem local microglia induce whisker map plasticity in the thalamus after peripheral nerve injury. *Cell Rep.* **2021**, *34*, 108823. [CrossRef] [PubMed]
23. Trevisan, G.; Benemei, S.; Materazzi, S.; De Logu, F.; De Siena, G.; Fusi, C.; Fortes Rossato, M.; Coppi, E.; Marone, I.M.; Ferreira, J.; et al. TRPA1 mediates trigeminal neuropathic pain in mice downstream of monocytes/macrophages and oxidative stress. *Brain* **2016**, *139*, 1361–1377. [CrossRef] [PubMed]
24. Correa, D.; Scheuber, M.I.; Shan, H.; Weinmann, O.W.; Baumgartner, Y.A.; Harten, A.; Wahl, A.S.; Skaar, K.L.; Schwab, M.E. Intranasal delivery of full-length anti-Nogo-A antibody: A potential alternative route for therapeutic antibodies to central nervous system targets. *Proc. Natl. Acad. Sci. USA* **2023**, *120*, e2200057120. [CrossRef]
25. Hanus, L.; Breuer, A.; Tchilibon, S.; Shiloah, S.; Goldenberg, D.; Horowitz, M.; Pertwee, R.G.; Ross, R.A.; Mechoulam, R.; Fride, E. HU-308: A specific agonist for CB(2), a peripheral cannabinoid receptor. *Proc. Natl. Acad. Sci. USA* **1999**, *96*, 14228–14233. [CrossRef]
26. Aw, E.; Zhang, Y.; Carroll, M. Microglial responses to peripheral type 1 interferon. *J. Neuroinflamm.* **2020**, *17*, 340. [CrossRef]
27. Stirling, D.R.; Swain-Bowden, M.J.; Lucas, A.M.; Carpenter, A.E.; Cimini, B.A.; Goodman, A. CellProfiler 4: Improvements in speed, utility and usability. *BMC Bioinform.* **2021**, *22*, 433. [CrossRef]
28. Zakrzewska, J.M. Differential diagnosis of facial pain and guidelines for management. *Br. J. Anaesth.* **2013**, *111*, 95–104. [CrossRef]
29. Cabral, G.A.; Raborn, E.S.; Griffin, L.; Dennis, J.; Marciano-Cabral, F. CB2 receptors in the brain: Role in central immune function. *Br. J. Pharmacol.* **2008**, *153*, 240–251. [CrossRef]
30. Romero-Sandoval, E.A.; Horvath, R.; Landry, R.P.; DeLeo, J.A. Cannabinoid receptor type 2 activation induces a microglial anti-inflammatory phenotype and reduces migration via MKP induction and ERK dephosphorylation. *Mol. Pain.* **2009**, *5*, 25. [CrossRef]
31. Ehrhart, J.; Obregon, D.; Mori, T.; Hou, H.; Sun, N.; Bai, Y.; Klein, T.; Fernandez, F.; Tan, J.; Shytle, R.D. Stimulation of cannabinoid receptor 2 (CB2) suppresses microglial activation. *J. Neuroinflamm.* **2005**, *2*, 29. [CrossRef] [PubMed]
32. Ramírez, B.G.; Blázquez, C.; Gómez del Pulgar, T.; Guzmán, M.; de Ceballos, M.L. Prevention of Alzheimer's disease pathology by cannabinoids: Neuroprotection mediated by blockade of microglial activation. *J. Neurosci.* **2005**, *25*, 1904–1913. [CrossRef] [PubMed]
33. Mecha, M.; Feliú, A.; Carrillo-Salinas, F.J.; Rueda-Zubiaurre, A.; Ortega-Gutiérrez, S.; de Sola, R.G.; Guaza, C. Endocannabinoids drive the acquisition of an alternative phenotype in microglia. *Brain Behav. Immun.* **2015**, *49*, 233–245. [CrossRef] [PubMed]
34. Komorowska-Müller, J.A.; Schmöle, A.C. CB2 Receptor in Microglia: The Guardian of Self-Control. *Int. J. Mol. Sci.* **2020**, *22*, 19. [CrossRef] [PubMed]

35. Sałat, K.; Furgała-Wojas, A.; Sałat, R. The Microglial Activation Inhibitor Minocycline, Used Alone and in Combination with Duloxetine, Attenuates Pain Caused by Oxaliplatin in Mice. *Molecules* **2021**, *26*, 3577. [CrossRef]
36. Soethoudt, M.; Grether, U.; Fingerle, J.; Grim, T.W.; Fezza, F.; de Petrocellis, L.; Ullmer, C.; Rothenhäusler, B.; Perret, C.; van Gils, N.; et al. Cannabinoid CB. *Nat. Commun.* **2017**, *8*, 13958. [CrossRef]
37. Batbold, D.; Shinoda, M.; Honda, K.; Furukawa, A.; Koizumi, M.; Akasaka, R.; Yamaguchi, S.; Iwata, K. Macrophages in trigeminal ganglion contribute to ectopic mechanical hypersensitivity following inferior alveolar nerve injury in rats. *J. Neuroinflamm.* **2017**, *14*, 249. [CrossRef]
38. Rajesh, M.; Mukhopadhyay, P.; Haskó, G.; Huffman, J.W.; Mackie, K.; Pacher, P. CB2 cannabinoid receptor agonists attenuate TNF-alpha-induced human vascular smooth muscle cell proliferation and migration. *Br. J. Pharmacol.* **2008**, *153*, 347–357. [CrossRef]
39. Wright, K.L.; Duncan, M.; Sharkey, K.A. Cannabinoid CB2 receptors in the gastrointestinal tract: A regulatory system in states of inflammation. *Br. J. Pharmacol.* **2008**, *153*, 263–270. [CrossRef]

cells

Article

Male-Dominant Spinal Microglia Contribute to Neuropathic Pain by Producing CC-Chemokine Ligand 4 Following Peripheral Nerve Injury

Fumihiro Saika [1,2,†], Tetsuya Sato [3,*,†,‡], Takeru Nakabayashi [3,†], Yohji Fukazawa [4], Shinjiro Hino [5], Kentaro Suzuki [6] and Norikazu Kiguchi [1,*]

1 Department of Physiological Sciences, School of Pharmaceutical Sciences, Wakayama Medical University, Wakayama 640-8156, Japan; f-saika@tumh.ac.jp
2 Faculty of Wakayama Health Care Sciences, Takarazuka University of Medical and Health Care, Wakayama 640-8392, Japan
3 H.U. Group Research Institute G.K., Tokyo 197-0833, Japan; takeru.nakabayashi@hugp.com
4 Department of Anatomy, Kansai University of Health Sciences, Osaka 590-0482, Japan; fukazawa@kansai.ac.jp
5 Department of Medical Cell Biology, Institute of Molecular Embryology and Genetics, Kumamoto University, Kumamoto 860-0811, Japan; s-hino@kumamoto-u.ac.jp
6 Faculty of Life and Environmental Sciences, University of Yamanashi, Yamanashi 400-8510, Japan; k-suzuki@yamanashi.ac.jp
* Correspondence: satote@saitama-med.ac.jp (T.S.); kiguchi@wakayama-med.ac.jp (N.K.); Tel.: +81-42-984-0318 (T.S.); +81-73-488-2581 (N.K.)
† These authors contributed equally to this work.
‡ Current address: Biomedical Research Center, Faculty of Medicine, Saitama Medical University, Saitama 350-1241, Japan.

Abstract: Recent studies have revealed marked sex differences in pathophysiological roles of spinal microglia in neuropathic pain, with microglia contributing to pain exacerbation exclusively in males. However, the characteristics of pain-enhancing microglia, which are more prominent in males, remain poorly understood. Here, we reanalyzed a previously published single-cell RNA sequencing dataset and identified a microglial subpopulation that significantly increases in the spinal dorsal horn (SDH) of male mice following peripheral nerve injury. CC-chemokine ligand 4 (CCL4) was highly expressed in this subpopulation and its mRNA levels were increased in the SDH after partial sciatic nerve ligation (PSL) only in male mice. Notably, CCL4 expression was reduced in male mice following microglial depletion, indicating that microglia are the primary source of CCL4. Intrathecal administration of maraviroc, an inhibitor of the CCL4–CC-chemokine receptor 5 (CCR5) signaling pathway, after PSL, significantly suppressed mechanical allodynia only in male mice. Furthermore, intrathecal administration of CCL4 induced mechanical allodynia in both sexes, accompanied by increased expression of c-fos, a neuronal excitation marker, in the SDH. These findings highlight a sex-biased difference in the gene expression profile of spinal microglia following peripheral nerve injury, with elevated CCL4 expression in male mice potentially contributing to pain exacerbation.

Keywords: allodynia; CCL4; CCR5; female; inflammation; sex; spinal cord

1. Introduction

Pain serves as an essential manifestation of abnormalities in the body and helps in initiating protective actions [1,2]. However, pain exceeding the normal physiological range requires medical evaluation and intervention. Neuropathic pain, resulting from nervous system damage, often does not respond to conventional analgesics, and has prompted extensive research into the underlying molecular mechanisms [3,4]. Recent studies have highlighted activation of microglia in the spinal dorsal horn (SDH) and the role of pain-related inflammatory mediators, such as cytokines and chemokines, in exacerbating and prolonging the pain [5,6]. Emerging evidence is also revealing sex-based differences in the involvement of microglia in neuropathic pain [7,8]. We and other research groups have demonstrated that administering microglial inhibitors (e.g., minocycline and pexidartinib) to rodent models of neuropathic pain suppresses pain in males, but not in females [9–11]. These findings suggest that, in males, microglia play a direct role in pain exacerbation under pathological conditions and contribute to neuropathic pain. Conversely, this mechanism does not appear to operate in microglia of females, implying the involvement of alternative pathways in them.

Microglia express receptors that respond to various stimuli from their environment and neighboring cells, and their activation has been implicated in pain induction [12,13]. Several recent studies have revealed significant sex-based differences in this process. For example, intrathecal (i.t.) administration of lipopolysaccharide, a toll-like receptor 4 agonist, or of colony-stimulating factor 1 (CSF1), which is essential for the survival and proliferation of microglia, induces morphological activation of microglia and reduces the pain threshold in rodent models [14–16]. This effect is pronounced in male mice but absent in female mice. Additionally, we demonstrated that selective activation of spinal microglia using Gq-DREADD (designer receptors exclusively activated by designer drugs) induces pain in male mice, but not in female mice [17]. These findings underscore sex-dependent differences in spinal microglial responsiveness to specific stimuli and their involvement in pain under physiological conditions. Despite growing evidence gathered employing various tools used to manipulate microglial activity, the key molecules that define male-dominant microglial involvement in neuropathic pain have not been identified definitively. Identifying these molecules is essential for unraveling the sex-biased mechanisms of pain regulation.

Microglia typically produce inflammatory factors that act on surrounding cells [18,19]. Therefore, identifying the factors uniquely expressed in pain-enhancing microglia and characterizing these cells is critical. A previous study using single-cell RNA sequencing (scRNA-seq) identified unique subpopulations of spinal microglia predominantly arising in male mice following peripheral nerve injury [20]. Based on these findings, we reanalyzed spinal microglial transcriptomes at single-cell resolution to identify a key soluble factor, the chemokine CC-chemokine ligand 4 (CCL4), which is predominantly expressed in male microglia under pathological conditions. Furthermore, we validated the functional significance of CCL4 via conventional biochemical and behavioral assays and elucidated the mechanisms underlying sex differences in microglia-mediated neuropathic pain.

2. Materials and Methods

2.1. Mice

All animal experiments were approved by the Animal Research Committee of Wakayama Medical University and were conducted in accordance with the in-house guidelines for the care and use of laboratory animals at Wakayama Medical University, as well as the Animal Research: Reporting of In Vivo Experiments (ARRIVE) guidelines. Male and female C57BL/6 mice (6–8-weeks-old) were purchased from SLC (Hamamatsu, Japan) and used for the experiments at 8–12-weeks of age. All mice were housed in groups of 5–6 in plastic cages under controlled temperature (23–24 °C), relative humidity (60–70%), and a 12 h dark/light cycle, with free access to food and water.

2.2. Processing of scRNA-Seq Data

We analyzed scRNA-seq data from male and female mouse microglia using publicly available data from the gene expression omnibus data repository (accession number: GSE162807) [20]. This study included 142,905 cells collected before and after peripheral nerve injury (at 3 days, 14 days, and 5 months), which were aligned to the mouse mm10 reference transcriptome. To identify the cell types, we analyzed all microglia samples and integrated them using the Harmony method [21] in the Seurat single-cell analysis package (Version 5.1.0) [22]. Initially, cells expressing more than 4000 genes (potential doublets) or fewer than 500 genes, as well as those with more than 5% mitochondrial genes, were filtered out. After normalizing, selecting variable features, and scaling each dataset separately for the filtered 135,806 cells, we performed principal component analysis for dimensional reduction. We then used the *IntegrateLayers* function to combine all the sample datasets. The integrated dataset was processed using the *FindNeighbors* function (reduction = "harmony", dims = 1:11) to obtain the nearest-neighbor graph, the *FindClusters* function (resolution = 0.3) to identify each cell population, and the *RunUMAP* function (reduction = "harmony", dims = 1:11) for visualization of the integrated dataset using Uniform Manifold Approximation and Projection (UMAP) dimension reduction. This process resulted in the identification of 11 cell type clusters.

2.3. Gene Set Enrichment Analysis

To functionally annotate the marker genes in each cluster, we conducted gene set enrichment analysis (GSEA) [23] using the R package fgsea (version 1.32.0). First, we identified marker genes in each cell cluster in Seurat using the *FindMarkers* function with the parameters test.use = wilcox, min.pct = 0.01 and logfc.threshold = 0.1. For each cluster, we generated a gene list sorted in decreasing order of $-\log_{10}$ p-values and used this list for GSEA. After performing GSEA with the molecular signatures database (MSigDB) hallmark dataset [24], we considered gene sets enriched at an FDR < 0.05.

2.4. Partial Sciatic Nerve Ligation (PSL) Model

The mice were subjected to PSL, as previously described [25,26]. Briefly, under isoflurane anesthesia, the left common sciatic nerve of each mouse was exposed at the mid-thigh level by making a small skin incision on one side, hereafter referred to as the ipsilateral side. Approximately one-third of the sciatic nerve was tightly ligated with a silk suture (Natsume Seisakusho, Tokyo, Japan), followed by suturing of the muscle and skin layers and sterilization of the surgical area with povidone–iodine. The untreated right limb was considered the contralateral limb.

2.5. Administration of Pexidartinib (PLX3397)

To deplete macrophages and microglia in vivo, PLX3397 (MedChemExpress, Monmouth Junction, NJ, USA), a CSF1 receptor (CSF1R) inhibitor, was formulated into the AIN-76A rodent diet (Research Diets, New Brunswick, NJ, USA) at 290 mg/kg. The PLX3397 dose was established based on a previous report [27]. The mice had free access to the PLX3397-formulated diet for two weeks instead of normal food, as PLX3397 is orally active. The AIN-76A rodent diet was used as the control.

2.6. Immunohistochemistry

The lumbar (L4–5) spinal cord was harvested from euthanized mice following transcardial perfusion with phosphate-buffered saline (PBS) and fixed in 4% paraformaldehyde/phosphate-buffer solution. The specimens were post-fixed in 4% paraformaldehyde and cryoprotected in 30% sucrose/PBS solution at 4 °C overnight. After embedding in a freezing compound (Sakura, Tokyo, Japan), frozen tissues were longitudinally sectioned at 30 μm thickness using a cryostat (Leica Microsystems, Wetzlar, Germany), and the sections were floated in PBS. The sections were treated with PBS containing 0.1% Triton X-100 (PBST) for 1 h and then blocked with 5% donkey serum at room temperature (15–25 °C) for 2 h. The sections were then incubated overnight at 4 °C with primary antibodies targeting IBA1 (rabbit polyclonal, 1:1000; Fujifilm Wako, Osaka, Japan), NeuN (mouse monoclonal, 1:500; Millipore, Billerica, MA, USA), and c-fos (rabbit polyclonal, 1:50; Santa Cruz Biotechnology, Dallas, TX, USA). The sections were rinsed in PBST and incubated with fluorescent dye-conjugated secondary antibodies (1:200; Thermo Fisher Scientific, Waltham, MA, USA) at room temperature for 2 h. Finally, the sections were washed with PBS, mounted on glass slides, and covered with coverslips using the DAPI-Fluoromount-G (Southern Biotechnology Associates, Birmingham, AL, USA). Fluorescent images were acquired using a confocal laser-scanning microscope (Olympus, Tokyo, Japan). The number of IBA1$^+$ cells within the lamina I-III of the SDH was measured in a square area (200 × 200 μm^2) using the FLUOVIEW software (Version 2.5.1.228).

2.7. Reverse Transcription-Quantitative Polymerase Chain Reaction (RT-qPCR)

Mice were euthanized using isoflurane, and fresh dorsal horns of the lumbar (L4–5) SDH samples were collected in RNAlater solution (Thermo Fisher Scientific). Total RNA was isolated from tissues using the TRIzol Plus RNA Purification Kit (Thermo Fisher Scientific) following the manufacturer's instructions. Briefly, tissues were placed in a 1.5 mL RNase-free tube and homogenized with TRIzol reagent. Chloroform was added to each sample, and the mixture was then centrifuged at 4 °C for 15 min. The aqueous phase containing RNA was transferred to a fresh tube, and RNA was isolated using a purification column. Total RNA extract (1 μg) was incubated with random primers (Promega, Madison, WI, USA) at 70 °C for 5 min and subsequently cooled on ice. Samples were converted into cDNA by incubation with M-MLV Reverse Transcriptase (Promega) and dNTP Mix (Promega). qPCR was performed using the AriaMx Real-Time PCR System (Agilent Technologies, Santa Clara, CA, USA) with template cDNA (10 ng), primers for each gene (Thermo Fisher Scientific), and SYBR Premix Ex Taq II (Takara Bio, Kusatsu, Japan). The reactions were performed under the following conditions: 3 min at 95 °C, followed by 45 cycles of step two comprising 10 s at 95 °C and 30 s at 60 °C. Fluorescence intensities were recorded and the data were normalized to β-actin expression (*Actb*). The primer sequences used were as follows: *Actb*, 5′-CAGCTGAGAGGGAAATCGTG-3′ and 5′-TCTCCAGGGAGGAAGAGGAT-3′; *Cd11b*, 5′-GTTTCTACTGTCCCCCAGCA-3′ and 5′-GTTGGAGCCGAACAAATAGC-3′; *Ccl4*, 5′-ATGAAGCTCTGCGTGTCTGC-3′ and 5′-GCCGGGAGGTGTAAGAGAAA-3′.

2.8. Drug Administration

A CC-chemokine receptor 5 (CCR5) antagonist (Maraviroc; Tocris Biosciences, Bristol, UK) was dissolved in dimethyl sulfoxide and diluted in sterile PBS for further use. Based on previous reports [9,28], 20 nmol maraviroc was administered intrathecally (i.t.) on day 7 after PSL. Recombinant CCL4 (BioLegend, San Diego, CA, USA) was dissolved in sterile PBS and i.t. administered at a dose of 1 or 10 pmol. Under isoflurane anesthesia, an i.t. injection was administered in the region between the spinal L5 and L6 vertebrae using a 30-gauge needle fitted with a Hamilton microsyringe [29].

2.9. Von Frey Test

The mechanical pain threshold was determined using the von Frey test, as previously described [17]. Briefly, mice were individually placed on a metal mesh grid floor (5 × 5 mm) and covered with an acrylic box. After a 2- to 3 h adaptation period, calibrated von Frey filaments (Neuroscience, Tokyo, Japan) were applied to the middle of the plantar surface of the hind paw through the mesh floor. The filament set used in this study consisted of nine calibrated von Frey filaments: 0.02, 0.04, 0.07, 0.16, 0.4, 0.6, 1.0, 1.4, and 2.0 g. Using the up–down method, testing began with the application of 0.4 g filament. Quick withdrawal, shaking, biting, or licking of the stimulated paw was considered a positive paw-withdrawal response. If no withdrawal response occurred, the next stronger stimulus was applied. Conversely, the next weaker stimulus was selected following paw withdrawal, in accordance with Chaplan's procedure [30]. Once the response threshold was crossed (two responses were straddling the threshold), the 50% paw-withdrawal threshold was calculated based on these responses.

2.10. Gene Expression Analysis

Expression profiling of CCL4 (*Ccl4*) gene across a diverse range of normal tissues, organs, and cell lines in mice was visualized using BioGPS (http://biogps.org/).

2.11. Statistical Analysis

All data are presented as the mean ± standard error of the mean (SEM). To compare differences between two groups, a two-tailed Student's *t*-test or Welch's *t*-test was used. To compare the differences between the four groups with two factors, two-way ANOVA followed by Tukey's multiple comparison test was used. Statistical analyses were performed using the GraphPad Prism software (GraphPad Software, Version 10.1.2, Boston, MA, USA), and statistical significance was set at $p < 0.05$.

3. Results

3.1. Expression of CCL4 by Male-Dominant Subpopulation of Spinal Microglia After Nerve Injury

To identify microglia-secreted pain-enhancing molecules in male-dominant subpopulations of activated microglia in the SDH after peripheral nerve injury (spared nerve injury: SNI), we reanalyzed a scRNA-seq dataset from a previously published study (Tansley et al.) [20]. Clustering analysis of microglia under the following conditions (day 3, day 14, 5 months after SNI and naïve controls for both sexes) revealed 11 distinct clusters (Figure 1A). Consistent with prior findings, canonical microglia genes, such as *Tmem119*, *Fcrls*, *P2ry12*, *Cx3cr1*, *Trem2*, and *C1qa*, were expressed in all cell type clusters, representing most of the analyzed microglia (Figure S1). In contrast, clusters 6 and 10, which accounted for a portion of the microglia, exhibited unique transcriptional profiles. The heatmap showed that the primary population in clusters 6 and 10 comprised microglia from male mice on day 3 after SNI, followed by female mice, whereas microglia from day 14 and 5 months after SNI, and naïve controls of both sexes were minimally represented (Figure 1B). Activated microglia play a crucial role in the development of neuropathic pain, highlighting the importance of early time points following nerve injury. Microglia categorized within clusters 6 and 10 showed an increase in both sexes on day 3 after SNI compared to naïve controls. However, the proportion of male microglia was greater than that of female microglia, suggesting that clusters 6 and 10 reflected sex-specific differences in activated microglia following nerve injury.

Given the pivotal role of inflammation-related molecules in the pathophysiology of neuropathic pain, we analyzed the gene expression patterns in these clusters according to sex. GSEA revealed activation of the TNFA_SIGNALING_VIA_NFKB pathway in cluster 6 microglia. Furthermore, 45 hallmark genes were identified as characteristic of male-dominant microglia after nerve injury. Although *Rhob* exhibited the most pronounced difference between males and females among all genes in cluster 6, we prioritized the soluble inflammatory molecule CCL4, which exhibited the second greatest sex difference. (Figure 1C). A violin plot for *Ccl4* expression levels across all microglia in cluster 6 confirmed that CCL4-expressing microglia were predominantly from male mice on day 3 after SNI (Figure 1D). Additionally, the number of CCL4-expressing microglia in males was greater than that in females in cluster 10 microglia (Figure S2), although the transcriptional profile of cluster 10 differed from that of cluster 6. These findings suggest that *Ccl4* expression is a defining feature of male-dominant microglia involved in neuropathic pain.

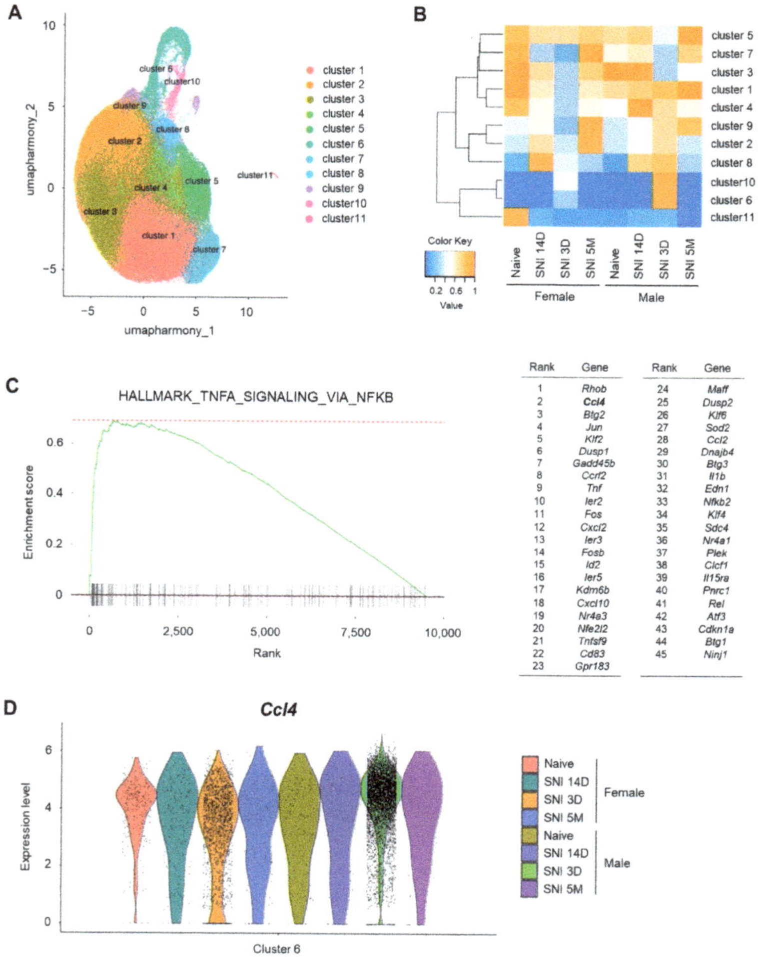

Figure 1. Male-dominant subpopulation of spinal microglia after nerve injury express CCL4. (**A**) Uniform manifold approximation and projection (UMAP) visualization of mouse microglia (n = 2 control samples and n = 6 injury samples from male and female mice), colored by cell type cluster. The control naive sample included 33,330 cells, whereas the spared nerve injury (SNI) model samples included 102,476 cells collected on day 3, day 14, and 5 months after SNI. (**B**) Heatmap based on cell number ratio. This heatmap indicates the ratio of cells classified into each cluster relative to the total number of cells in each sample. The vertical axis represents the cell type cluster numbers, whereas the horizontal axis represents the sample types. (**C**) Result of gene set enrichment analysis (GSEA) of marker genes in males compared with those in females for the "HALLMARK_TNFA_SIGNALING_VIA_NFKB" gene set. List of 45 leading edge genes are shown. (**D**) Violin plot for *Ccl4* genes, markers of cluster 6 cells.

3.2. Male Microglia-Dominant Upregulation of CCL4 in the Spinal Dorsal Horn

Using immunohistochemistry, we evaluated whether microglial activation differed in the SDH of male and female mice following PSL. On day 7 after PSL, the number of IBA1$^+$ microglia was similarly increased on the ipsilateral side of the SDH in both male and female mice (Figure 2A,B). Next, we used RT-qPCR to assess the time course of *Ccl4* expression in the SDH after PSL. *Ccl4* was significantly upregulated on days 7 and 14 in male mice, but not in female mice, with expression levels being markedly higher in males at both time points (Figure 2C), whereas other pain-related genes, such as *Ccl3*, were similarly upregulated in both sexes [9]. As previously reported, treatment with PLX3397, an inhibitor of the CSF1 receptor, substantially reduced the expression of the microglial marker *Cd11b* in the SDH on day 7 after PSL in a sex-dependent manner. Consistently, *Ccl4* expression was significantly decreased upon PLX3397 treatment in male mice but remained unchanged in female mice (Figure 2D). A database search using BioGPS confirmed that *Ccl4* is expressed not only by peripheral macrophages but also by microglia (Figure S3). These findings indicate that *Ccl4* is upregulated in SDH microglia after PSL in males but not in females.

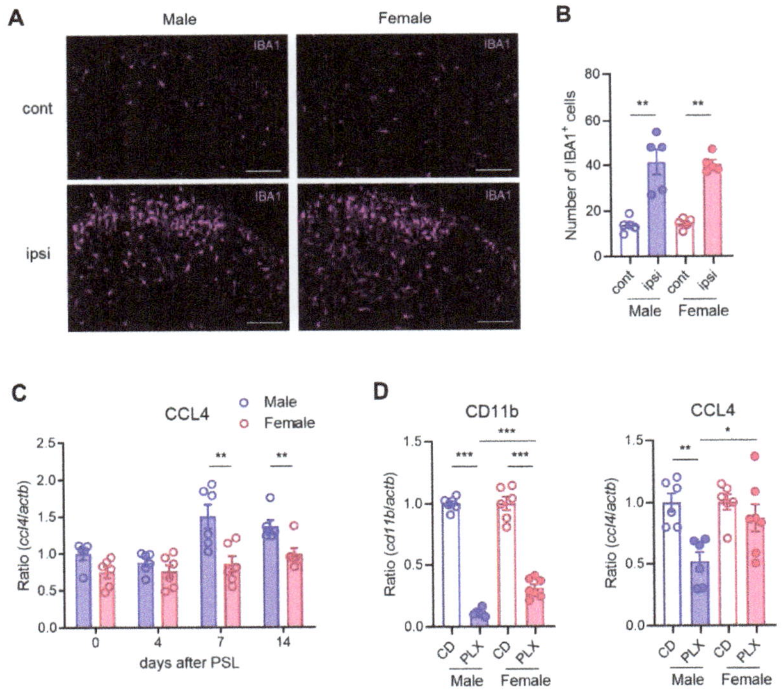

Figure 2. Male microglia-dominant upregulation of CCL4 in the spinal dorsal horn. Male and female mice were subjected to partial sciatic nerve ligation (PSL), and the lumbar spinal dorsal horn (SDH) was collected. (**A**) IBA1$^+$ microglia in the SDH on day 7 after PSL were visualized using immunohistochemistry. Scale bars = 100 µm. (**B**) Quantitative analysis of the number of IBA1$^+$ cells in the lamina I-III in the SDH ($n = 5$, Welch's t-test, ** $p < 0.01$). (**C**) mRNA levels of *Ccl4* on days 0, 4, 7, and 14 after PSL were analyzed using RT-qPCR (n = 5–6, Student's t-test, ** $p < 0.01$). (**D**) Mice were fed a control diet (CD) or PLX3397 diet for 7 days before PSL and subsequently subjected to PSL. The lumbar ipsilateral SDH was collected on day 7 after PSL. Fold changes in mRNA levels of *Cd11b* and *Ccl4* in PLX3397-fed mice compared with those in control-fed mice. (n = 6–7, Student's t-test, *** $p < 0.001$, ** $p < 0.01$, * $p < 0.05$).

3.3. Sexually Dimorphic Effect of CCR5 Antagonist in Relieving Neuropathic Pain

CCR5 is the principal receptor for CCL4, despite the complexity of the chemokine ligand–receptor system [31,32]. To assess whether the CCL4–CCR5 axis plays a sex-dependent pathophysiological role in neuropathic pain in the SDH, we investigated the effects of CCR5 blockade. Maraviroc, a CCR5 antagonist [33], was evaluated for its suppressive effects on PSL-induced neuropathic pain. Mechanical pain thresholds, assessed using the von Frey test, were significantly reduced on the ipsilateral side on day 7 after PSL in both male and female mice, confirming the development of mechanical allodynia. The i.t. administration of maraviroc (20 nmol) on day 7 transiently, but markedly, relieved PSL-induced mechanical allodynia 3 h after administration in male mice. In contrast, maraviroc did not suppress allodynia in female mice (Figure 3). These findings suggest that enhancement of the CCL4–CCR5 axis in the SDH plays a critical role in neuropathic pain in male mice.

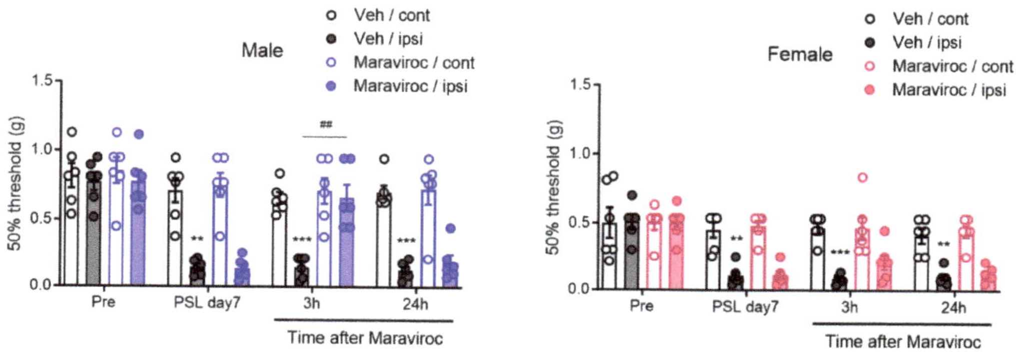

Figure 3. Sexually dimorphic effect of CCR5 antagonist maraviroc in relieving neuropathic pain. Male and female mice were subjected to PSL, and maraviroc (20 nmol) was intrathecally (i.t.) administered once on day 7 after PSL. The 50% mechanical threshold on days 0 (pre) and 7 after PSL, and 3 and 24 h after i.t. administration of maraviroc in male and female mice were assessed employing the up–down method using the von Frey test ($n = 6$, two-way ANOVA followed by Tukey's multiple comparison test, *** $p < 0.001$, ** $p < 0.01$ vs. Veh/cont, ## $p < 0.01$).

3.4. Sex-Independent Allodynic Effects of CCL4 in the Spinal Dorsal Horn

To determine whether an increase in CCL4 is sufficient to induce mechanical allodynia, we evaluated the effects of exogenous CCL4 on pain sensitivity in both male and female mice. A single i.t. administration of CCL4 (1 or 10 pmol) significantly reduced the mechanical pain threshold, indicating mechanical allodynia 6 h after administration in both sexes. The allodynic effects persisted for at least 72 h in a dose-dependent manner (Figure 4A). Further analysis showed that c-fos protein expression in the SDH was significantly increased in both male and female mice on day 1 after i.t. administration of CCL4 (10 pmol). These c-fos signals colocalized with NeuN, a marker of neuronal nuclei, indicating the activation of pain-processing neurons in the SDH (Figure 4B,C). These results indicate that CCL4 exerts a potent allodynic effect by reducing the mechanical pain thresholds in a sex-independent manner.

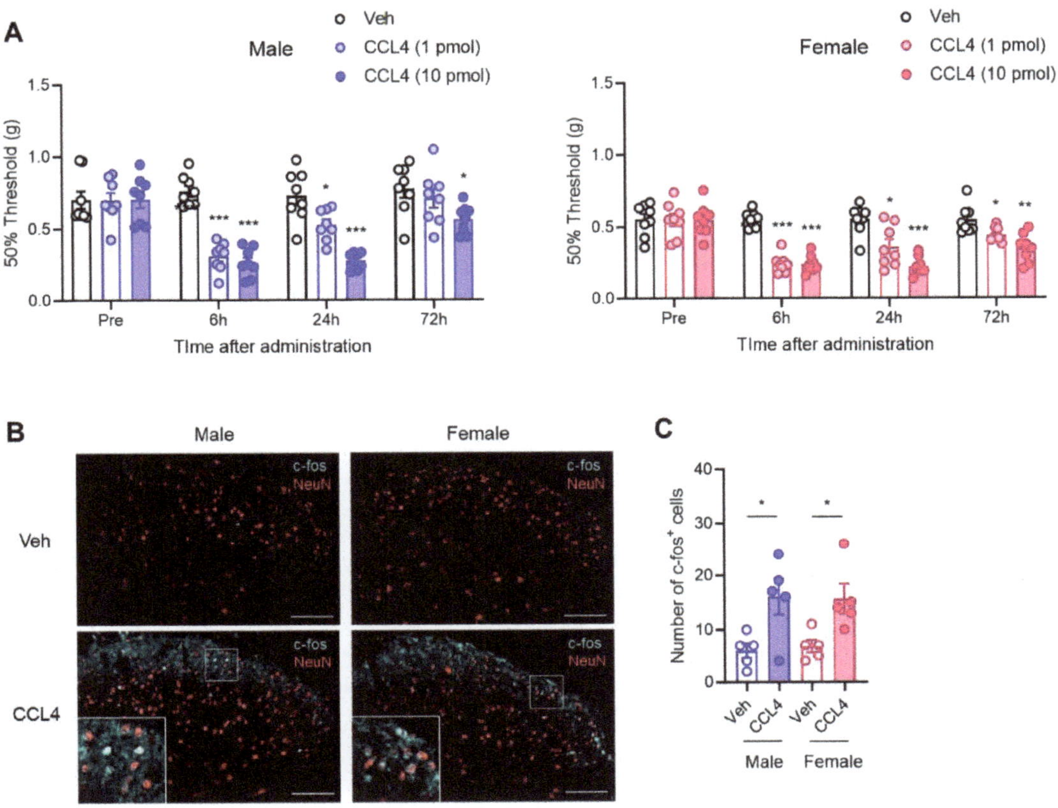

Figure 4. Sex-independent allodynic effects of intrathecally administered CCL4. Recombinant CCL4 (1 or 10 pmol) was intrathecally administered to naïve male and female mice. (**A**) The 50% mechanical threshold before and 6, 24, and 72 h after administration in male and female mice were assessed employing the up–down method using the von Frey test ($n = 8$, two-way ANOVA followed by Tukey's multiple comparison test, *** $p < 0.001$, ** $p < 0.01$, * $p < 0.05$ vs. Veh). (**B**) Expression of c-fos protein in the SDH on day 1 after administration of CCL4 (10 pmol) was visualized via immunohistochemistry. Scale bars = 100 µm. (**C**) Quantitative analysis of the number of c-fos$^+$ cells in the lamina I–III of the SDH ($n = 5$, Student's t-test, * $p < 0.05$).

4. Discussion

In this study, we identified sex-biased characteristics of spinal microglia that contribute to the etiology of neuropathic pain. Bioinformatic analysis of published scRNA-seq datasets for spinal microglia revealed male-dominant expression of the chemokine *Ccl4* in specific microglial subpopulations following SNI. We further confirmed that *Ccl4* mRNA was upregulated in the SDH of males, but not females, after PSL, despite both sexes exhibiting similar morphological activation of spinal microglia. The i.t. administration of maraviroc, an inhibitor of the CCL4-CCR5 signaling pathway, significantly suppressed PSL-induced mechanical allodynia in male mice. However, the anti-allodynic effects of maraviroc were not significant in female mice. Notably, the i.t. administration of recombinant CCL4 induced mechanical allodynia in both sexes, indicating that the sex-dependent pathophysiological roles of microglia-derived CCL4 in neuropathic pain are mediated via its expression following peripheral nerve injury.

Following the discovery of a causal link between microglial activation and neuropathic pain [34], numerous studies have highlighted the roles of soluble factors derived from activated microglia in pain hypersensitivity. For example, inflammatory cytokines (e.g., interleukin-1β and tumor necrosis factor-α), chemokines (e.g., CCL2 and CCL3), growth factors, and lipid mediators have been shown to play critical roles in neuropathic pain by enhancing neuronal excitability and/or amplifying neuroinflammation [5,13,35,36]. This process is accompanied by the activation of microglia and astrocytes, ultimately leading to hyperexcitation of pain-processing pathways [6,37]. Although accumulating evidence suggests pronounced sexually dimorphic characteristics of microglia in the etiology of neuropathic pain [7,8,38], the existence of sex differences in the pathophysiological roles of these molecules remains underexplored. Elucidating the mechanisms underlying sex differences in microglia-driven neuropathic pain requires identification of critical molecules that define male-dominant microglial contributions to neuropathic pain.

In a previous study, Tansley et al. classified spinal microglia from naïve, sham, and SNI mice into 11 distinct clusters [20]. Among these, unique subpopulations (original clusters 7–9) were predominantly composed of microglia from SNI day 3 in both sexes. On SNI day 3, morphological activation and proliferation of microglia in the SDH were significantly increased in both sexes, and these microglia exhibited distinct transcriptome profiles compared to other subpopulations. Notably, the number of proliferative microglia in male mice was greater than that in female mice, suggesting sex differences in microglial characteristics [20]. We hypothesized that further analysis of their transcriptome data would help identify novel soluble molecules that drive functional sex differences in microglia. Our findings revealed that the TNFα signaling via the NF-κB pathway was upregulated in cluster 6, aligning with the established role of neuroinflammatory process in the etiology of neuropathic pain. Importantly, the proportion of microglia in male SNI day 3 was greater than that in female SNI day 3 within cluster 6, and CCL4 expressing microglia were primarily observed in males. Given that cluster 6 in our analysis aligns with previous findings [20], CCL4 emerges as a key molecule associated with sex differences in activated microglia after nerve injury.

Several neuropathic pain models have been established to investigate the etiology of neuropathic pain and to evaluate the therapeutic potential of drugs [39,40]. Among the four main nerve injury models, spinal microglial activation is a common feature, and the inhibition of activated microglia has been shown to alleviate neuropathic pain across these models [11,25,41,42]. However, the temporal dynamics of microglial activation and expression profiles of microglia-derived molecules may vary between models. To elucidate the common pathophysiological mechanisms underlying neuropathic pain, it is essential to examine key phenomena across different neuropathic pain models. Several studies have demonstrated significant microglial activation and upregulation of inflammatory molecules, sustained for at least two weeks after nerve injury, in neuropathic pain models [9,36,43]. Therefore, although Tansley et al. employed the SNI model [20], their scRNA-seq transcriptome data can be used to investigate the mechanisms underlying PSL-induced neuropathic pain. Indeed, the male-dominant upregulation of CCL4 after PSL in our study indicates the existence of shared mechanisms between the SNI and PSL models.

CCL4, also known as macrophage inflammatory protein-1β, exerts diverse effects on various cell types via its interaction with CCR5. Several lines of evidence suggest that the CCL4–CCR5 axis facilitates inflammation that underlies intractable diseases, such as osteoarthritis and diabetes, and promotes tumor development and progression by modulating lymphocytes, macrophages, and tissue-resident cells [44–47]. Additionally, CCL4 is upregulated under pathological neurogenic conditions following nerve injury, and inhibition of macrophage- or Schwann cell-derived CCL4 has been shown to attenuate neuropathic pain [48]. CCL4 also plays a crucial role in several neuroinflammatory diseases of the central nervous system, including traumatic brain injury and Alzheimer's disease [47,49,50]. Thus, CCL4 upregulation likely contributes to pain hypersensitivity at the spinal level. A previous study reported that CCL4 was upregulated in the SDH following chronic constriction injury and that i.t. administration of maraviroc attenuated neuropathic pain in male rats [28]. However, the study did not address sex-related differences. In the present study, we demonstrated that CCL4 was upregulated exclusively in male mice, and that the CCL4–CCR5 signaling pathway was enhanced in the SDH following peripheral nerve injury in a sex-dependent manner.

Given that perineurally administered maraviroc at the site surrounding the injured sciatic nerve exhibited anti-allodynic effects in both male and female PSL models, as previously reported [9], it is surprising that i.t. administered maraviroc showed anti-allodynic effects exclusively in male PSL models. Despite the sexually dimorphic anti-allodynic effects of i.t. administered maraviroc, the i.t. administration of CCL4 induced robust allodynia in both sexes, whereas a previous report presented an allodynic effect of i.t. administered CCL4 only in males [51]. These findings suggest that activation of CCR5 in the SDH leads to pain hypersensitivity in both males and females. CCR5 is highly expressed in immune cells, including spinal microglia, and is also known to be expressed in neurons [51–54]. Indeed, the induction of c-fos expression following CCL4 administration supports the excitation of pain-processing neurons. Interestingly, CCL3, another CCR5 ligand, was similarly upregulated in the SDH of both sexes after nerve injury, as previously reported [9]. However, CCL4 upregulation showed clear sex differences, being elevated only in males. This transcriptional disparity may underlie sex-biased characteristics of microglia. Understanding the mechanisms underlying these sex-related differences is crucial. We previously demonstrated that the sexually dimorphic characteristics of spinal microglia under neuropathic pain conditions are influenced by circulating androgens [9]. Further studies are needed to determine whether androgens modulate CCL4 expression and to elucidate the transcriptional mechanisms responsible for this regulation.

Overall, we demonstrated that the upregulation of CCL4 in the SDH following peripheral nerve injury occurs exclusively in males. Importantly, CCL4 may serve as a marker for male-dominant microglia involved in neuropathic pain. Although pharmacological inhibition of the CCL4–CCR5 pathway with i.t. maraviroc showed sexually dimorphic alleviation of neuropathic pain in the PSL model, activation of CCR5 via i.t. administration of exogenous CCL4 induced allodynia in both sexes. This suggests that the induction of CCL4 in activated microglia within the SDH may represent a crucial mechanism underlying sex differences in microglia-driven neuropathic pain. Considering that this sexual dimorphism in microglia is androgen-dependent, elucidating the regulatory mechanisms of spinal microglia mediated by androgen signaling can provide deeper insights into sex differences in pain processing.

Supplementary Materials: The following supporting information can be downloaded at: https://www.mdpi.com/article/10.3390/cells14070484/s1, Figure S1: Expression of canonical microglia genes in all cell type clusters; Figure S2: Expression of *Ccl4* in male-dominant subpopulation; Figure S3: Expression profiling of *Ccl4* in mice.

Author Contributions: Conceptualization, T.S., S.H., K.S. and N.K.; methodology, T.S. and N.K.; software, T.S. and T.N.; validation, T.S., S.H., K.S. and N.K.; formal analysis, F.S., T.S., T.N. and N.K.; investigation, F.S., T.S., T.N. and N.K.; resources, T.S. and N.K.; data curation, F.S., T.S., T.N., Y.F.,

S.H., K.S. and N.K.; writing—original draft preparation, F.S., T.S. and N.K.; writing—review and editing, T.S., T.N., Y.F., S.H., K.S. and N.K.; visualization, T.S. and N.K.; supervision, N.K.; project administration, N.K.; funding acquisition, T.S., Y.F., K.S. and N.K. All authors have read and agreed to the published version of the manuscript.

Funding: This work was supported by JSPS KAKENHI (Grant Number: 23K08371 to Y.F.), the Japan Agency for Medical Research and Development (Grant Number: JP21gk0210029 to N.K.), the Smoking Research Foundation (to N.K.), the Takeda Science Foundation (to N.K.), the Mochida Memorial Foundation (to N.K.), the Naito Foundation (to N.K.), the Daiichi Sankyo Foundation of Life Science (to N.K.), and the Program of the Joint Usage/Research Center for Developmental Medicine and High Depth Omics, IMEG, Kumamoto University (to T.S., K.S. and N.K.).

Institutional Review Board Statement: This study was approved by the Animal Research Committee of Wakayama Medical University (approval No. 781, 828, Tora 37, Tora90).

Informed Consent Statement: Not applicable.

Data Availability Statement: The original contributions presented in this study are included in the article/Supplementary Materials. Further inquiries can be directed to the corresponding authors.

Acknowledgments: We are grateful to Mayumi Shibutani, Haruna Mano, Akari Minami, and Momo Kubota for their technical assistance. The authors thank the Medical Research Center organizers at Saitama Medical University for providing continuous support throughout this study.

Conflicts of Interest: The authors declare no conflicts of interest.

Abbreviations

The following abbreviations are used in this manuscript:

CCL4	CC-chemokine ligand 4
CCR5	CC-chemokine receptor 5
CSF1	colony-stimulating factor 1
GSEA	gene set enrichment analysis
PSL	partial sciatic nerve ligation
RT-qPCR	reverse transcription-quantitative polymerase chain reaction
scRNA-seq	single-cell RNA sequencing
SDH	spinal dorsal horn
SNI	spared nerve injury
UMAP	uniform manifold approximation and projection

References

1. Julius, D.; Basbaum, A.I. Molecular mechanisms of nociception. *Nature* **2001**, *413*, 203–210. [CrossRef] [PubMed]
2. Basbaum, A.I.; Bautista, D.M.; Scherrer, G.; Julius, D. Cellular and molecular mechanisms of pain. *Cell* **2009**, *139*, 267–284. [CrossRef] [PubMed]
3. CoHen, S.P.; Vase, L.; Hooten, W.M. Chronic pain: An update on burden, best practices, and new advances. *Lancet* **2021**, *397*, 2082–2097. [CrossRef]
4. Colloca, L.; Ludman, T.; Bouhassira, D.; Baron, R.; Dickenson, A.H.; Yarnitsky, D.; Freeman, R.; Truini, A.; Attal, N.; Finnerup, N.B.; et al. Neuropathic pain. *Nat. Rev. Dis. Primers* **2017**, *3*, 17002. [CrossRef] [PubMed]
5. Ji, R.R.; Chamessian, A.; Zhang, Y.Q. Pain regulation by non-neuronal cells and inflammation. *Science* **2016**, *354*, 572–577. [CrossRef]
6. Tsuda, M.; Masuda, T.; Kohno, K. Microglial diversity in neuropathic pain. *Trends Neurosci.* **2023**, *46*, 597–610. [CrossRef]
7. Midavaine, E.; Cote, J.; Marchand, S.; Sarret, P. Glial and neuroimmune cell choreography in sexually dimorphic pain signaling. *Neurosci. Biobehav. Rev.* **2021**, *125*, 168–192. [CrossRef]
8. Mapplebeck, J.C.S.; Beggs, S.; Salter, M.W. Sex differences in pain: A tale of two immune cells. *Pain* **2016**, *157* (Suppl. S1), S2–S6. [CrossRef]

9. Saika, F.; Fukazawa, Y.; Hatano, Y.; Kishioka, S.; Hino, Y.; Hino, S.; Suzuki, K.; Kiguchi, N. Sexually dimorphic effects of pexidartinib on nerve injury-induced neuropathic pain in mice. *Glia* **2024**, *72*, 1402–1417. [CrossRef]
10. Sorge, R.E.; Mapplebeck, J.C.; Rosen, S.; Beggs, S.; Taves, S.; Alexander, J.K.; Martin, L.J.; Austin, J.S.; Sotocinal, S.G.; Chen, D.; et al. Different immune cells mediate mechanical pain hypersensitivity in male and female mice. *Nat. Neurosci.* **2015**, *18*, 1081–1083. [CrossRef]
11. Taves, S.; Berta, T.; Liu, D.L.; Gan, S.; Chen, G.; Kim, Y.H.; Van de Ven, T.; Laufer, S.; Ji, R.R. Spinal inhibition of p38 MAP kinase reduces inflammatory and neuropathic pain in male but not female mice: Sex-dependent microglial signaling in the spinal cord. *Brain Behav. Immun.* **2016**, *55*, 70–81. [CrossRef] [PubMed]
12. Butovsky, O.; Weiner, H.L. Microglial signatures and their role in health and disease. *Nat. Rev. Neurosci.* **2018**, *19*, 622–635. [CrossRef]
13. Inoue, K.; Tsuda, M. Microglia in neuropathic pain: Cellular and molecular mechanisms and therapeutic potential. *Nat. Rev. Neurosci.* **2018**, *19*, 138–152. [CrossRef]
14. Kuhn, J.A.; Vainchtein, I.D.; Braz, J.; Hamel, K.; Bernstein, M.; Craik, V.; Dahlgren, M.W.; Ortiz-Carpena, J.; Molofsky, A.B.; Molofsky, A.V.; et al. Regulatory T-cells inhibit microglia-induced pain hypersensitivity in female mice. *Elife* **2021**, *10*, e69056. [CrossRef]
15. Sorge, R.E.; LaCroix-Fralish, M.L.; Tuttle, A.H.; Sotocinal, S.G.; Austin, J.S.; Ritchie, J.; Chanda, M.L.; Graham, A.C.; Topham, L.; Beggs, S.; et al. Spinal cord Toll-like receptor 4 mediates inflammatory and neuropathic hypersensitivity in male but not female mice. *J. Neurosci.* **2011**, *31*, 15450–15454. [CrossRef]
16. Lee, J.; Hwang, H.; Lee, S.J. Distinct roles of GT1b and CSF-1 in microglia activation in nerve injury-induced neuropathic pain. *Mol. Pain.* **2021**, *17*, 17448069211020918. [CrossRef] [PubMed]
17. Saika, F.; Matsuzaki, S.; Kishioka, S.; Kiguchi, N. Chemogenetic Activation of CX3CR1-Expressing Spinal Microglia Using Gq-DREADD Elicits Mechanical Allodynia in Male Mice. *Cells* **2021**, *10*, 874. [CrossRef] [PubMed]
18. Distefano-Gagne, F.; Bitarafan, S.; Lacroix, S.; Gosselin, D. Roles and regulation of microglia activity in multiple sclerosis: Insights from animal models. *Nat. Rev. Neurosci.* **2023**, *24*, 397–415. [CrossRef]
19. Paolicelli, R.C.; Sierra, A.; Stevens, B.; Tremblay, M.E.; Aguzzi, A.; Ajami, B.; Amit, I.; Audinat, E.; Bechmann, I.; Bennett, M.; et al. Microglia states and nomenclature: A field at its crossroads. *Neuron* **2022**, *110*, 3458–3483. [CrossRef]
20. Tansley, S.; Uttam, S.; Urena Guzman, A.; Yaqubi, M.; Pacis, A.; Parisien, M.; Deamond, H.; Wong, C.; Rabau, O.; Brown, N.; et al. Single-cell RNA sequencing reveals time- and sex-specific responses of mouse spinal cord microglia to peripheral nerve injury and links ApoE to chronic pain. *Nat. Commun.* **2022**, *13*, 843. [CrossRef]
21. Korsunsky, I.; Millard, N.; Fan, J.; Slowikowski, K.; Zhang, F.; Wei, K.; Baglaenko, Y.; Brenner, M.; Loh, P.R.; Raychaudhuri, S. Fast, sensitive and accurate integration of single-cell data with Harmony. *Nat. Methods* **2019**, *16*, 1289–1296. [CrossRef] [PubMed]
22. Hao, Y.; Stuart, T.; Kowalski, M.H.; Choudhary, S.; Hoffman, P.; Hartman, A.; Srivastava, A.; Molla, G.; Madad, S.; Fernandez-Granda, C.; et al. Dictionary learning for integrative, multimodal and scalable single-cell analysis. *Nat. Biotechnol.* **2024**, *42*, 293–304. [CrossRef] [PubMed]
23. Subramanian, A.; Tamayo, P.; Mootha, V.K.; Mukherjee, S.; Ebert, B.L.; Gillette, M.A.; Paulovich, A.; Pomeroy, S.L.; Golub, T.R.; Lander, E.S.; et al. Gene set enrichment analysis: A knowledge-based approach for interpreting genome-wide expression profiles. *Proc. Natl. Acad. Sci. USA* **2005**, *102*, 15545–15550. [CrossRef] [PubMed]
24. Liberzon, A.; Birger, C.; Thorvaldsdottir, H.; Ghandi, M.; Mesirov, J.P.; Tamayo, P. The Molecular Signatures Database (MSigDB) hallmark gene set collection. *Cell Syst.* **2015**, *1*, 417–425. [CrossRef] [PubMed]
25. Kiguchi, N.; Kobayashi, D.; Saika, F.; Matsuzaki, S.; Kishioka, S. Inhibition of peripheral macrophages by nicotinic acetylcholine receptor agonists suppresses spinal microglial activation and neuropathic pain in mice with peripheral nerve injury. *J. Neuroinflamm.* **2018**, *15*, 96. [CrossRef]
26. Seltzer, Z.; Dubner, R.; Shir, Y. A novel behavioral model of neuropathic pain disorders produced in rats by partial sciatic nerve injury. *Pain* **1990**, *43*, 205–218. [CrossRef]
27. Elmore, M.R.; Najafi, A.R.; Koike, M.A.; Dagher, N.N.; Spangenberg, E.E.; Rice, R.A.; Kitazawa, M.; Matusow, B.; Nguyen, H.; West, B.L.; et al. Colony-stimulating factor 1 receptor signaling is necessary for microglia viability, unmasking a microglia progenitor cell in the adult brain. *Neuron* **2014**, *82*, 380–397. [CrossRef]
28. Kwiatkowski, K.; Piotrowska, A.; Rojewska, E.; Makuch, W.; Jurga, A.; Slusarczyk, J.; Trojan, E.; Basta-Kaim, A.; Mika, J. Beneficial properties of maraviroc on neuropathic pain development and opioid effectiveness in rats. *Prog. Neuropsychopharmacol. Biol. Psychiatry* **2016**, *64*, 68–78. [CrossRef]

29. Kiguchi, N.; Uta, D.; Ding, H.; Uchida, H.; Saika, F.; Matsuzaki, S.; Fukazawa, Y.; Abe, M.; Sakimura, K.; Ko, M.C.; et al. GRP receptor and AMPA receptor cooperatively regulate itch-responsive neurons in the spinal dorsal horn. *Neuropharmacology* **2020**, *170*, 108025. [CrossRef]
30. Chaplan, S.R.; Bach, F.W.; Pogrel, J.W.; Chung, J.M.; Yaksh, T.L. Quantitative assessment of tactile allodynia in the rat paw. *J. Neurosci. Methods* **1994**, *53*, 55–63. [CrossRef]
31. Comerford, I.; McColl, S.R. Atypical chemokine receptors in the immune system. *Nat. Rev. Immunol.* **2024**, *24*, 753–769. [CrossRef]
32. Proudfoot, A.E. Chemokine receptors: Multifaceted therapeutic targets. *Nat. Rev. Immunol.* **2002**, *2*, 106–115. [CrossRef]
33. Dorr, P.; Westby, M.; Dobbs, S.; Griffin, P.; Irvine, B.; Macartney, M.; Mori, J.; Rickett, G.; Smith-Burchnell, C.; Napier, C.; et al. Maraviroc (UK-427,857), a potent, orally bioavailable, and selective small-molecule inhibitor of chemokine receptor CCR5 with broad-spectrum anti-human immunodeficiency virus type 1 activity. *Antimicrob. Agents Chemother.* **2005**, *49*, 4721–4732. [CrossRef] [PubMed]
34. Tsuda, M.; Shigemoto-Mogami, Y.; Koizumi, S.; Mizokoshi, A.; Kohsaka, S.; Salter, M.W.; Inoue, K. P2X4 receptors induced in spinal microglia gate tactile allodynia after nerve injury. *Nature* **2003**, *424*, 778–783. [CrossRef] [PubMed]
35. Milligan, E.D.; Watkins, L.R. Pathological and protective roles of glia in chronic pain. *Nat. Rev. Neurosci.* **2009**, *10*, 23–36. [CrossRef] [PubMed]
36. Matsushita, K.; Tozaki-Saitoh, H.; Kojima, C.; Masuda, T.; Tsuda, M.; Inoue, K.; Hoka, S. Chemokine (C-C motif) receptor 5 is an important pathological regulator in the development and maintenance of neuropathic pain. *Anesthesiology* **2014**, *120*, 1491–1503. [CrossRef]
37. Ji, R.R.; Xu, Z.Z.; Gao, Y.J. Emerging targets in neuroinflammation-driven chronic pain. *Nat. Rev. Drug Discov.* **2014**, *13*, 533–548. [CrossRef]
38. Mogil, J.S. Qualitative sex differences in pain processing: Emerging evidence of a biased literature. *Nat. Rev. Neurosci.* **2020**, *21*, 353–365. [CrossRef]
39. Mogil, J.S. Animal models of pain: Progress and challenges. *Nat. Rev. Neurosci.* **2009**, *10*, 283–294. [CrossRef]
40. Calvo, M.; Dawes, J.M.; Bennett, D.L. The role of the immune system in the generation of neuropathic pain. *Lancet Neurol.* **2012**, *11*, 629–642. [CrossRef]
41. Masuda, T.; Tsuda, M.; Yoshinaga, R.; Tozaki-Saitoh, H.; Ozato, K.; Tamura, T.; Inoue, K. IRF8 is a critical transcription factor for transforming microglia into a reactive phenotype. *Cell Rep.* **2012**, *1*, 334–340. [CrossRef]
42. Guan, Z.; Kuhn, J.A.; Wang, X.; Colquitt, B.; Solorzano, C.; Vaman, S.; Guan, A.K.; Evans-Reinsch, Z.; Braz, J.; Devor, M.; et al. Injured sensory neuron-derived CSF1 induces microglial proliferation and DAP12-dependent pain. *Nat. Neurosci.* **2016**, *19*, 94–101. [CrossRef] [PubMed]
43. Yang, H.; Wu, L.; Deng, H.; Chen, Y.; Zhou, H.; Liu, M.; Wang, S.; Zheng, L.; Zhu, L.; Lv, X. Anti-inflammatory protein TSG-6 secreted by bone marrow mesenchymal stem cells attenuates neuropathic pain by inhibiting the TLR2/MyD88/NF-kappaB signaling pathway in spinal microglia. *J. Neuroinflamm.* **2020**, *17*, 154. [CrossRef] [PubMed]
44. Koch, A.E.; Kunkel, S.L.; Shah, M.R.; Fu, R.; Mazarakis, D.D.; Haines, G.K.; Burdick, M.D.; Pope, R.M.; Strieter, R.M. Macrophage inflammatory protein-1 beta: A C-C chemokine in osteoarthritis. *Clin. Immunol. Immunopathol.* **1995**, *77*, 307–314. [CrossRef]
45. Chang, T.T.; Chen, J.W. Emerging role of chemokine CC motif ligand 4 related mechanisms in diabetes mellitus and cardiovascular disease: Friends or foes? *Cardiovasc. Diabetol.* **2016**, *15*, 117. [CrossRef] [PubMed]
46. Mukaida, N.; Sasaki, S.I.; Baba, T. CCL4 Signaling in the Tumor Microenvironment. *Adv. Exp. Med. Biol.* **2020**, *1231*, 23–32.
47. Ciechanowska, A.; Mika, J. CC Chemokine Family Members' Modulation as a Novel Approach for Treating Central Nervous System and Peripheral Nervous System Injury-A Review of Clinical and Experimental Findings. *Int. J. Mol. Sci.* **2024**, *25*, 3788. [CrossRef]
48. Saika, F.; Kiguchi, N.; Kobayashi, Y.; Fukazawa, Y.; Kishioka, S. CC-chemokine ligand 4/macrophage inflammatory protein-1beta participates in the induction of neuropathic pain after peripheral nerve injury. *Eur. J. Pain.* **2012**, *16*, 1271–1280. [CrossRef]
49. Xia, M.Q.; Qin, S.X.; Wu, L.J.; Mackay, C.R.; Hyman, B.T. Immunohistochemical study of the beta-chemokine receptors CCR3 and CCR5 and their ligands in normal and Alzheimer's disease brains. *Am. J. Pathol.* **1998**, *153*, 31–37. [CrossRef]
50. Ciechanowska, A.; Popiolek-Barczyk, K.; Pawlik, K.; Ciapala, K.; Oggioni, M.; Mercurio, D.; De Simoni, M.G.; Mika, J. Changes in macrophage inflammatory protein-1 (MIP-1) family members expression induced by traumatic brain injury in mice. *Immunobiology* **2020**, *225*, 151911. [CrossRef]
51. Rojewska, E.; Zychowska, M.; Piotrowska, A.; Kreiner, G.; Nalepa, I.; Mika, J. Involvement of Macrophage Inflammatory Protein-1 Family Members in the Development of Diabetic Neuropathy and Their Contribution to Effectiveness of Morphine. *Front. Immunol.* **2018**, *9*, 494. [CrossRef] [PubMed]
52. Sorce, S.; Myburgh, R.; Krause, K.H. The chemokine receptor CCR5 in the central nervous system. *Prog. Neurobiol.* **2011**, *93*, 297–311. [CrossRef] [PubMed]

53. Festa, B.P.; Siddiqi, F.H.; Jimenez-Sanchez, M.; Won, H.; Rob, M.; Djajadikerta, A.; Stamatakou, E.; Rubinsztein, D.C. Microglial-to-neuronal CCR5 signaling regulates autophagy in neurodegeneration. *Neuron* **2023**, *111*, 2021–2037.e12. [CrossRef]
54. Dong, F.L.; Yu, L.; Feng, P.D.; Ren, J.X.; Bai, X.H.; Lin, J.Q.; Cao, D.L.; Deng, Y.T.; Zhang, Y.; Shen, H.H.; et al. An atlas of neuropathic pain-associated molecular pathological characteristics in the mouse spinal cord. *Commun. Biol.* **2025**, *8*, 70. [CrossRef] [PubMed]

cells

Article

Microgliosis in the Spinal Dorsal Horn Early After Peripheral Nerve Injury Is Associated with Damage to Primary Afferent Aβ-Fibers

Yuto Shibata [1,†], Yuki Matsumoto [1,2,†], Keita Kohno [1], Yasuharu Nakashima [2] and Makoto Tsuda [1,3,*]

1 Department of Molecular and System Pharmacology, Graduate School of Pharmaceutical Sciences, Kyushu University, 3-1-1 Maidashi, Higashi-ku, Fukuoka 812-8582, Japan
2 Department of Orthopaedic Surgery, Graduate School of Medical Sciences, Kyushu University, 3-1-1 Maidashi, Higashi-ku, Fukuoka 812-8582, Japan
3 Kyushu University Institute for Advanced Study, 744 Motooka Nishi-ku, Fukuoka 819-0395, Japan
* Correspondence: tsuda@phar.kyushu-u.ac.jp
† These authors contributed equally to this work.

Abstract: Neuropathic pain results from a lesion or disease affecting the somatosensory nervous system. Injury to primary afferent nerves leads to microgliosis in the spinal dorsal horn (SDH), which plays a crucial role in developing neuropathic pain. Within the SDH, primary afferent fibers broadly project, and microglia are nearly ubiquitously distributed under normal conditions. However, not all microglia react to injuries affecting primary afferent fibers, resulting in spatially heterogeneous microgliosis within the SDH. The mechanisms underlying this phenomenon remain elusive. In this study, the spatial relationship between microgliosis and the projections of injured nerves was investigated by generating mice that had expressed tdTomato in the fourth lumbar dorsal root ganglion (L4-DRG) neurons via intra-L4-spinal nerve (SpN) injection of adeno-associated viral vectors. After transection of the L4-SpN, we found that microgliosis in the SDH selectively occurred in the innervation territories of the injured primary afferent fibers. However, denervating transient receptor potential vanilloid 1 (TRPV1)-expressing primary afferent fibers in the SDH through intrathecal injection of capsaicin did not trigger microgliosis, nor did it influence the microgliosis induced by L4-SpN injury. Conversely, pharmacological damage to myelinated DRG neurons, including Aβ-fibers, was sufficient to induce microgliosis. Furthermore, L4-SpN injury also induced microgliosis in the gracile nucleus, which primarily receives innervation from Aβ-fibers. These findings suggest that microgliosis in the SDH shortly after peripheral nerve injury is predominantly associated with damage to primary afferent Aβ-fibers.

Keywords: microglia; primary afferents; Aβ-fibers; C-fibers; peripheral nerve injury; spinal dorsal horn

1. Introduction

Neuropathic pain arises from lesions or diseases of the somatosensory nervous system. Increasing evidence from studies on neuropathic pain models shows that nerve damage causes significant changes not only in neurons [1–3] but also in non-neuronal cells [4,5], particularly microglia (tissue-resident macrophages), in the central nervous system (CNS). Following peripheral nerve damage, microglia in the spinal dorsal horn (SDH) respond rapidly, undergoing extensive changes in morphology, cell number, marker expression, transcriptional and translational activities, and function [4,6]. These reactive microglia in the SDH, evident shortly after nerve damage, play a critical role in the subsequent pathological changes in CNS function and the development of neuropathic pain [4,6]. Therefore, understanding the reactive processes of microglia triggered by nerve damage is crucial for elucidating the onset of neuropathic pain and developing therapeutic strategies.

Nerve damage-induced reactive microglia have been widely documented across various experimental models. A traditional model involves transection of peripheral nerves, such as the sciatic nerve, which leads to molecular and cellular changes in microglia within the spinal cord (both ventral and dorsal horns) [4,7–9]. Such alterations commonly follow various types of injuries (e.g., partial/complete ligation, or compression) to different peripheral nerve sites [sciatic nerve, tibial nerve, common peroneal nerve, and spinal nerve (SpN)] [4,10–14]. Unlike peripheral tissue inflammation, nerve injury clearly induces molecular and cellular changes in the microglia of the SDH [8,13,15,16]. Damage to primary afferent fibers plays a critical role in these reactive processes. In the SDH, primary afferent fibers are widely innervated, and under normal conditions, microglia are almost ubiquitously distributed. Interestingly, not all microglia respond to injuries affecting primary afferent nerves, leading to spatial heterogeneity in reactive microglia within the SDH [17]. In particular, areas in the SDH where microglia exhibit morphological changes and high expression of ionized calcium-binding adapter molecule 1 (IBA1, a microglial marker) are preferentially observed [18]. However, the mechanisms underlying this spatial heterogeneity of microglial responses in the SDH following nerve injury remain unclear.

The aim of this study was to investigate this unsolved issue and also to provide a clue to elucidate the role of injured neurons in microglial responses in the SDH and its molecular mechanisms. In this study, the spatial correlation between reactive microglia and injured nerve fiber projections in the SDH was analyzed. For this analysis, a genetic technique was developed to specifically label injured nerve fibers using adeno-associated viral (AAV) vectors to express fluorescent proteins in certain segments of dorsal root ganglion (DRG) neurons, specifically the fourth lumbar (L4). Using mice with labeled L4-DRG and pharmacological methods to damage either myelinated primary afferent fibers (particularly Aβ-fibers) or unmyelinated nociceptors, this study showed that the spatial heterogeneity of microglial responses [upregulation of IBA1 expression and an increase in microglial cell numbers (referred to as microgliosis in this study)] within the SDH following nerve injury correlates with the projection territories of damaged primary afferent Aβ-fibers.

2. Materials and Methods

2.1. Animals

Male C57BL/6J mice (CLEA Japan, Tokyo Japan) and male B6.Cg-*Gt(ROSA)26Sor*$^{tm14(CAG\text{-}tdTomato)Hze}$/J (*ROSA26*^{tdTomato}) mice (Stock No: 007914, The Jackson Laboratory, Bar Harbor, ME, USA) were used. All mice used were aged 8–10 weeks at the start of each experiment and were housed individually or in groups at a temperature of 22 ± 1 °C with a 12 h light–dark cycle, and were fed food and water ad libitum. All animal experiments were conducted according to relevant national and international guidelines contained in the "Act on Welfare and Management of Animals" (Ministry of Environment of Japan) and the "Regulation of Laboratory Animals" (Kyushu University) and under the protocols approved by the Institutional Animal Care and Use committee review panels at Kyushu University.

2.2. Recombinant AAV Vector Production and L4-SpN Injection

Viral vector production was performed according to our previous method [19]. The gene-encoding Cre and tdTomato were subcloned into the pENTR plasmid (Thermo Fisher Scientific, Waltham, MA, USA). To produce AAV vectors, Cre and tdTomato were inserted into pZac2.1-enhanced synapsin promoter (ESYN)-WPRE. rAAV vectors were produced from human embryonic kidney 293T (HEK293T) cells with triple transfection [pZac, cis plasmid; pAAV2/9 (University of Pennsylvania Gene Therapy Program Vector Core), trans plasmid; pAd DeltaF6, adenoviral helper plasmid (University of Pennsylvania Geno Therapy Program Vector Core)] and purified by two cesium chloride density gradient purification steps. The vector was dialyzed against phosphate-buffered saline (PBS) containing 0.001% (v/v) Pluronic-F68 (Thermo Fisher Scientific, Waltham, MA, USA) or 0.001% (v/v) poloxamer 188 Non-ionic Surfactant (#24040032; Thermo Fisher Scientific, Waltham, MA, USA) using Vivaspin Turbo 15 100,000 MWCO (#VS15T41; Sartorius, Gottingen, Germany). The genome titer of rAAV was determined by Pico Green fluorometric reagent (#P7589; Thermo Fisher Scientific, Waltham, MA, USA) following denaturation of the AAV particles. Vectors were stored at −80 °C until use.

Viral injections were performed according to our previously described method [19]. Mice were deeply anesthetized via intraperitoneal (i.p.) injection using a mixture of anesthetic agents: medetomidine hydrochloride (0.15 mg/kg), midazolam (2 mg/kg), and butorphanol tartrate (2.5 mg/kg). The skin was incised at L3–S1, and the paraspinal muscles and fat were removed to expose the L5-traverse process, revealing the parallel-lying L3- and L4-SpNs. A glass microcapillary, filled with rAAV solution (AAV-ESYN-tdTomato, or AAV-ESYN-Cre), was inserted into the L4-SpN to a depth of 150 μm from the surface of the nerve. Microinjection of 200 nL of rAAV solution into wild-type (WT) and *ROSA26*^{tdTomato} mice, respectively, was performed using a Micro4 Micro Syringe Pump Controller (World Precision Instruments, Sarasota, FL, USA). After microinjection, the glass microcapillary was removed, the skin was sutured with 5-0 silk, and the mice were kept under a heating light until they recovered. According to our previous method [19], the virus-injected mice were used for further analyses 3 weeks after injection. The viral titers used were AAV2/9-ESYN-tdTomato and AAV2/9-ESYN-Cre, both at 1.0×10^{13} GC/mL.

WT mice were injected with saporins into the L4-SpN using a Micro4 Micro Syringe Pump Controller, administering 400 nL of saporin solution [saporin conjugated with cholera toxin B subunit (CTB-SAP; IT-14, Advanced Targeting Systems, Carlsbad, CA, USA), isolectin B4 (IB4-SAP; IT-10, Advanced Targeting Systems, Carlsbad, CA, USA), and unconjugated saporin (Ctrl-SAP; PR-01, Advanced Targeting Systems, Carlsbad, CA, USA)] and PBS as controls. CTB-SAP [20] and IB4-SAP [21] are known to be internalized by myelinated and unmyelinated afferents, respectively, and induce neuronal death through ribosomal inactivation. Fourteen days after injection, the presence of IBA$^+$ microglia in the L3/4-SDH and neurofilament 200 (NF200)$^+$ or IB4$^+$ neurons in the L4-DRG was analyzed by immunohistochemistry. To achieve a dose of 0.6 pmol/400 nL/mouse, saporin concentrations were adjusted in PBS as follows: Ctrl-SAP, 0.05 mg/mL; CTB-SAP, 0.2 mg/mL; IB4-SAP, 0.3 mg/mL.

2.3. Peripheral Nerve Injury

An SpN injury model was used [22] with some modifications, as described previously [23]. Briefly, under isoflurane (2%) anesthesia, a small incision at L3–S1 was made. The paraspinal muscle and fat were removed from the L5 traverse process, which exposed the parallel-lying L3- and L4-SpNs. The L4-SpN was then carefully isolated and cut. The wound and the surrounding skin were sutured with 5-0 silk. In experiments of a combination with AAV microinjection, the injection site was exposed again, and the L4-SpN was carefully cut.

2.4. Hot Plate Test

To assess thermosensory behaviors, mice were placed on a metal surface (25 × 20 cm) maintained at 55 °C within a 25 cm-high Plexiglass box. The latency to either lick the hind paw or jump was recorded as a nocifensive end point [24].

2.5. Intrathecal Injection

Under 2% isoflurane anesthesia, a 30 G needle attached to a 25 μL Hamilton syringe was inserted into the intervertebral space between the L5 and L6 spinal vertebrae in mice, as previously described [23,25]. To ablate transient receptor potential vanilloid 1 ($TRPV1^+$) afferent fibers in the SDH, WT mice were injected intrathecally with capsaicin [10 μg/5 μL; #M2028, Sigma, Saint Louis, MO, USA; capsaicin was first dissolved in ethanol and diluted by PBS with Tween-80 (the final concentration of ethanol and Tween-80 was 10%)] [26] or vehicle (10% ethanol/10% Tween 80/PBS). Denervation of $TRPV1^+$ nerve fibers in the SDH was assessed 3 and 7 days after capsaicin injection by immunostaining. Microglial response in the SDH was observed 7 days after L4-SpN injury was performed at 7 days post-capsaicin treatment.

2.6. Immunohistochemistry

Mice were deeply anesthetized with i.p. injection of pentobarbital and transcardially perfused with PBS followed by ice-cold 4% paraformaldehyde/PBS. The transverse L4 segments of the spinal cord were removed and postfixed in the same fixative for 3 h at 4 °C. According to our previous method [23], DRG sections (15 μm) and transverse spinal sections (30 μm) were incubated for 48 h at 4 °C with primary antibodies for anti-glial fibrillary acidic protein (GFAP) (rabbit monoclonal; 1:2000; 13-0300, Invitrogen, Waltham, MA, USA), anti-IBA1 (guinea pig polyclonal; 1:2000; 234 00, Synaptic Systems, Goettingen, Germany), Alexa Fluor 647-conjugated anti- myelin basic protein (MBP) (1:1000; SMI99, BioLegend, San Diego, CA, USA), anti-NF200 (rabbit polyclonal; 1:2000; N4142, Sigma-Aldrich, Saint Louis, MO, USA), biotin-conjugated IB4 (1:1000; I21414, Invitrogen), and anti-TRPV1 (guinea pig polyclonal; 1:1000; GP14100, Neuromics, Waltham, MA, USA). Tissue sections were incubated with secondary antibodies conjugated to Alexa Fluor 488 (1:1000; A-21208, Thermo Fisher, Waltham, MA, USA), Alexa Fluor 546 (1:1000; A-11056, Thermo Fisher, Waltham, MA, USA), or streptavidin Alexa Fluor 488 (1:1000; S11223, Invitrogen, Waltham, MA, USA). For Nissl staining, DRG sections were stained for 30 min at room temperature. The samples were mounted with Vectashield hardmount (Vector Laboratories, Newark, CA, USA) or ProLong Glass Antifade Mountant (Invitrogen, Waltham, MA, USA). Three to five sections from one tissue were randomly selected, and images were taken using a confocal laser microscope imaging system (LSM700/900, Carl Zeiss, Oberkochen, Germany) and analyzed using Image J. The number of $tdTomato^+$ cells and $NF200^+$ and $IB4^+$ cells was counted manually using the cell counter plugin of Image J [19]. For quantification of the number of microglia in the SDH, $IBA1^+$ cells with clear cell bodies and with an S/N ratio of 2.0 or more were counted [18]. The numbers of particles was counted using the Analyze Particles function of Image J. The region of interest (ROI) was determined by drawing the boundary between gray and white matter of the SDH based on the differential interference contrast (DIC) image.

2.7. Statistical Analysis

Quantitative data are shown as mean ± SEM. Statistical significance was determined using the unpaired *t*-test and one-way analysis of variance (ANOVA) with the post hoc Tukey's multiple-comparisons test using Prism 7 (GraphPad, San Diego, CA, USA). Values were considered significantly different at $p < 0.05$.

3. Results

3.1. New Methods to Visualize Injured Primary Afferent Fibers in the SDH Using AAV Vectors

Initially, the spatial relationship between reactive microglia and injured nerve fiber projections in the SDH following nerve injury were examined. To visualize injured primary afferent fibers, a method for specifically and genetically visualizing neurons in the L4-DRG and their nerve fiber projections was developed using an AAV vector expressing the fluorescent protein tdTomato under the control of the ESYN promoter (AAV-ESYN-tdTomato) (Figure 1A). This AAV was unilaterally injected into the L4-SpN of WT mice. Four weeks later, tdTomato fluorescence was specifically expressed in neurons in the L4-DRG ipsilateral to the injection but not contralateral to it (Figure 1B,C). Additionally, tdTomato was not expressed in the neighboring L3 or L5-DRGs. In the L4-DRG, approximately 70% of the total neurons were positive for tdTomato (Figure 1C). In the spinal cord of the WT mice injected with AAV-ESYN-tdTomato, the dorsal root of the L4-DRG neurons at its entry zone and their nerve fibers in the parenchyma of the SDH were clearly visualized (Figure 1D,E). tdTomato expression was not observed in ventral horn neurons (Figure 1D), confirming the specific labeling of L4-DRG neurons. As a complementary approach, *ROSA26*^{tdTomato} mice in which AAV-ESYN-Cre had been injected into the L4-SpN to induce tdTomato expression were used (Figure 1F). As anticipated, this strategy resulted in expression of tdTomato in L4-DRG neurons (Figure 1G,H); however, unexpectedly, tdTomato expression was also observed in satellite glial cells surrounding the DRG neurons (Figure 1I). Nevertheless, tdTomato$^+$ fibers of L4-DRG neurons in the SDH were also clearly observed (Figure 1J,K). Thus, we established two methods for genetically and specifically visualizing L4-DRG nerve fiber projections in the SDH.

3.2. Spatial Correlation Between Injured Nerve Fiber Projection and Reactive Microglia

Using the established methods, we explored the relationship between the projection territories of injured L4-DRG nerve fibers and the areas of reactive microglia following L4-SpN transection (Figure 2A). L4-SpN transection markedly increased the immunofluorescence levels of IBA1 and the number of microglia (referred to as microgliosis) 7 days post-injury (Figure 2B). Notably, the L4-SpN injury-induced microgliosis was not spatially uniform within the L4-SDH but was selectively observed in the innervation territories of tdTomato$^+$ injured L4-DRG nerve fibers (Figure 2B,C). Additionally, consistent with the rostro-caudal projection pattern of L4-DRG neurons in the SDH, the nerve injury-induced microgliosis was observed in the neighboring segments (L2, L3, and L5) (Figure 2B). This spatial pattern also correlated with the projection territories of tdTomato$^+$ fibers. These findings provide compelling evidence that microgliosis in the SDH after nerve injury is localized to the projection territories of injured primary afferent nerve fibers.

3.3. TRPV1$^+$ Primary Afferent Fibers Are Dispensable for Microgliosis in the SDH After Peripheral Nerve Injury

DRG neurons are divided into several classes, such as myelinated and unmyelinated, the latter of which are positive for TRPV1 and IB4 in mice [27]. Their spatial projection patterns in the SDH also differ [27–29]. To assess the impact of damage to TRPV1$^+$ primary afferent fibers on microglial reactivity in the SDH, mice were treated with an intrathecal injection of capsaicin (10 μg/5 μL), a dose that was sufficient to denervate TRPV1$^+$ fibers in the SDH [26]. Seven days post-injection, TRPV1 immunofluorescence in primary afferent fibers in the SDH was absent (Figure 3A), and nocifensive behaviors in response to noxious heat were not observed (the latency to display nocifensive behaviors: vehicle-treated mice, 11.0 ± 0.4 s; capsaicin-treated mice, 30 ± 0.0 s, n = 3–4), confirming the denervation of

TRPV1⁺ primary afferent fibers in the SDH. In these mice, ablation of spinal TRPV1⁺ nerve fibers had no effect on the L4-SpN transection-induced microgliosis (increased IBA1 immunofluorescence levels and microglial cell count) in the L4-SDH 7 days post-injury (Figure 3B,C). Moreover, at an early time point (3 days post-capsaicin injection), spinal TRPV1⁺ nerve fiber ablation alone did not increase IBA1 immunofluorescence or microglial cell number (Figure 3D). These results indicate that TRPV1⁺ primary afferent fibers are dispensable for microgliosis in response to peripheral nerve injury.

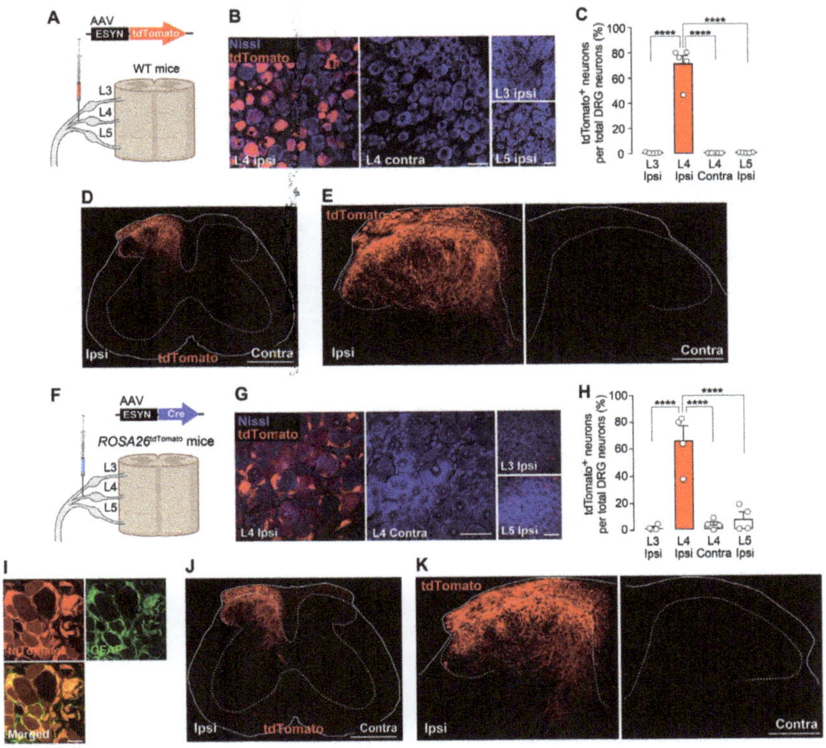

Figure 1. Specific labeling of L4-DRG neurons and their projections in the SDH by injecting AAV vectors into the L4-SpN. (**A**) Schematic illustration of L4-SpN microinjection of AAV-ESYN-tdTomato in WT mice. (**B**) Representative fluorescence images of tdTomato⁺ cells (red) and DRG neurons stained by Nissl (blue) in the L3/4/5 segments. Scale bars, 50 μm. (**C**) Percentage of tdTomato⁺ cells in total Nissl⁺ DRG neurons in each segment [n = 5 mice (3 or 4 slices per mouse)]. **** $p < 0.0001$, one-way ANOVA with post hoc Tukey's multiple-comparisons test. (**D**,**E**) tdTomato expression in the L4-SDH at 3 weeks (**D**) and 4 weeks (**E**) after microinjection of AAV-ESYN-tdTomato. Scale bars; 500 μm (**D**) and 200 μm (**E**). (**F**) Schematic illustration of L4-SpN injection of AAV-ESYN-Cre in *ROSA26*^{tdTomato} mice. (**G**) Representative fluorescence images of tdTomato⁺ cells (red) and Nissl⁺ DRG neurons (blue). Scale bars, 50 μm. (**H**) Percentage of tdTomato⁺ Nissl⁺ neurons per total Nissl⁺ DRG neurons in each segment [n = 4 mice tested (4 slices per mouse)]. **** $p < 0.0001$, one-way ANOVA with post hoc Tukey's multiple-comparisons test. (**I**) Immunolabeling of tdTomato⁺ cells (red) with GFAP (green), a satellite glial marker, in the L4-DRG. Scale bars, 20 μm. (**J**,**K**) tdTomato⁺ nerve fibers in the L4-SDH at 3 weeks (**J**) and 4 weeks (**K**) after microinjection of AAV-ESYN-Cre. Scale bars, 500 μm (**J**) and 200 μm (**K**). Data are shown as the mean ± SEM. Panels A and F were created with BioRender.com.

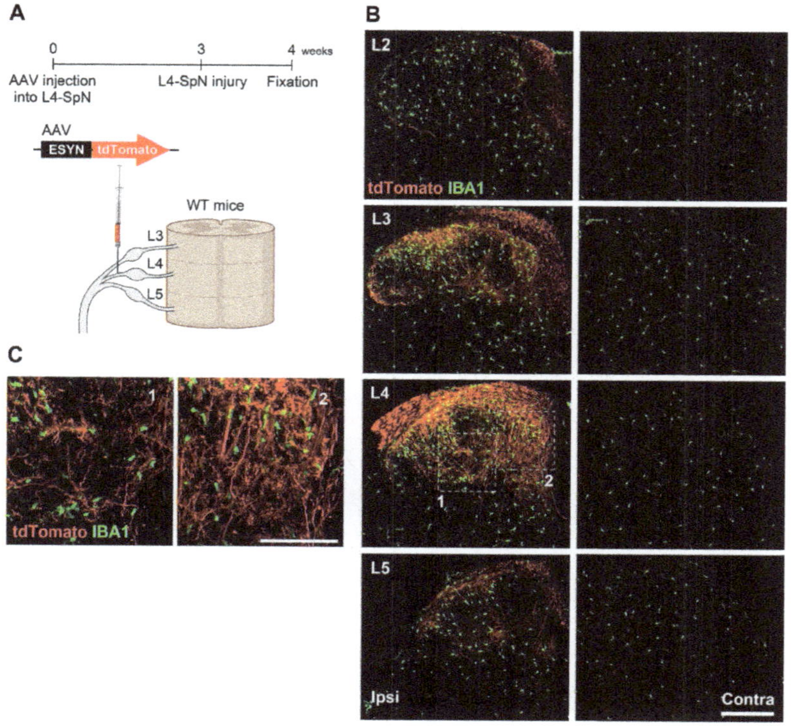

Figure 2. Similarity in spatial patterns of injured L4-DRG spinal nerve projection and reactive microglia in the SDH. (**A**) Schematic timeline and illustration of AAV injection, L4-SpN injury, and fixation. (**B**,**C**) Immunolabeling of IBA1$^+$ cells (green) and tdTomato$^+$ nerve fibers (red) in each SDH segment and two selected areas (1 and 2) 7 days after L4-SpN injury. Scale bars, 200 μm (**B**) and 100 μm (**C**). Panel A was created with BioRender.com.

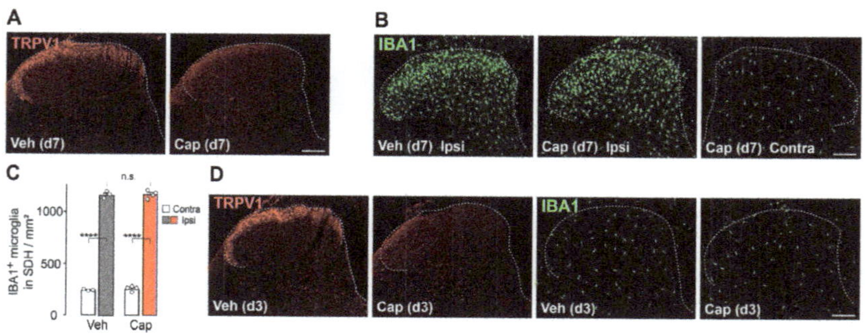

Figure 3. Minimal contribution of TRPV1$^+$ primary afferent fibers in the SDH to microgliosis after L4-SpN injury. (**A**) Representative immunofluorescence images of TRPV1$^+$ nerve fibers (red) in the SDH 7 days after intrathecal injection of vehicle (Veh) or capsaicin (Cap; 10 μg) to WT mice. Scale bar, 100 μm. (**B**) Representative images of IBA1$^+$ cells (green) in the SDH 7 days after L4-SpN injury in WT mice with intrathecal injection of vehicle or capsaicin. Scale bar, 100 μm. (**C**) Quantification of IBA1$^+$ cells in the SDH ipsilateral and contralateral to the injury (n = 3 to 4 mice). **** $p < 0.0001$, unpaired t-test. n.s., not significant. (**D**) Representative immunofluorescence images of TRPV1$^+$ nerve fibers (red) and IBA1$^+$ cells (green) in the SDH 3 days after intrathecal capsaicin injection. Scale bar, 100 μm. Data are shown as mean ± SEM.

3.4. Microgliosis in the SDH Involves Damage to Primary Afferent Aβ-Fibers

Given the minimal involvement of TRPV1^{+} nerve fibers in nerve injury-induced microgliosis in the SDH, the role of myelinated primary afferent fibers, such as Aβ-fibers, was investigated. Aβ-fibers from the lower spinal segments innervate not only the SDH but also the gracile nucleus (GN) in the brainstem [30]. Consistently, tdTomato^{+} fibers of L4-DRG neurons in the GN were observed (Figure 4A), most of which were ensheathed in MBP (Figure 4B), confirming the projection of myelinated primary afferent fibers from the L4-DRG. Furthermore, following L4-SpN injury in these mice, the number of IBA1^{+} cells increased in the GN, a spatial pattern closely matching the injured myelinated nerve fiber projection territories (Figure 4C,D). As the GN selectively receives primary afferent Aβ-fibers [30], these data suggest that primary afferent Aβ-fibers are involved in nerve injury-induced microgliosis.

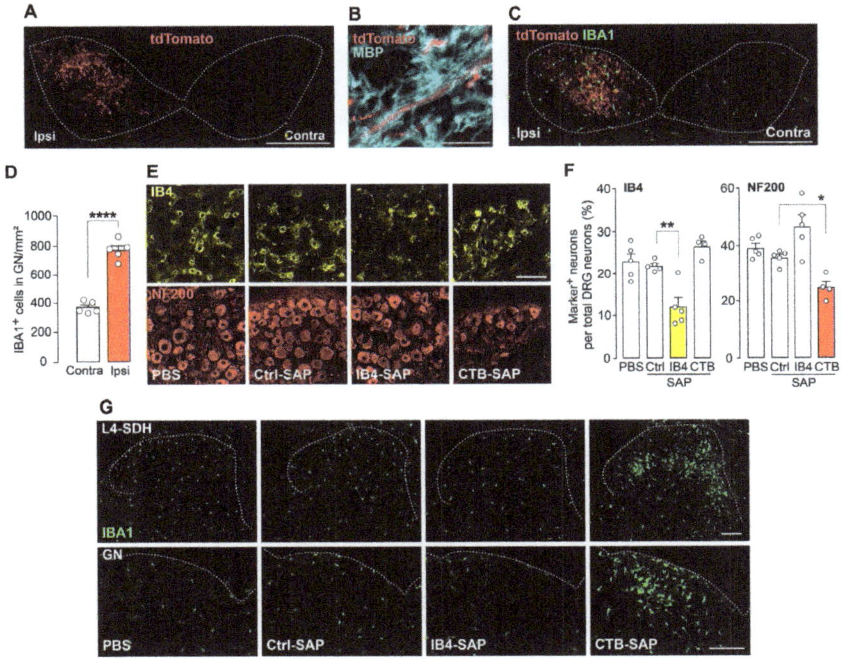

Figure 4. Damage to primary afferent Aβ-fibers is important for L4-SpN injury-induced microgliosis. (**A**) tdTomato expression (red) of primary afferent nerve fibers in the GN. Scale bar, 200 μm. (**B**) Immunostaining of tdTomato^{+} nerve fibers (red) and MBP (cyan) in the GN. Scale bar, 10 μm. (**C**) tdTomato^{+} nerve fibers (red) and IBA1^{+} cells (green) in the GN of WT mice 7 days after L4-SpN injury. Scale bar, 200 μm. (**D**) The number of IBA1^{+} cells after L4-SpN injury (day 7) in the ipsilateral or contralateral GN (n = 5 mice). **** $p < 0.0001$, unpaired t-test. (**E**) Representative immunofluorescence images of IB4^{+} (yellow) and NF200^{+} (red) neurons in the L4-DRG (ipsilateral and contralateral sides) 14 days after injection of PBS, Ctrl-SAP, IB4-SAP, and CTB-SAP. Scale bar, 100 μm. (**F**) Quantification of the number of IB4^{+} or NF200^{+} DRG neurons (Nissl^{+}) after injection of each saporin (Ctrl-SAP, IB4-SAP, and CTB-SAP) and PBS (n = 4–6 mice). Scale bar, 100 μm. * $p < 0.05$ and ** $p < 0.01$, one-way ANOVA with post hoc Tukey's multiple-comparisons test. (**G**) Representative immunofluorescence images of IBA^{+} cells (green) in the SDH and GN after PBS and each saporin injection. Scale bars, 100 μm. Data are shown as mean ± SEM.

To determine whether damage to primary afferent Aβ-fibers induces microgliosis, a pharmacological ablation approach with saporin-conjugated reagents was used. CTB-SAP was injected into the L4-SpN to target primary afferent Aβ-fibers, which express GM1 ganglioside, the target of CTB [20]. Intra-L4-SpN injection of CTB-SAP reduced the number of L4-DRG neurons expressing NF200 (a marker for large-diameter myelinated fibers, including Aβ-fibers) [31] compared with control saporin (Ctrl-SAP) (Figure 4E,F). Notably, CTB-SAP injection into the L4-SpN significantly increased IBA1 immunofluorescence and microglial cell number in the L4-SDH (Figure 4G). Similar microgliosis was also observed in the GN. However, Ctrl-SAP did not induce changes in microglia in the SDH, indicating that the microgliosis caused by CTB-SAP was not due to the non-specific effects of saporin itself. Additionally, intra-L4-SpN injection of IB4-SAP significantly decreased the number of IB4$^+$ DRG neurons (Figure 4E,F) but did not cause microgliosis, as seen in CTB-SAP-treated mice (Figure 4G). Overall, these results demonstrate that pharmacological damage to primary afferent Aβ-fibers, but not C-fibers, is sufficient to induce microgliosis in the SDH.

4. Discussion

Since it was first reported in the late 1970s that microglia in the SDH respond to sciatic nerve axotomy [7,32], numerous studies have been published on the cellular and molecular alterations in SDH microglia following nerve injury [4,6]. However, the mechanism underlying the spatial heterogeneity of microgliosis in the SDH after nerve injury has not been fully elucidated. To shed light on this unresolved issue, this study developed a method for genetically labeling L4-DRG neurons by injecting an AAV into the L4-SpN and transecting the L4-SpN, thus specifically visualizing injured nerve fibers in the SDH and other regions. Using these nerve-labeled mice, we demonstrated for the first time that the spatial heterogeneity of nerve injury-induced microgliosis in the SDH is associated with the projection territories of injured myelinated primary afferent fibers, particularly Aβ-fibers. Indeed, reactive microglia after L4-SpN injury were preferentially localized along the projection pathways of myelinated fibers, including Aβ-fibers [33,34]. Additionally, microgliosis was observed in the GN and was highly restricted to the projection territory of injured myelinated fibers from L4-DRG neurons. Furthermore, pharmacological damage to NF200$^+$ DRG neurons and their myelinated fibers using CTB-SAP was sufficient to induce microgliosis in both the SDH and the GN. Aδ-fibers and proprioceptor projections are also found in the GN [30], but, in contrast to Aβ-fibers, these projections in the GN were very limited. In addition, a spatial correlation between primary afferent fibers labeled with CTB and microgliosis resulting from peripheral nerve injury has been demonstrated in another model [14]. Therefore, Aβ-fibers would have a major contribution to nerve injury-induced microgliosis in the SDH and GN, although the possible involvement of other fibers (e.g., Aδ-fibers and proprioceptors) cannot be excluded.

In stark contrast, we found that unmyelinated C-fibers play a minimal role in microglial reaction. This differential role provides a crucial clue for elucidating the mechanisms underlying microgliosis after nerve injury. However, it should be noted that there are also contradictory findings regarding the role of unmyelinated C-fibers. Electrical stimulation of C-fibers in normal mice has been reported to cause microglia to become reactive in the SDH [35]. Additionally, it has recently been shown that conditional knockout of genes (GPR151 [36] and MyD88 [37]) in primary afferent C-fibers suppresses nerve injury-induced microgliosis in the SDH. In contrast, it has also been reported that nerve injury-induced microgliosis is not affected in rats whose TRPV1$^+$ neurons have been ablated by systemic treatment with capsaicin during the neonatal period [38]. Although compensatory changes after the loss of TRPV1$^+$ neurons could be considered, our study clearly demonstrated that acute denervation of TRPV1$^+$ primary afferent fibers in the adult SDH does not influence SpN injury-induced SDH microgliosis, nor does the denervation itself cause microgliosis.

These findings are supported by a previous study indicating that functional blockade of TRPV1$^+$ C-fibers has no effect on nerve injury-induced microgliosis [39]. Moreover, pharmacological damage to IB4$^+$ DRG neurons also did not result in microgliosis. Given that TRPV1$^+$ and IB4$^+$ fibers constitute most unmyelinated C-fibers [31], it appears that unmyelinated C-fibers are neither necessary nor sufficient to induce microgliosis in the SDH after peripheral nerve injury, emphasizing the significance of myelinated Aβ-fiber damage.

Several studies have reported the effect of selective gene loss in primary afferent neurons on nerve injury-induced microgliosis in the SDH [4,40]. Among these molecules, colony-stimulating factor 1 (CSF1) is currently considered the most promising candidate responsible for microgliosis following nerve injury [17,41]. CSF1 expression is rapidly induced in injured DRG neurons, and conditional knockout of CSF1 in primary afferent neurons significantly suppresses nerve injury-induced microgliosis in the SDH [17]. However, CSF1 induction is observed in injured DRG neurons with almost all soma sizes [41] (e.g., positive to NF200, CGRP, and IB4 [42]), suggesting that CSF1 upregulation occurs in various types of DRG neurons, including both myelinated Aβ-fibers and unmyelinated C-fibers [17,41], and that CSF1 in each DRG population may play a distinct role in nerve injury-induced microgliosis. Additionally, the role of demyelination of Aβ-fibers and its related factors (e.g., lysophosphatidic acid (LPA)) should also be considered [43–46]. Demyelination and LPA are known to induce microgliosis in various models of CNS disease [46,47]. To identify DRG neuron subtypes involved in microgliosis and to understand its molecular mechanisms (e.g., CSF1 and demyelination-related factors), further investigations using our established methods, which enable manipulation of gene expression in injured DRG neurons, are required.

It should also be noted that there are limitations to our methods. For example, intra-SpN injection of AAV vectors did not induce gene expression in all L4-DRG neurons, and the levels of gene expression varied among these neurons. This variability may explain why microgliosis in the SDH and the areas with tdTomato$^+$ damaged myelinated fibers did not completely overlap. Technical improvements that enable gene expression in a larger number of DRG neurons are necessary for the future. Nevertheless, this is the first paper to establish a method, in conjunction with a neuropathic pain model developed by L4-SpN injury, that allows for gene expression specifically in injured primary afferent fibers. By incorporating genes used for optogenetics and chemogenetics, this technique could also be useful for functionally manipulating injured neurons and for elucidating the role of injured neurons in nerve injury-induced alterations in central nervous system function and pain behavior.

5. Conclusions

In this study, we established a novel method to visualize injured nerves using an AAV-mediated gene expression approach and demonstrated the spatial correlation between microgliosis and injured myelinated Aβ-fiber projections in the SDH and GN. Pharmacological damage to primary afferent Aβ-fibers was sufficient to induce microgliosis in both the SDH and the GN. Given that microgliosis in the SDH shortly after peripheral nerve injury is a critical step in the development of neuropathic pain [4], this study paves the way to identifying the molecular and cellular mechanisms underlying microgliosis following nerve injury and, consequently, to better understanding neuropathic pain.

Author Contributions: Conceptualization, M.T.; methodology, Y.S., Y.M., K.K. and M.T.; validation, Y.S. and Y.M.; formal analysis, Y.S. and Y.M.; investigation, Y.S. and Y.M.; data curation, Y.S. and Y.M.; writing—original draft preparation, Y.S. and Y.M.; writing—review and editing, Y.S., Y.M., K.K.

and M.T.; visualization, Y.S., Y.M., K.K. and M.T.; supervision, Y.N. and M.T.; project administration, M.T.; funding acquisition, M.T. All authors have read and agreed to the published version of the manuscript.

Funding: This work was supported by the Japan Society for the Promotion of Science (JSPS) KAKENHI grants JP24H00067 (M.T.) and JP20H05900 (M.T.), by the Core Research for Evolutional Science and Technology (CREST) program from the Japan Agency for Medical Research and Development (AMED) under grant number 25gm1510013h (M.T.), and by the Research Support Project for Life Science and Drug Discovery (Basis for Supporting Innovative Drug Discovery and Life Science Research (BINDS)) from AMED under grant number JP25ama121031 (M.T.).

Institutional Review Board Statement: All animal experiments were conducted according to the national and international guidelines contained in the "Act on Welfare and Management of Animals" (Ministry of Environment of Japan) and the "Regulation of Laboratory Animals" (Kyushu University), and under the protocols approved by the Institutional Animal Care and Use committee review panels at Kyushu University [the approval code for this study is A23-302-1 (27 September 2024)].

Informed Consent Statement: Not applicable.

Data Availability Statement: All data generated or analyzed during this study are included in the paper.

Acknowledgments: We would like to thank Editage for editing a draft of this manuscript, and the University of Pennsylvania vector core for providing pZac2.1, pAAV2/9, and pAd DeltaF6 plasmid. Some figure panels were created with BioRender.com.

Conflicts of Interest: The authors declare no conflicts of interest.

Abbreviations

SDH	Spinal dorsal horn
L4	Fourth lumbar
DRG	Dorsal root ganglion
SpN	Spinal nerve
TRPV1	Transient receptor potential vanilloid 1
CNS	Central nervous system
IBA1	Ionized calcium-binding adapter molecule 1
AAV	Adeno-associated virus
ESYN	Enhanced synapsin
PBS	Phosphate-buffered saline
i.p.	Intraperitoneal
WT	Wild type
CTB	Cholera toxin B subunit
IB4	Isolectin B4
Ctrl	Control
SAP	Saporin
NF200	Neurofilament 200
MBP	Myelin basic protein
ROI	Region of interest
DIC	Differential interference contrast
n.s.	Not significant
GN	Gracile nucleus
CSF1	Colony-stimulating factor 1
LPA	Lysophosphatidic acid

References

1. Kuner, R.; Flor, H. Structural plasticity and reorganisation in chronic pain. *Nat. Rev. Neurosci.* **2016**, *18*, 20–30. [CrossRef] [PubMed]
2. Moehring, F.; Halder, P.; Seal, R.P.; Stucky, C.L. Uncovering the cells and circuits of touch in normal and pathological settings. *Neuron* **2018**, *100*, 349–360. [CrossRef] [PubMed]
3. Finnerup, N.B.; Kuner, R.; Jensen, T.S. Neuropathic pain: From mechanisms to treatment. *Physiol. Rev.* **2021**, *101*, 259–301. [CrossRef]
4. Inoue, K.; Tsuda, M. Microglia in neuropathic pain: Cellular and molecular mechanisms and therapeutic potential. *Nat. Rev. Neurosci.* **2018**, *19*, 138–152. [CrossRef]
5. Ji, R.R.; Donnelly, C.R.; Nedergaard, M. Astrocytes in chronic pain and itch. *Nat. Rev. Neurosci.* **2019**, *20*, 667–685. [CrossRef]
6. Tsuda, M.; Masuda, T.; Kohno, K. Microglial diversity in neuropathic pain. *Trends Neurosci.* **2023**, *46*, 597–610. [CrossRef]
7. Gilmore, S.A. Proliferation of non-neuronal cells in spinal cords of irradiated, immature rats following transection of the sciatic nerve. *Anat. Rec.* **1975**, *181*, 799–811. [CrossRef]
8. Masuda, T.; Iwamoto, S.; Yoshinaga, R.; Tozaki-Saitoh, H.; Nishiyama, A.; Mak, T.W.; Tamura, T.; Tsuda, M.; Inoue, K. Transcription factor irf5 drives p2x4r+-reactive microglia gating neuropathic pain. *Nat. Commun.* **2014**, *5*, 3771. [CrossRef] [PubMed]
9. Denk, F.; Crow, M.; Didangelos, A.; Lopes, D.M.; McMahon, S.B. Persistent alterations in microglial enhancers in a model of chronic pain. *Cell Rep.* **2016**, *15*, 1771–1781. [CrossRef]
10. Coyle, D.E. Partial peripheral nerve injury leads to activation of astroglia and microglia which parallels the development of allodynic behavior. *Glia* **1998**, *23*, 75–83. [CrossRef]
11. Colburn, R.W.; Rickman, A.J.; DeLeo, J.A. The effect of site and type of nerve injury on spinal glial activation and neuropathic pain behavior. *Exp. Neurol.* **1999**, *157*, 289–304. [CrossRef] [PubMed]
12. Stuesse, S.L.; Cruce, W.L.; Lovell, J.A.; McBurney, D.L.; Crisp, T. Microglial proliferation in the spinal cord of aged rats with a sciatic nerve injury. *Neurosci. Lett.* **2000**, *287*, 121–124. [CrossRef] [PubMed]
13. Tsuda, M.; Shigemoto-Mogami, Y.; Koizumi, S.; Mizokoshi, A.; Kohsaka, S.; Salter, M.W.; Inoue, K. P2x4 receptors induced in spinal microglia gate tactile allodynia after nerve injury. *Nature* **2003**, *424*, 778–783. [CrossRef] [PubMed]
14. Beggs, S.; Salter, M.W. Stereological and somatotopic analysis of the spinal microglial response to peripheral nerve injury. *Brain Behav. Immun.* **2007**, *21*, 624–633. [CrossRef]
15. Honore, P.; Rogers, S.D.; Schwei, M.J.; Salak-Johnson, J.L.; Luger, N.M.; Sabino, M.C.; Clohisy, D.R.; Mantyh, P.W. Murine models of inflammatory, neuropathic and cancer pain each generates a unique set of neurochemical changes in the spinal cord and sensory neurons. *Neuroscience* **2000**, *98*, 585–598. [CrossRef]
16. Masuda, T.; Tsuda, M.; Yoshinaga, R.; Tozaki-Saitoh, H.; Ozato, K.; Tamura, T.; Inoue, K. Irf8 is a critical transcription factor for transforming microglia into a reactive phenotype. *Cell Rep.* **2012**, *1*, 334–340. [CrossRef]
17. Guan, Z.; Kuhn, J.A.; Wang, X.; Colquitt, B.; Solorzano, C.; Vaman, S.; Guan, A.K.; Evans-Reinsch, Z.; Braz, J.; Devor, M.; et al. Injured sensory neuron-derived csf1 induces microglial proliferation and dap12-dependent pain. *Nat. Neurosci.* **2016**, *19*, 94–101. [CrossRef]
18. Kohno, K.; Kitano, J.; Kohro, Y.; Tozaki-Saitoh, H.; Inoue, K.; Tsuda, M. Temporal kinetics of microgliosis in the spinal dorsal horn after peripheral nerve injury in rodents. *Biol. Pharm. Bull.* **2018**, *41*, 1096–1102. [CrossRef]
19. Kanehisa, K.; Koga, K.; Maejima, S.; Shiraishi, Y.; Asai, K.; Shiratori-Hayashi, M.; Xiao, M.F.; Sakamoto, H.; Worley, P.F.; Tsuda, M. Neuronal pentraxin 2 is required for facilitating excitatory synaptic inputs onto spinal neurons involved in pruriceptive transmission in a model of chronic itch. *Nat. Commun.* **2022**, *13*, 2367. [CrossRef]
20. Llewellyn-Smith, I.J.; Martin, C.L.; Arnolda, L.F.; Minson, J.B. Tracer-toxins: Cholera toxin b-saporin as a model. *J. Neurosci. Methods* **2000**, *103*, 83–90. [CrossRef]
21. Tarpley, J.W.; Kohler, M.G.; Martin, W.J. The behavioral and neuroanatomical effects of ib4-saporin treatment in rat models of nociceptive and neuropathic pain. *Brain Res.* **2004**, *1029*, 65–76. [CrossRef] [PubMed]
22. Ho Kim, S.; Mo Chung, J. An experimental model for peripheral neuropathy produced by segmental spinal nerve ligation in the rat. *Pain.* **1992**, *50*, 355–363. [CrossRef]
23. Kohno, K.; Shirasaka, R.; Yoshihara, K.; Mikuriya, S.; Tanaka, K.; Takanami, K.; Inoue, K.; Sakamoto, H.; Ohkawa, Y.; Masuda, T.; et al. A spinal microglia population involved in remitting and relapsing neuropathic pain. *Science* **2022**, *376*, 86–90. [CrossRef]
24. Kohro, Y.; Matsuda, T.; Yoshihara, K.; Kohno, K.; Koga, K.; Katsuragi, R.; Oka, T.; Tashima, R.; Muneta, S.; Yamane, T.; et al. Spinal astrocytes in superficial laminae gate brainstem descending control of mechanosensory hypersensitivity. *Nat. Neurosci.* **2020**, *23*, 1376–1387. [CrossRef] [PubMed]

25. Hylden, J.L.; Wilcox, G.L. Intrathecal morphine in mice: A new technique. *Eur. J. Pharmacol.* **1980**, *67*, 313–316. [CrossRef]
26. Cavanaugh, D.J.; Lee, H.; Lo, L.; Shields, S.D.; Zylka, M.J.; Basbaum, A.I.; Anderson, D.J. Distinct subsets of unmyelinated primary sensory fibers mediate behavioral responses to noxious thermal and mechanical stimuli. *Proc. Natl. Acad. Sci. USA* **2009**, *106*, 9075–9080. [CrossRef]
27. Zwick, M.; Davis, B.M.; Woodbury, C.J.; Burkett, J.N.; Koerber, H.R.; Simpson, J.F.; Albers, K.M. Glial cell line-derived neurotrophic factor is a survival factor for isolectin b4-positive, but not vanilloid receptor 1-positive, neurons in the mouse. *J. Neurosci.* **2002**, *22*, 4057–4065. [CrossRef]
28. Shehab, S.A.; Hughes, D.I. Simultaneous identification of unmyelinated and myelinated primary somatic afferents by co-injection of isolectin b4 and cholera toxin subunit b into the sciatic nerve of the rat. *J. Neurosci. Methods* **2011**, *198*, 213–221. [CrossRef] [PubMed]
29. Qi, L.; Iskols, M.; Shi, D.; Reddy, P.; Walker, C.; Lezgiyeva, K.; Voisin, T.; Pawlak, M.; Kuchroo, V.K.; Chiu, I.M.; et al. A mouse drg genetic toolkit reveals morphological and physiological diversity of somatosensory neuron subtypes. *Cell* **2024**, *187*, 1508–1526.e16. [CrossRef]
30. Upadhyay, A.; Gradwell, M.A.; Vajtay, T.J.; Conner, J.; Sanyal, A.A.; Azadegan, C.; Patel, K.R.; Thackray, J.K.; Bohic, M.; Imai, F.; et al. The dorsal column nuclei scale mechanical sensitivity in naive and neuropathic pain states. *Cell Rep.* **2025**, *44*, 115556. [CrossRef]
31. Peirs, C.; Seal, R.P. Neural circuits for pain: Recent advances and current views. *Science* **2016**, *354*, 578–584. [CrossRef] [PubMed]
32. Gilmore, S.A.; Skinner, R.D. Intraspinal non-neuronal cellular responses to peripheral nerve injury. *Anat. Rec.* **1979**, *194*, 369–387. [CrossRef]
33. Li, L.; Rutlin, M.; Abraira, V.E.; Cassidy, C.; Kus, L.; Gong, S.; Jankowski, M.P.; Luo, W.; Heintz, N.; Koerber, H.R.; et al. The functional organization of cutaneous low-threshold mechanosensory neurons. *Cell* **2011**, *147*, 1615–1627. [CrossRef]
34. Bai, L.; Lehnert, B.P.; Liu, J.; Neubarth, N.L.; Dickendesher, T.L.; Nwe, P.H.; Cassidy, C.; Woodbury, C.J.; Ginty, D.D. Genetic identification of an expansive mechanoreceptor sensitive to skin stroking. *Cell* **2015**, *163*, 1783–1795. [CrossRef]
35. Hathway, G.J.; Vega-Avelaira, D.; Moss, A.; Ingram, R.; Fitzgerald, M. Brief, low frequency stimulation of rat peripheral c-fibres evokes prolonged microglial-induced central sensitization in adults but not in neonates. *Pain* **2009**, *144*, 110–118. [CrossRef] [PubMed]
36. Xia, L.P.; Luo, H.; Ma, Q.; Xie, Y.K.; Li, W.; Hu, H.; Xu, Z.Z. Gpr151 in nociceptors modulates neuropathic pain via regulating p2x3 function and microglial activation. *Brain* **2021**, *144*, 3405–3420. [CrossRef]
37. Liu, X.J.; Liu, T.; Chen, G.; Wang, B.; Yu, X.L.; Yin, C.; Ji, R.R. Tlr signaling adaptor protein myd88 in primary sensory neurons contributes to persistent inflammatory and neuropathic pain and neuroinflammation. *Sci. Rep.* **2016**, *6*, 28188. [CrossRef] [PubMed]
38. Szeredi, I.D.; Jancso, G.; Oszlacs, O.; Santha, P. Prior perineural or neonatal treatment with capsaicin does not alter the development of spinal microgliosis induced by peripheral nerve injury. *Cell Tissue Res.* **2021**, *383*, 677–692. [CrossRef]
39. Suter, M.R.; Berta, T.; Gao, Y.J.; Decosterd, I.; Ji, R.R. Large a-fiber activity is required for microglial proliferation and p38 mapk activation in the spinal cord: Different effects of resiniferatoxin and bupivacaine on spinal microglial changes after spared nerve injury. *Mol. Pain.* **2009**, *5*, 53. [CrossRef]
40. Chen, G.; Zhang, Y.Q.; Qadri, Y.J.; Serhan, C.N.; Ji, R.R. Microglia in pain: Detrimental and protective roles in pathogenesis and resolution of pain. *Neuron* **2018**, *100*, 1292–1311. [CrossRef]
41. Okubo, M.; Yamanaka, H.; Kobayashi, K.; Dai, Y.; Kanda, H.; Yagi, H.; Noguchi, K. Macrophage-colony stimulating factor derived from injured primary afferent induces proliferation of spinal microglia and neuropathic pain in rats. *PLoS ONE* **2016**, *11*, e0153375. [CrossRef]
42. Grosu, A.V.; Gheorghe, R.O.; Filippi, A.; Deftu, A.F.; Isler, M.; Suter, M.; Ristoiu, V. Dorsal root ganglia csf1(+) neuronal subtypes have different impact on macrophages and microglia after spared nerve injury. *J. Peripher. Nerv. Syst.* **2024**, *29*, 514–527. [CrossRef]
43. Inoue, M.; Rashid, M.H.; Fujita, R.; Contos, J.J.; Chun, J.; Ueda, H. Initiation of neuropathic pain requires lysophosphatidic acid receptor signaling. *Nat. Med.* **2004**, *10*, 712–718. [CrossRef] [PubMed]
44. Ma, L.; Uchida, H.; Nagai, J.; Inoue, M.; Aoki, J.; Ueda, H. Evidence for de novo synthesis of lysophosphatidic acid in the spinal cord through phospholipase a2 and autotaxin in nerve injury-induced neuropathic pain. *J. Pharmacol. Exp. Ther.* **2010**, *333*, 540–546. [CrossRef] [PubMed]
45. Gritsch, S.; Lu, J.; Thilemann, S.; Wortge, S.; Mobius, W.; Bruttger, J.; Karram, K.; Ruhwedel, T.; Blanfeld, M.; Vardeh, D.; et al. Oligodendrocyte ablation triggers central pain independently of innate or adaptive immune responses in mice. *Nat. Commun.* **2014**, *5*, 5472. [CrossRef] [PubMed]

46. Kent, S.A.; Miron, V.E. Microglia regulation of central nervous system myelin health and regeneration. *Nat. Rev. Immunol.* **2024**, *24*, 49–63. [CrossRef]
47. Ueda, H. Lysophosphatidic acid signaling is the definitive mechanism underlying neuropathic pain. *Pain* **2017**, *158* (Suppl. S1), S55–S65. [CrossRef]

cells

MDPI

Article

Association Between Synovial NTN4 Expression and Pain Scores, and Its Effects on Fibroblasts and Sensory Neurons in End-Stage Knee Osteoarthritis

Ayumi Tsukada [1,†], Yui Uekusa [1,†], Etsuro Ohta [2,3,4], Akito Hattori [2], Manabu Mukai [1], Dai Iwase [1], Jun Aikawa [1], Yoshihisa Ohashi [1], Gen Inoue [1], Masashi Takaso [1] and Kentaro Uchida [1,5,*]

1 Department of Orthopaedic Surgery, Kitasato University School of Medicine, Sagamihara City 252-0374, Kanagawa, Japan; amidesutarere9010@yahoo.co.jp (A.T.); uekusa18y@gmail.com (Y.U.); m.manabu0829@hotmail.co.jp (M.M.); daiiwase19760601@yahoo.co.jp (D.I.); jun43814@gmail.com (J.A.); 44134413oo@gmail.com (Y.O.); ginoue@kitasato-u.ac.jp (G.I.); mtakaso@kitasato-u.ac.jp (M.T.)
2 Division of Blood Transfusion and Transplantation, Kitasato University School of Health Sciences, Minamiuonuma 949-7241, Nigata, Japan; eohta@kitasato-u.ac.jp (E.O.); hattori.akito@kitasato-u.ac.jp (A.H.)
3 Program in Cellular Immunology, Graduate School of Medical Science, Kitasato, Sagamihara City 252-0375, Kanagawa, Japan
4 Center for Cell Design, Institute for Regenerative Medicine and Cell Design, Kitasato University School of Allied Health Sciences, Sagamihara City 252-0374, Kanagawa, Japan
5 Medical Sciences Research Institute, Shonan University, Chigasaki City 253-0083, Kanagawa, Japan
* Correspondence: kuchida@med.kitasato-u.ac.jp
† These authors contributed equally to this work.

Abstract: Osteoarthritis (OA) is a chronic joint disease marked by synovial inflammation, cartilage degradation, and persistent pain. Although Netrin-4 (NTN4) has been implicated in pain modulation in rheumatoid arthritis (RA), its role in OA pain remains less understood. Previous research has documented that NTN4 promotes axonal growth in rodent-derived neurons; however, its effects on human sensory neurons are yet to be fully explored. NTN4 also plays a multifactorial role in various non-neuronal cells, such as endothelial cells, tumor cells, and stromal cells. Nevertheless, its specific impact on synovial fibroblasts, which are key components of the synovium and have been linked to OA pain, is still unclear. This study examined the correlation between NTN4 expression levels and pain severity in OA, specifically investigating its effects on human iPSC-derived sensory neurons (iPSC-SNs) and synovial fibroblasts from OA patients. Our findings indicate a positive correlation between synovial *NTN4* expression and pain severity. Recombinant human Netrin-4 (rh-NTN4) was also shown to enhance neurite outgrowth in human iPSC-SNs, suggesting a potential role in neuronal sensitization. Additionally, rh-NTN4 stimulated the production of pro-inflammatory cytokines (IL-6, IL-8) and chemokines (CXCL1, CXCL6, CXCL8) in synovium-derived fibroblastic cells, implicating it in synovial inflammation. Collectively, these results suggest that NTN4 may contribute to KOA pathology by promoting synovial inflammation and potentially sensitizing sensory neurons, thereby influencing the mechanisms of underlying pain.

Keywords: osteoarthritis; netrin-4; synovium; sensory neuron; fibroblast

1. Introduction

Knee osteoarthritis (KOA) is a chronic degenerative condition characterized by knee pain, commonly affecting middle-aged and elderly individuals. Pain management has long been a cornerstone of KOA treatment according to established guidelines. Indeed, the 2019 guidelines from the Osteoarthritis Research Society International (OARSI) emphasize the significance of addressing pain as a primary pathological factor in KOA [1]. Therefore, effectively addressing peripheral pain sensitivity is crucial for alleviating KOA-related pain and enhancing the quality of life for patients with this condition.

In human studies on osteoarthritis (OA), inflammation in the synovium contributes to pain through various inflammatory mediators, with evidence linking the severity of synovitis in the knee to pain levels [2,3]. The synovium is richly innervated with nociceptive nerve fibers, which respond to harmful mechanical stimuli and pain-inducing substances. These nociceptors are further sensitized by inflammatory mediators released during OA progression, thereby amplifying the perception of pain [4–8]. This dynamic interplay between inflammation and neuronal sensitization underscores the critical role of sensory innervation in OA pain mechanisms.

Sensory innervation in synovium plays a pivotal role in pain perception, with nerve fibers and their associated receptors directly transmitting pain signals to the central nervous system. The sensitization of nociceptors and their heightened response to pro-inflammatory stimuli contribute significantly to the chronic pain experienced in KOA [9]. Recent advancements in neuroscience have highlighted the involvement of the Netrin family in modulating neuronal sensitization and synaptic plasticity, two processes fundamental to pain transmission and persistence [10–13]. The Netrin family consists of netrin-1, -3, -4, -5, G1, and G2, which are secreted proteins with diverse biological roles in neural development, axonal guidance, and angiogenesis. Secreted Netrins interact with specific receptors such as DCC (deleted in colorectal cancer), Neogenin, and members of the Unc5 homolog family, while the Netrin-G subfamily binds to unique receptor types [14].

Among the Netrins, Netrin-4 (NTN4), also known as β-netrin, is a laminin-like secreted protein involved in the development of various tissues, including the central nervous system, bones, kidneys, and blood vessels. NTN4 has been extensively studied in the context of axonal guidance, angiogenesis, and tumor biology, with its crystal structure elucidated to reveal insights into its molecular interactions [15,16]. A recent study reported a possible association between NTN4 and pain in patients with rheumatoid arthritis, noting that it augments the branching of sensory neurons [17]. However, the correlation between NTN4 and pain in OA patients remains unclear. Additionally, a previous study evaluated the effect of human recombinant NTN4 on rodent-derived neurons, demonstrating enhanced branching of sensory neurons in vitro [17,18]. Although the homology between human and mouse *NTN4* is 87%, it is still uncertain whether human NTN4 has a similar effect on human-derived sensory neurons. Advances in induced pluripotent stem cell (iPSC) technology have opened new avenues for researching human sensory neurons [19,20]. iPSC-derived human sensory neurons have proven useful for evaluating the function of NTN4 on human sensory neurons.

Recent studies have highlighted the crucial role of synovial fibroblasts in OA pain [21,22]. In healthy joints, these fibroblasts are quiescent, and primarily tasked with producing extracellular matrix proteins to maintain tissue structure and synovial fluid composition. However, in OA, the synovium becomes inflamed, and the fibroblasts within are activated. This activation contributes to pain by sensitizing nociceptive nerve endings in the synovium. Synovial fibroblasts can augment the growth of nociceptive axons and neurites, extending into synovial papillary processes that position nociceptors close to the lining fibroblasts, thereby promoting pain perception.

Previous research has demonstrated that fibroblast-secreted NTN4 facilitates axonal and neurite growth, supporting a mechanistic link between NTN4 and pain propagation in rheumatoid arthritis [17]. In contrast, NTN4 has been shown to influence non-neuronal cells [23–25]; for example, our prior studies have reported that NTN4 stimulates the expression of pro-inflammatory cytokine mRNA in stromal cells derived from the infrapatellar fat pad (IPFP) of OA patients [23]. This suggests that NTN4 might also activate synovial fibroblasts through autocrine or paracrine actions. Synovitis mechanisms, including the release of pro-inflammatory mediators such as tumor necrosis factor alpha (TNFα), interleukin 1 beta (IL1β), monocyte chemoattractant protein 1 (MCP1), and IL6, are known to sensitize and activate nociceptors, lowering the threshold required for pain activation and leading to pain sensitization [26–28]. These mediators are positively correlated with patient-reported pain and are thought to increase the excitability of sensory neurons [29]. Thus, the potential role of NTN4 in promoting synovial inflammation and activating synovial fibroblasts, directly or indirectly, could be significant in understanding OA pain mechanisms. However, the specific impacts of NTN4 on synovial fibroblasts themselves remain to be fully elucidated.

Our study aims to investigate the correlation between *NTN4* expression and pain levels in OA patients, examining the effects of NTN4 on synovial fibroblasts and human iPSC-SNs.

2. Materials and Methods

2.1. Study Participants

This study adhered to the Declaration of Helsinki guidelines and received approval from the Kitasato University Institutional Review Board (protocol code: B19-259; approved on 27 January 2020). The sample size was calculated based on the correlation coefficient (ρ = 0.377) observed in the initial 10 cases between NTN4 expression and pain score. After determining the effect size, it was estimated that 50 cases would be required with a significance level of 0.05 and a statistical power of 0.8. Patients diagnosed with KOA based on clinical and radiographic evaluations were recruited for this research. Exclusion criteria included the presence of rheumatoid arthritis, autoimmune diseases, inflammatory arthropathies, systemic joint-affecting diseases, or a history of joint replacement surgery. Synovial tissue samples were collected from KOA patients confirmed by radiography who underwent total knee arthroplasty at our institution. During surgery, tissue samples were obtained from the affected knee. Fifty synovial specimens were rapidly frozen in liquid nitrogen at −196 °C and stored at −80 °C for RNA extraction.

2.2. Culture of iPSC-Derived Sensory Neurons and Measurement of Neurite Length

Human induced pluripotent stem cell (iPSC)-derived sensory neurons (iPSC-SNs) were sourced from ReproCELL, Inc. (Yokohama, Japan, Catalog No. RCDN001N). Neurons were plated on 10-mm coverslips coated with poly-L-ornithine and fibronectin in 24-well plates. Cultures were maintained at 37 °C in a humidified atmosphere with 5% CO_2, and the medium was refreshed every 3 to 4 days. Prior to recombinant human Netrin-4 (rh-NTN4; Cat. no. 1254-N4-025, R&D Systems, Minne-apolis, MN, USA) stimulation, RNA was extracted from the cells to assess the expression of endogenous NTN4 and its receptors using quantitative PCR (qPCR). After 24 h, the neurons were treated with rh-NTN4 at concentrations of 50 and 500 ng/mL. Neuronal morphology images were captured after 4 days of treatment using an Olympus CKX53 inverted microscope. Immunofluorescence analysis was performed on day 14. As described by Ohta et al. [30], neurons were fixed and stained for analysis. Cells were incubated with primary antibodies against TUBB3 (T8660, Sigma-Aldrich, St. Louis, MO, USA) at 4 °C overnight. Following PBS rinsing, the cells were incubated at room temperature for 1 h with Alexa Fluor 647-conjugated secondary antibodies (A-21242, Invitrogen, Carlsbad, CA, USA), with nuclei counterstained using Hoechst 33342 (Sigma-Aldrich). Fluorescence images were acquired using a Keyence

BZ-X810 microscope, equipped with a 20× objective lens (NA 0.45; Keyence), a 1.0× optical zoom, and capturing five random fields of view from a single cover glass.

The neurite length of iPSC-SNs was measured in vitro on days 4 and 14. Measurements were performed using the ImageJ (version 1.4.3) plug-in NeuronJ, which was specifically downloaded for this purpose. Neurites extending from the cell body of each neuron were traced, and branched neurites were measured as the total length of all neurites originating from a single cell body. Measurements were taken in two wells for each concentration, with 10 neurons analyzed per well. This process was repeated three times to ensure reproducibility.

2.3. Isolation and Culture of Synovium-Derived Fibroblasts

Eight synovial tissue samples were finely minced into small fragments and enzymatically digested using 0.2% collagenase type I (Sigma-Aldrich) in α-minimal essential medium (α-MEM; Nacalai Tesque Inc., Kyoto, Japan) at 37 °C for 2 h with gentle agitation to facilitate cell dissociation. The resulting suspension was filtered through a 100-μm cell strainer (pluriSelect, Leipzig, Germany) to remove undigested debris. The filtrate was centrifuged at 300× *g* for 5 min, and the resulting cell pellet was resuspended in phosphate-buffered saline (PBS). To deplete non-fibroblast populations, the cells were incubated at 4 °C for 30 min with biotin-labeled anti-CD45 (Cat. No. 304004) and CD31 (Cat. No. 536604) antibodies (BioLegend, San Diego, CA, USA), which target hematopoietic lineage cells and endothelial cells, respectively. After antibody incubation, unbound antibodies were removed by washing the cells twice with PBS. Magnetic separation was performed using streptavidin-conjugated magnetic particles (BD Biosciences, Franklin Lakes, NJ, USA) in combination with the IMag magnetic separation system (BD Biosciences). For the separation process, the cell suspension was placed on a magnetic board at room temperature for 8 min to allow binding of antibody-labeled cells to the magnetic particles. The negative fraction, enriched in fibroblasts, was collected. To enhance fibroblast purity, the collected negative fraction was subjected to a second magnetic separation under the same conditions. This sequential separation ensured the effective depletion of CD45-positive and CD31-positive cells, resulting in a fibroblast-enriched population. The isolated fibroblasts were plated in 75 cm^2 tissue culture flasks (Thermo Fisher Scientific, Waltham, MA, USA) and maintained at 37 °C in a humidified incubator with 5% CO_2. Cells were cultured in α-MEM supplemented with 10% fetal bovine serum (FBS; Gibco, Thermo Fisher Scientific) and 1% penicillin-streptomycin (Nacalai Tesque). The culture medium was replaced every 3 to 4 days to remove non-adherent cells and support fibroblast proliferation. Prior to rh-NTN4 stimulation, RNA was extracted from the fibroblasts to assess the expression of endogenous NTN4 and its receptors using qPCR. The results were then compared to those obtained from iPSC-SNs. Once cells reached approximately 80% subconfluency, they were detached using 0.25% trypsin/EDTA (Nacalai Tesque) and reseeded into 6-well plates at a density of 2×10^5 cells per well for experimental treatments. After a 3-day incubation period to allow cells to adhere and stabilize, fibroblasts were treated with rh-NTN4 or vehicle control (α-MEM supplemented with 10% FBS). rh-NTN4 was administered at final concentrations of 50 ng/mL or 500 ng/mL. The treatments were performed for 3, 6, or 24 h to evaluate both short-term and long-term effects on fibroblast gene and protein expression. At the conclusion of each treatment period, cells were lysed directly in TRIzol reagent (Invitrogen, Carlsbad, CA, USA) to extract total RNA for quantitative PCR (qPCR) analysis. For protein analysis, cell supernatants were collected after 24 h of treatment and stored at −80 °C for subsequent quantification using enzyme-linked immunosorbent assay (ELISA). In addition, flow cytometry was performed after 24 h of treatment to evaluate VCAM-1 expression (n = 8). All experiments were performed with biological replicates, and each experiment was conducted in duplicate to ensure reproducibility.

2.4. qPCR

Total RNA was extracted from synovial tissue using MaXtract high-density tubes (Qiagen, Valencia, CA, USA) in combination with the phenol/chloroform method, ensuring high RNA yield and purity. For cultured cell samples, RNA was extracted using the Direct-zol MicroPrep kit (Zymo Research, Orange, CA, USA), following the manufacturer's protocol, which includes an in-column DNase I treatment step to remove genomic DNA contamination. RNA purity and concentration were assessed spectrophotometrically (Denovix, Tokyo, Japan), ensuring an A260/A280 ratio greater than 1.8 for all samples to confirm RNA integrity. Complementary DNA (cDNA) was synthesized from 0.5 µg of total RNA using Superscript III Reverse Transcriptase (Invitrogen, Carlsbad, CA, USA). Oligo(dT) primers and dNTPs were purchased from Takara Bio (Shiga, Japan). The reaction was carried out in a thermal cycler under the following conditions: primer annealing at 65 °C for 5 min, reverse transcription at 50 °C for 50 min, and enzyme inactivation at 70 °C for 15 min. Primers for the target genes were designed using Primer-BLAST (NCBI) to ensure specificity and efficiency. All primers were synthesized by Hokkaido System Science (Hokkaido, Japan). Primer sequences are listed in Table 1.

Table 1. Primer sequences.

Gene		Sequence	bp
CXCL1	sense	GCT TGC CTC AAT CCT GCA TC	73
	antisense	AGT TGG ATT TGT CAC TGT TCA GC	
CXCL6	sense	GGT CCT TCG GGC TCC TTG TG	125
	antisense	ACG CGT AAA CAA GTG CAA CG	
IL6	sense	GAG GAG ACT TGC CTG GTG AAA	199
	antisense	TGG CAT TTG TGG TTG GGT CA	
IL8	sense	ACA CTG CGC CAA CAC AGA AA	89
	antisense	CAA CCC TCT GCA CCC AGT TT	
MMP1	sense	ACT TAC ATC GTG TTG CGG CT	164
	antisense	CGA TGG GCT GGA CAG GAT TT	
MMP3	sense	GTG GAG TTC CTG ACG TTG GT	164
	antisense	TGG AGT CAC CTC TTC CCA GA	
MMP13	sense	TGA CTG AGA GGC TCC GAG AA	111
	antisense	CAT CAG GAA CCC CGC ATC TT	
NEO1	sense	GGGCATGAGTCAGAGGACAG	127
	antisense	CGAGGGAATGGATGGGATGG	
NTN4	sense	TGT TGT CAA GAA GGG CGC TA	159
	antisense	ACG CGA AGG TTG GTG ATCT T	
UNC5B	sense	CAGAACGACCACGTCACACA	121
	antisense	ACCAGTAATCCTCCAGCCCA	
VCAM1	sense	CCA TCC ACA AAG CTG CAA GA	70
	antisense	CTG GAG CTG GTA GAC CCT CG	

Quantitative PCR (qPCR) was performed in a 25 µL reaction volume containing 2 µL of cDNA, 12.5 µL of TB Green Premix Ex Taq II (Tli RNaseH Plus, Takara Bio, Shiga, Japan), 2 µL of each primer, and 8.5 µL of nuclease-free water. Reactions were run in triplicate using the CFX-96 Real-Time PCR Detection System (Bio-Rad, Hercules, CA, USA) under the following cycling conditions: initial denaturation at 95 °C for 10 min, followed by 40 cycles of denaturation at 95 °C for 15 s, and annealing/extension at 60 °C for 1 min. Gene expression levels were normalized to GAPDH as an internal control, and relative expression levels were calculated using the delta-delta CT method. Amplification specificity was confirmed by melt curve analysis.

2.5. ELISA

rh-NTN4 was administered to fibroblast cultures at final concentrations of 50 ng/mL or 500 ng/mL. The treatments were performed for 24 h to evaluate protein concentrations in the supernatants. MMP-1 was measured using an ELISA kit (R&D Systems, Minneapolis, MN, USA, Catalog No. DY901B). MMP-3 was also measured using an ELISA kit (R&D Systems, Minneapolis, MN, USA, Catalog No. DY513-05). Additionally, MMP-13 levels were determined using an ELISA kit (R&D Systems, Minneapolis, MN, USA, Catalog No. DY511). Interleukin-6 (IL-6) was measured using an ELISA kit (BioLegend, San Diego, CA, USA, Cat. No. 430515). Interleukin-8 (IL-8) was also measured using another ELISA kit (BioLegend, Cat. No. 431504). Furthermore, CXCL1 was determined using an ELISA kit (ProteinTech, Chicago, IL, USA, Cat. No. KE00133), and CXCL6 was measured with another ELISA kit (ProteinTech, Cat. No. KE00274). All measurements were performed according to the manufacturer's instructions.

2.6. Flow Cytometric Analysis

rh-NTN4 was administered to fibroblast cultures at final concentrations of 50 ng/mL or 500 ng/mL for 24 h. To evaluate VCAM-1 expression, flow cytometric analysis was performed after 24 h of treatment. Cultured fibroblasts were detached using a 0.25% Trypsin/EDTA solution. After removing trypsin, the cells were stained with the following antibodies: APC-conjugated anti-VCAM-1 (BioLegend, Cat. No. 305810), PE-Cy7-conjugated anti-CD90 (BioLegend, Cat. No.328124), and FITC-conjugated anti-CD45 (BioLegend, Cat. No. 304006). The stained cells were analyzed using a FACSVerse flow cytometer (BD Biosciences). Data were processed and analyzed using FlowJo software to calculate the mean fluorescence intensity (MFI) of VCAM-1 expression.

2.7. Statistical Analysis

The sample size and statistical power were determined using G*Power 3 software to ensure adequate sensitivity for detecting significant differences and correlations. Data were analyzed using SPSS software (Version 28.0; IBM Corp., Armonk, NY, USA). The Shapiro–Wilk test was used to evaluate data normality within each group. Since the data did not follow a normal distribution ($p < 0.05$), non-parametric tests were applied. The Kruskal–Wallis test, a non-parametric alternative to one-way ANOVA that does not require normality assumptions, was used to evaluate the effects of NTN4 on gene expression in fibroblasts under different treatment time points and concentrations. Post-hoc analysis was performed using the Dunn test to account for multiple comparisons. In addition, Spearman's rank correlation coefficient was calculated to determine associations between NTN4 expression and VAS scores for active pain. Linear regression analysis was also conducted to examine relationships between NTN4 expression and clinical factors such as age, BMI, K/L grade, and VAS scores. Statistical significance was defined as a p-value of less than 0.05.

3. Results

3.1. NTN4 Expression and KOA Pathology

Table 2 outlines the demographic and clinical characteristics of the 50 study participants, which includes patients across different K/L grades, reflecting various levels of OA severity. Key clinical data recorded for each patient included age, BMI, and pain scores assessed via the Visual Analog Scale (VAS) (Table 2). No significant differences were found in age, BMI, or VAS pain scores between male and female participants (Table 2).

Table 2. Clinical characteristics of knee osteoarthritis patients.

	All (n = 50)	Male (n = 13)	Female (n = 37)	p-Value
Age (years)	75.0 ± 7.5	71.9 ± 6.4	75.8 ± 7.6	0.174
BMI (kg/m^2)	26.6 ± 4.1	26.6 ± 3.2	26.6 ± 4.4	0.805
KL grade (2/3/4), n	2/8/40	0/0/13	2/8/4	0.648
VAS (cm)	7.0 ± 2.4	7.5± 1.8	6.9± 2.6	0.311

BMI, body mass index; KL grade, Kellgren–Lawrence grade; VAS., Visual Analog Scale.

To investigate the relationship between *NTN4* expression and pain in OA patients, we analyzed *NTN4* expression levels in synovial tissue samples using qPCR. A positive correlation was observed between *NTN4* expression levels and pain scores, as measured by the VAS (ρ = 0.362, p = 0.010, Figure 1).

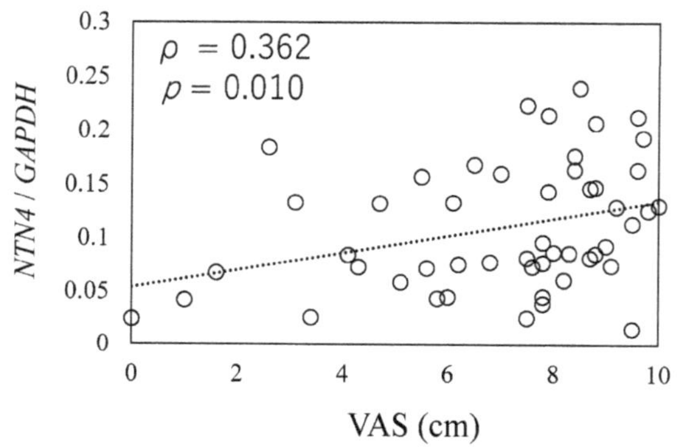

Figure 1. Relationship between *NTN4* expression and osteoarthritis pathology.

The correlation between *NTN4* expression and pain score was measured by the Visual Analog Scale (VAS). X-axis: indicates the scores on the Visual Analog Scale (VAS) for pain during movement, as reported by patients. Y-axis: Represents the expression levels of *NTN4* in synovial tissue, normalized to the housekeeping gene *GAPDH*.

Further, a linear regression analysis was conducted to investigate associations between *NTN4* expression and key clinical variables, such as K/L grade, age, BMI, and VAS scores. This analysis revealed a significant association between *NTN4* expression and VAS scores (β = 0.397, p = 0.005). In contrast, other variables, including age (β = −0.250, p = 0.060), gender (β = −0.250, p = 0.079), BMI (β = 0.507, p = 0.670), and K/L grade (β = 0.020, p = 0.805), did not show significant associations with *NTN4* expression (Table 3).

Table 3. Linear regression analysis of variables associated with *NTN4* expression.

Variable	β	p-Value
Age (years)	−0.266	0.060
Gender, male/female	−0.250	0.079
BMI (kg/m^2)	0.089	0.507
KL grade	0.020	0.884
VAS (mm)	0.397	0.005

3.2. The Effect of rh-NTN4 on Neurite Outgrowth in Human iPSC-SNs

qPCR analysis of human iPSC-SNs showed a significantly lower expression of *NTN4* (p = 0.010, Figure 2A) compared to the fibroblasts, while no significant differences were observed in the expression levels of *UNC5B* ($p = 0.065$, Figure 2B). However, a higher expression of *NEO1* encoding neogenin was observed in iPSC-SNs ($p < 0.001$, Figure 2C).

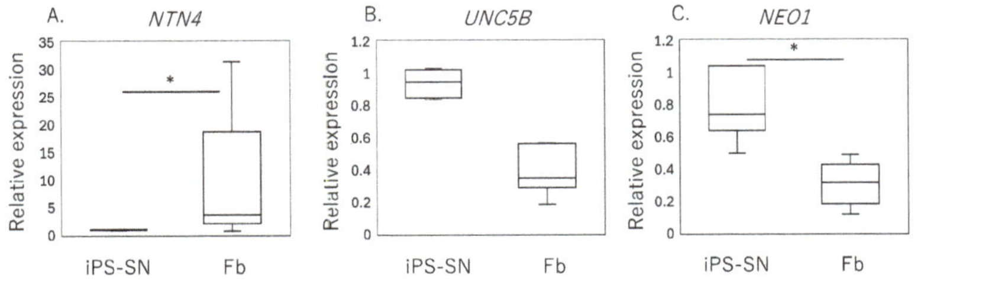

Figure 2. Expression of *NTN4* and its receptors in iPSC-derived sensory neurons (iPSC-SNs) and synovial fibroblasts. Box-and-whisker plots showing the relative expression levels of (**A**) *NTN4*, (**B**) *UNC5B*, and (**C**) *NEO1* in fibroblasts (Fb), normalized to the expression levels in iPSC-derived sensory neurons (iPSC-SNs), which are set as 1. An asterisk (*) indicates statistical significance with p-values less than 0.05, demonstrating differences between fibroblast and iPSC-SN expression levels.

To assess the effect of rh-NTN4 on neurite outgrowth, the length of the neurites in human iPSC-SNs was measured at 4 and 14 days in vitro (Figure 3A–K). In the short-term culture (4 days), rh-NTN4 treatment significantly enhanced neurite outgrowth compared to the control group ($p = 0.013$, Figure 3A–D). Similarly, in the 14-day culture, sensory neurons treated with rh-NTN4 displayed markedly longer neurites than the untreated controls ($p = 0.027$, Figure 3E–H).

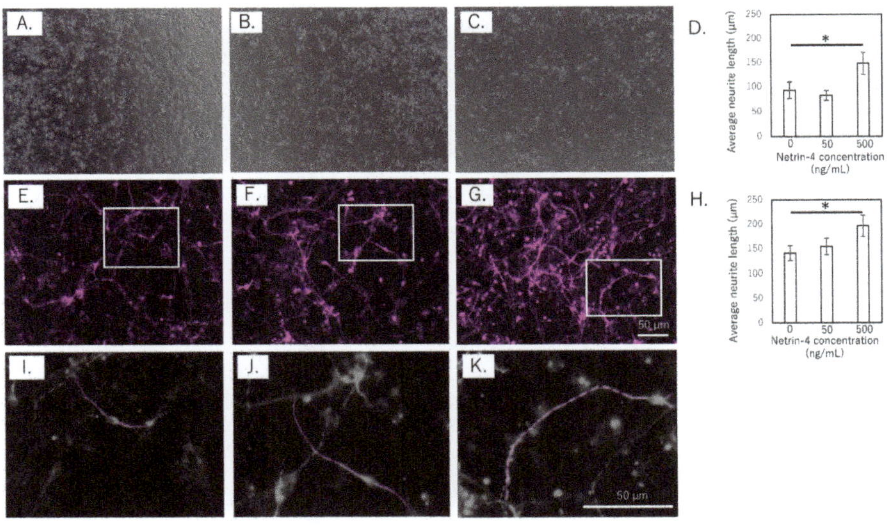

Figure 3. Effect of rh-NTN4 on neurite outgrowth in human iPSC-derived sensory neurons (**A–C**). Optical microscopy images at day 4 after rh-NTN4 stimulation: (**A**) vehicle-treated cells, (**B**) cells treated with 50 ng/mL rh-NTN4, (**C**) cells treated with 500 ng/mL rh-NTN4. (**D**) Quantification of neurite length in vehicle-treated, 50 ng/mL rh-NTN4-treated, and 500 ng/mL rh-NTN4-treated cells at day 4. Data are presented as mean ± SE. (**E–G**) Fluorescence microscopy images at day 14 after rh-NTN4 stimulation: (**E**) vehicle-treated cells, (**F**) cells treated with 50 ng/mL rh-NTN4, (**G**) cells treated with 500 ng/mL rh-NTN4. (**H**) Quantification of neurite length in vehicle-treated, 50 ng/mL rh-NTN4-treated, and 500 ng/mL rh-NTN4-treated cells at day 14. Data are presented as mean ± SE. * indicates significant differences between groups ($p < 0.05$). (**I–K**) Neurite length per neuron in the (**E–G**) images was measured using NeuronJ, an ImageJ plug-in.

3.3. Effect of rh-NTN4 on the Expression of Inflammatory Cytokines, Chemokines, Matrix Metalloproteinases (MMPs), and VCAM1 in Synovial Fibroblasts

qPCR analysis revealed that exogenous rh-NTN4 stimulation did not alter the expression of endogenous *NTN4*, *UNC5B*, or *NEO1* (Figure 4A–C). Following the initial analysis, we evaluated the effects of rh-NTN4 on synovial fibroblasts, drawing on insights from our previous RNA-Seq analysis of infrapatellar fat pad-derived stromal cells treated with rh-NTN4 [23]. *MMP1* and *MMP3* exhibited significant differences in expression between the 0 ng/mL and 500 ng/mL groups at all time points (*MMP1*: 3 h, $p = 0.001$; 6 h, $p < 0.001$; 24 h, $p < 0.001$; *MMP3*: 3 h, $p = 0.011$; 6 h, $p = 0.001$; 24 h, $p < 0.001$) (Figure 4D,E). Similarly, *MMP13* showed significant differences at 6 h ($p < 0.001$) and 24 h ($p = 0.003$) (Figure 4F). *VCAM1* expression was significantly different between the 0 ng/mL and 500 ng/mL groups at 3 h ($p = 0.016$), 6 h ($p = 0.001$), and 24 h ($p < 0.001$), and between the 0 ng/mL and 50 ng/mL groups at 24 h ($p = 0.046$) (Figure 4G).

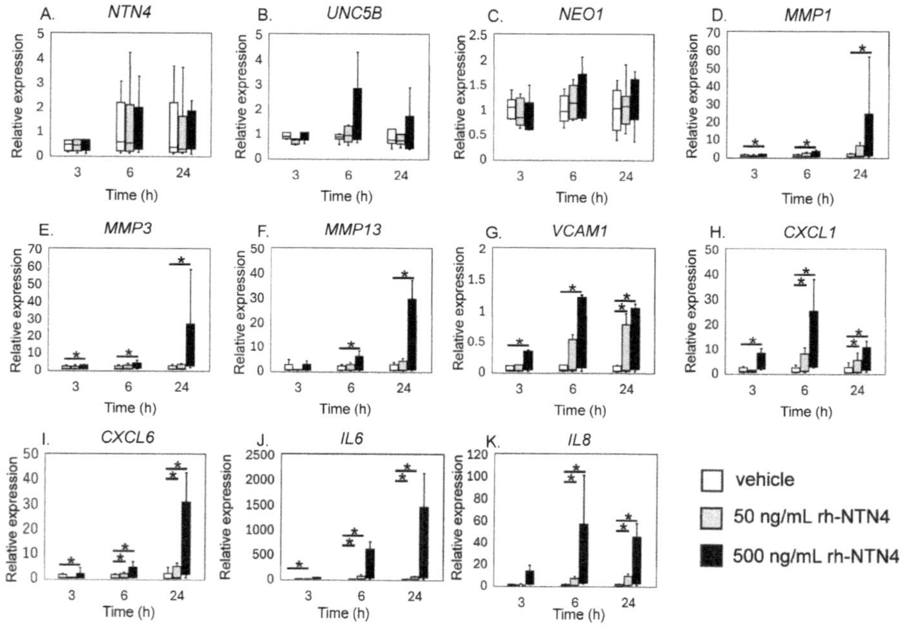

Figure 4. qPCR analysis of vehicle and recombinant Netrin-4 treated fibroblastic cells derived from the synovium. Relative expression levels of (**A**) *NTN4*, (**B**) *UNC5B*, (**C**) *NEO1*, (**D**) *MMP1*, (**E**) *MMP3*, (**F**) *MMP13*, (**G**) *VCAM1*, (**H**) *CXCL1*, (**I**) *CXCL6*, (**J**) *IL6*, and (**K**) *IL8* following rh-NTN4 treatment compared to vehicle control is presented using box-and-whisker plots. These plots depict the median, quartiles, and range of each dataset. Statistical significances between groups are clearly indicated with lines, and asterisks (*) denote *p*-values less than 0.05, highlighting statistically significant differences.

CXCL1 and *CXCL6* both displayed significant differences between the 0 ng/mL and 500 ng/mL groups at 3, 6, and 24 h (*CXCL1*: all $p < 0.001$; *CXCL6*: 3 h, $p = 0.024$; 6 h, $p < 0.001$; 24 h, $p < 0.001$), with additional differences observed between the 0 ng/mL and 50 ng/mL groups at 6 h and 24 h ($p = 0.046$ for both markers) (Figure 4H,I). The *IL6* and *IL8* expression levels were also significantly different between the 0 ng/mL and 500 ng/mL groups at all time points ($p < 0.001$), and between the 0 ng/mL and 50 ng/mL groups at 6 h and 24 h ($p = 0.046$) (Figure 4J,K).

To validate the qPCR data, ELISA was performed on culture supernatants. MMP-1 levels were significantly higher with 50 ng/mL rh-NTN4 treatment ($p = 0.024$) and further increased at 500 ng/mL rh-NTN4 ($p < 0.001$) (Figure 5A). MMP-3 levels demonstrated a significant rise at 500 ng/mL rh-NTN4 ($p < 0.001$) (Figure 5B). Similarly, MMP-13 levels were significantly elevated at 500 ng/mL compared to the vehicle control ($p = 0.037$) (Figure 5C). Furthermore, CXCL1 levels increased significantly with 50 ng/mL rh-NTN4 ($p = 0.046$) and were further elevated at 500 ng/mL ($p < 0.001$) (Figure 5D). Likewise, CXCL6 demonstrated a significant elevation at both 50 ng/mL ($p = 0.046$) and 500 ng/mL rh-NTN4 ($p < 0.001$) (Figure 5E). Both IL-6 and IL-8 levels were also significantly elevated in the presence of 50 ng/mL rh-NTN4 (IL-8: $p = 0.046$) and 500 ng/mL rh-NTN4 (IL-6: $p < 0.001$; IL-8: $p < 0.001$) compared to vehicle control (Figure 5F,G). To validate the qPCR result of VCAM1 expression, flow cytometric analysis was also performed (Figure 6A–D). The MFI in 500 ng/mL rh-NTN4 was significantly higher than that in vehicle control ($p = 0.040$) (Figure 6D).

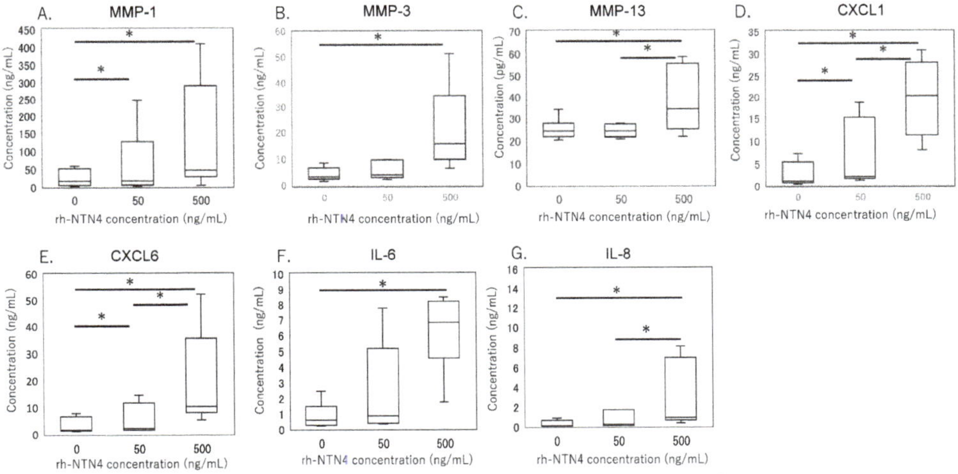

Figure 5. ELISA analysis of cell supernatant in vehicle- and recombinant Netrin-4-treated fibroblastic cells derived from the synovium. Concentrations of (**A**) MMP-1, (**B**) MMP-3, (**C**) MMP-13, (**D**) CXCL1, (**E**) CXCL6, (**F**) IL-6, and (**G**) IL-8 in cell supernatant are presented in box-and-whisker plots, showing the median, 25th and 75th percentiles, and range. * $p < 0.05$.

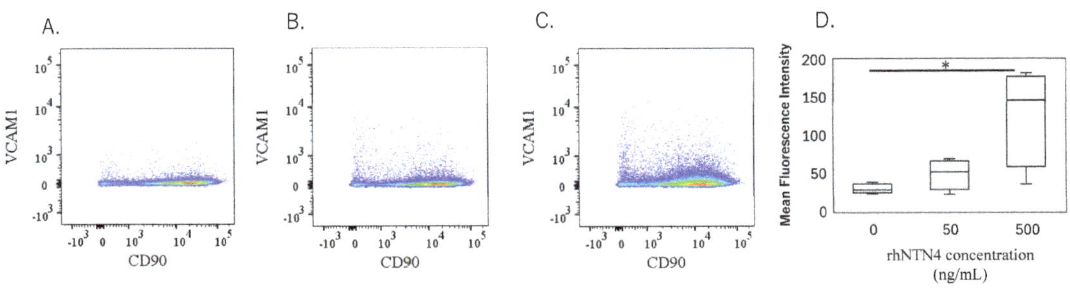

Figure 6. Flow cytometric analysis of vehicle- and recombinant Netrin-4-treated fibroblastic cells derived from the synovium. (**A–C**): Dot plot analysis of fibroblastic cells treated with vehicle (**A**), 50 ng/mL (**B**), and 500 ng/mL (**C**) recombinant human Netrin-4 (rh-NTN4). (**D**) Mean fluorescence intensity of vehicle- and rhNTN4-treated fibroblastic cells displayed as box-and-whisker plots. * indicates significant differences ($p < 0.05$) vs. vehicle.

4. Discussion

This study advances our understanding of the role of NTN4 in KOA by exploring its potential links to pain severity and its effects on synovial fibroblasts and human iPS-SNs. Our results demonstrate a significant correlation between NTN4 levels in the synovium and pain scores. Furthermore, NTN4 was found to promote axonal growth in iPS-SNs and to stimulate the production of inflammatory mediators in synovial fibroblasts. These findings suggest that NTN4 plays a complex role in KOA, potentially influencing both neuronal sensitization and inflammatory responses. Further investigation into NTN4-mediated neuronal and inflammatory pathways could provide deeper insights into how NTN4 contributes to OA pain mechanisms.

Previous research has shown that synovium, a richly innervated tissue, plays a pivotal role in pain perception in KOA [13,31,32] NTN4's influence on sensory neurons, particularly in promoting neurite outgrowth, underscores its potential role in sensitization(ref. This study also demonstrated that NTN4 can affect neuronal growth and branching via the UNC5B receptor in rats, impacting pain sensitivity [18]. While Netrin-4's impact on sensory neurons was previously evaluated using mouse DRG neurons [17], its effects on human cells have not been clarified. In this study, we utilized human iPSC-SNs to further explore NTN4's biological roles. Our findings confirm the expression of NTN4 receptors, *NEO1* and *UNC5B*, in human-derived neurons, and we observed enhanced axonal elongation similar to that seen in mouse cells [17]. These observations suggest that NTN4 could play a similar role in promoting sensory nerve outgrowth within the human synovium, potentially increasing the density of pain-transmitting fibers. However, it is important to note that while these results are promising, they primarily provide a foundation for hypothesizing about NTN4's function in human OA. The actual impact on pain perception in KOA patients remains to be directly demonstrated. The hypothesis that NTN4 may amplify pain signals by increasing sensory nerve innervation aligns with broader research linking increased sensory innervation to heightened pain sensitivity in OA-affected joints [33]. Future studies are necessary to elucidate the precise mechanisms by which NTN4 may influence pain pathways in human OA.

Previous studies have shown that the severity of synovitis correlates with increased knee pain, highlighting a connection between inflammation and nociceptive activity [34–38]. IL-6 plays a critical role in the development and persistence of hyperalgesia across various pain models [39]. Elevated IL-6 levels in the spinal cord have been shown to induce mechanical hyperalgesia in rats and are linked to nociceptive sensory processes [40,41]. Dysregulation of IL-6 leads to the production and release of various inflammatory mediators, which can activate neurocytes and potentially contribute to neuropathic pain [24,39]. Targeting IL-6 signaling has been shown to result in significant clinical improvements in inflammatory arthritis [42]. Similarly, elevated plasma IL-8 concentrations, which reflect higher levels of peripheral inflammation, have been associated with lower pressure pain thresholds in OA patients [43]. Our findings indicate that NTN4 could potentially play a role in these processes by promoting axonal growth in sensory neurons and upregulating pro-inflammatory cytokines and chemokines in synovial fibroblasts. However, while these results are promising, they should be interpreted with caution. The direct causal relationships between NTN4 action, inflammatory mediator production, and pain perception in KOA have not been definitively established. This study suggests that NTN4 might contribute to local inflammation and neuronal sensitization, potentially exacerbating pain perception in KOA, but further research is necessary to confirm these preliminary findings and understand the underlying mechanisms.

In our study, NTN4 was identified as playing a role in promoting the expression of MMP1, MMP3, and MMP13, all of which are involved in extracellular matrix degradation [44]. These MMPs play a pivotal role in the breakdown of collagen and proteoglycans, leading to the progressive deterioration of synovial tissue integrity in OA. By driving the expression of these catabolic enzymes, NTN4 may contribute to the pathological remodeling of the synovial microenvironment observed in OA. Targeting NTN4 could, therefore, offer a novel therapeutic strategy to mitigate matrix degradation and preserve synovial and joint health. This lack of variation may be attributed to this study's focus on end-stage OA, where synovial inflammation and ECM degradation are already at advanced levels, potentially obscuring differences that might be more evident in earlier stages of disease progression. The role of NTN4 in the earlier phases of OA, where ECM turnover is initiated and inflammatory responses are developing, remains unexplored and could provide critical insights into its involvement in the initiation and progression of OA pathology. Examining NTN4 activity in normal or early-stage OA synovium could clarify whether its upregulation serves as an early trigger for pathological processes, such as the recruitment of inflammatory cells, nociceptor sensitization, and ECM breakdown. Furthermore, early-stage studies might reveal whether NTN4 inhibition has the potential to slow or even prevent OA progression by preserving synovial integrity and dampening inflammatory responses before irreversible joint damage occurs.

Several limitations warrant further attention in future studies. First, our study did not assess whether iPSC-SNs express nociceptive neuropeptides such as Substance P, nor did we explore how NTN4 stimulation might alter such expression. This is crucial for fully understanding NTN4's role in neuronal sensitization and pain mechanisms in OA. Second, this study was limited in exploring the interactions between NTN4 and other inflammatory mediators, which could provide deeper insights into its regulatory mechanisms. Additionally, examining NTN4's effects on fibroblast behavior, matrix degradation, and crosstalk with cartilage cells was not conducted, which would offer a more comprehensive understanding of its role in OA pathology.

Finally, further research involving larger, diverse patient cohorts and healthy controls is essential to enhance our understanding of NTN4's systemic effects on joint tissues and its specific role in modulating pain mechanisms in OA. This should include studies using a variety of tissue samples, such as those obtained from synovial fluid punctures, and utilizing a broad range of patient samples along with the Osteoarthritis Research Society International (OARSI) staging system [45] to elucidate the complex dynamics of NTN4 in relation to the inflammatory milieu and pain pathways at different stages of OA. This could elucidate the complex dynamics of NTN4 and its relationship with the inflammatory milieu and pain pathways at different stages of OA.

5. Conclusions

This study suggests a potential correlation between NTN4 levels in the synovium and pain severity in KOA, indicating a role for NTN4 in pain modulation. Our findings show that NTN4 promotes axonal growth in human iPSC-derived sensory neurons and stimulates the release of pro-inflammatory cytokines in synovial fibroblasts. These results may imply that NTN4 could exacerbate pain by enhancing neuronal growth and inflammatory mediator release. Further research is essential to fully explore how NTN4-mediated effects contribute to the mechanisms of OA pain.

Author Contributions: Conceptualization, K.U. and M.T.; methodology, K.U., E.O. and Y.O.; formal analysis, A.T., Y.U., E.O., A.H. and G.I.; investigation, A.T., M.M., D.I., J.A. and G.I.; resources,

M.M., D.I. and J.A.; data curation, Y.O, M.T. and K.U.; writing—original draft preparation, A.T. and K.U.; writing—review and editing, K.U.; visualization, A.T., Y.O. and K.U.; supervision, M.T; project administration, M.M. and K.U.; funding acquisition, M.M. and K.U. All authors have read and agreed to the published version of the manuscript.

Funding: This research was funded by the Grant-in-Aid for Research Activity Start-up, grant number 23K19662, the Patients' Association of Kitasato University School of Medicine, Kitasato University Research Grant for Young Researchers, a research grant for young physicians and health professionals from SRL, Inc., and the All Kitasato Project Study, all of which do not have grant numbers. The APC was funded by Kitasato University School of Medicine.

Institutional Review Board Statement: This study was approved by the IRB of Kitasato University (reference number: B19-259; approval date: 27 January 2020).

Informed Consent Statement: We applied the opt-out method to obtain consent for this study.

Data Availability Statement: The original contributions presented in this study are included in the article. Further inquiries can be directed to the corresponding author(s).

Acknowledgments: We sincerely appreciate the valuable support provided by Junko Ohashi, Yuko Onuki, and Yuka Ito throughout the course of this study.

Conflicts of Interest: Yui Uekusa has received research grants from SRL Inc. (grant number: N/A). The remaining authors declare that the research was conducted in the absence of any commercial or financial relationships that could be construed as potential conflicts of interest. The authors declare that this study received funding from Grant-in-Aid for Research Activity Start-up grant number 23K19662, Patients' Association of Kitasato University School of Medicine, Kitasato University Research Grant for Young Researchers, and All Kitasato Project Study. The funder was not involved in the study design, collection, analysis, interpretation of data, the writing of this article, or the decision to submit it for publication.

References

1. Bannuru, R.R.; Osani, M.C.; Vaysbrot, E.E.; Arden, N.K.; Bennell, K.; Bierma-Zeinstra, S.M.A.; Kraus, V.B.; Lohmander, L.S.; Abbott, J.H.; Bhandari, M.; et al. OARSI guidelines for the non-surgical management of knee, hip, and polyarticular osteoarthritis. *Osteoarthr. Cartil.* **2019**, *27*, 1578–1589. [CrossRef]
2. Hill, C.L.; Hunter, D.J.; Niu, J.; Clancy, M.; Guermazi, A.; Genant, H.; Gale, D.; Grainger, A.; Conaghan, P.; Felson, D.T. Synovitis detected on magnetic resonance imaging and its relation to pain and cartilage loss in knee osteoarthritis. *Ann. Rheum. Dis.* **2007**, *66*, 1599–1603. [CrossRef]
3. Torres, L.; Dunlop, D.; Peterfy, C.; Guermazi, A.; Prasad, P.; Hayes, K.; Song, J.; Cahue, S.; Chang, A.; Marshall, M.; et al. The relationship between specific tissue lesions and pain severity in persons with knee osteoarthritis. *Osteoarthr. Cartil.* **2006**, *14*, 1033–1040. [CrossRef]
4. Chu, K.L.; Chandran, P.; Joshi, S.K.; Jarvis, M.F.; Kym, P.R.; McGaraughty, S. TRPV1-related modulation of spinal neuronal activity and behavior in a rat model of osteoarthritic pain. *Brain Res.* **2011**, *1369*, 158–166. [CrossRef] [PubMed]
5. Miller, R.E.; Malfait, A.M. Osteoarthritis pain: What are we learning from animal models? *Best Pract. Res. Clin. Rheumatol.* **2017**, *31*, 676–687. [CrossRef] [PubMed]
6. Miranda, J.A.; Stanley, P.; Gore, K.; Turner, J.; Dias, R.; Rees, H. A Preclinical Physiological Assay to Test Modulation of Knee Joint Pain in the Spinal Cord: Effects of Oxycodone and Naproxen. *PLoS ONE* **2014**, *9*, e106108. [CrossRef] [PubMed]
7. Philpott, H.T.; O'Brien, M.; McDougall, J.J. Attenuation of early phase inflammation by cannabidiol prevents pain and nerve damage in rat osteoarthritis. *PAIN®* **2017**, *158*, 2442–2451. [CrossRef]
8. Sagar, D.; Nwosu, L.; Walsh, D.; Chapman, V. Dissecting the contribution of knee joint NGF to spinal nociceptive sensitization in a model of OA pain in the rat. *Osteoarthr. Cartil.* **2015**, *23*, 906–913. [CrossRef]
9. Lee, Y.C.; Lu, B.; Bathon, J.M.; Haythornthwaite, J.A.; Smith, M.T.; Page, G.G.; Edwards, R.R. Pain sensitivity and pain reactivity in osteoarthritis. *Arthritis Care Res.* **2011**, *63*, 320–327. [CrossRef]
10. Cirulli, V.; Yebra, M. Netrins: Beyond the brain. *Nat. Rev. Mol. Cell. Biol.* **2007**, *8*, 296–306. [CrossRef]
11. Kennedy, T.E.; Serafini, T.; de la Torre, J.; Tessier-Lavigne, M. Netrins are diffusible chemotropic factors for commissural axons in the embryonic spinal cord. *Cell* **1994**, *78*, 425–435. [CrossRef]

12. Nakashiba, T.; Ikeda, T.; Nishimura, S.; Tashiro, K.; Honjo, T.; Culotti, J.G.; Itohara, S. Netrin-G1: A Novel Glycosyl Phosphatidylinositol-Linked Mammalian Netrin That Is Functionally Divergent from Classical Netrins. *J. Neurosci.* **2000**, *20*, 6540–6550. [CrossRef]
13. Stoppiello, L.A.; Mapp, P.I.; Wilson, D.; Hill, R.; Scammell, B.E.; Walsh, D.A. Structural Associations of Symptomatic Knee Osteoarthritis. *Arthritis Rheumatol.* **2014**, *66*, 3018–3027. [CrossRef] [PubMed]
14. Sun, K.L.W.; Correia, J.P.; Kennedy, T.E. Netrins: Versatile extracellular cues with diverse functions. *Development* **2011**, *138*, 2153–2169. [CrossRef]
15. Vennela, J.; Pottakkat, B.; Vairappan, B.S.; Verma, S.K.; Mukherjee, V. Hepatic Expression of NTN4 and Its Receptors in Patients with Hepatocellular Carcinoma. *Asian Pac. J. Cancer Prev.* **2023**, *24*, 4285–4292. [CrossRef]
16. Villanueva, A.A.; Puvogel, S.; Lois, P.; Munoz-Palma, E.; Ramirez Orellana, M.; Lubieniecki, F.; Casco Claro, F.; Gallegos, I.; Garcia-Castro, J.; Sanchez-Gomez, P.; et al. The Netrin-4/Laminin gamma1/Neogenin-1 complex mediates migration in SK-N-SH neuroblastoma cells. *Cell Adh. Migr.* **2019**, *13*, 33–40. [CrossRef] [PubMed]
17. Bai, Z.; Bartelo, N.; Aslam, M.; Murphy, E.A.; Hale, C.R.; Blachere, N.E.; Parveen, S.; Spolaore, E.; DiCarlo, E.; Gravallese, E.M.; et al. Synovial fibroblast gene expression is associated with sensory nerve growth and pain in rheumatoid arthritis. *Sci. Transl. Med.* **2024**, *16*, eadk3506. [CrossRef] [PubMed]
18. Hayano, Y.; Takasu, K.; Koyama, Y.; Yamada, M.; Ogawa, K.; Minami, K.; Asaki, T.; Kitada, K.; Kuwabara, S.; Yamashita, T. Dorsal horn interneuron-derived Netrin-4 contributes to spinal sensitization in chronic pain via Unc5B. *J. Exp. Med.* **2016**, *213*, 2949–2966. [CrossRef]
19. Liu, Y.; Balaji, R.; de Toledo, M.A.S.; Ernst, S.; Hautvast, P.; Kesdoğan, A.B.; Körner, J.; Zenke, M.; Neureiter, A.; Lampert, A. The pain target NaV1.7 is expressed late during human iPS cell differentiation into sensory neurons as determined in high-resolution imaging. *Pflügers Arch. Eur. J. Physiol.* **2024**, *476*, 975–992. [CrossRef]
20. Neureiter, A.; Eberhardt, E.; Lampert, A. Differentiation of iPS-Cells into Peripheral Sensory Neurons. *Methods Mol. Biol.* **2022**, *2429*, 175–188.
21. E Nanus, D.; Badoume, A.; Wijesinghe, S.N.; Halsey, A.M.; Hurley, P.; Ahmed, Z.; Botchu, R.; Davis, E.T.; A Lindsay, M.; Jones, S.W. Synovial tissue from sites of joint pain in knee osteoarthritis patients exhibits a differential phenotype with distinct fibroblast subsets. *EBioMedicine* **2021**, *72*, 103618. [CrossRef] [PubMed]
22. Tsuchiya, M.; Ohashi, Y.; Kodera, Y.; Satoh, M.; Matsui, T.; Fukushima, K.; Iwase, D.; Aikawa, J.; Mukai, M.; Inoue, G.; et al. CD39+CD55− Fb Subset Exhibits Myofibroblast-Like Phenotype and Is Associated with Pain in Osteoarthritis of the Knee. *Biomedicines* **2023**, *11*, 3047. [CrossRef] [PubMed]
23. Uekusa, Y.; Mukai, M.; Tsukada, A.; Iwase, D.; Aikawa, J.; Shibata, N.; Ohashi, Y.; Inoue, G.; Takaso, M.; Uchida, K. Elevated Netrin-4 Expression and Its Action in Infrapatellar Fat Pad. *Int. J. Mol. Sci.* **2024**, *25*, 11369. [CrossRef]
24. Vezzani, A.; Viviani, B. Neuromodulatory properties of inflammatory cytokines and their impact on neuronal excitability. *Neuropharmacology* **2015**, *96*, 70–82. [CrossRef]
25. Zhang, D.; Zhu, Z.; Wen, K.; Zhang, S.; Liu, J. Netrin-4 promotes VE-cadherin expression in endothelial cells through the NF-kappaB signaling pathway. *Exp. Ther. Med.* **2024**, *28*, 351. [CrossRef] [PubMed]
26. Ji, R.-R.; Xu, Z.-Z.; Gao, Y.-J. Emerging targets in neuroinflammation-driven chronic pain. *Nat. Rev. Drug Discov.* **2014**, *13*, 533–548. [CrossRef]
27. Julius, D.; Basbaum, A.I. Molecular mechanisms of nociception. *Nature* **2001**, *413*, 203–210. [CrossRef]
28. White, F.A.; Bhangoo, S.K.; Miller, R.J. Chemokines: Integrators of Pain and Inflammation. *Nat. Rev. Drug Discov.* **2005**, *4*, 834–844. [CrossRef]
29. Matsuda, M.; Huh, Y.; Ji, R.-R. Roles of inflammation, neurogenic inflammation, and neuroinflammation in pain. *J. Anesthesia* **2018**, *33*, 131–139. [CrossRef]
30. Ohta, E.; Nihira, T.; Uchino, A.; Imaizumi, Y.; Okada, Y.; Akamatsu, W.; Takahashi, K.; Hayakawa, H.; Nagai, M.; Ohyama, M.; et al. I2020T mutant LRRK2 iPSC-derived neurons in the Sagamihara family exhibit increased Tau phosphorylation through the AKT/GSK-3beta signaling pathway. *Hum. Mol. Genet.* **2015**, *24*, 4879–4900. [CrossRef]
31. Baker, K.; Grainger, A.; Niu, J.; Clancy, M.; Guermazi, A.; Crema, M.; Hughes, L.; Buckwalter, J.; Wooley, A.; Nevitt, M.; et al. Relation of synovitis to knee pain using contrast-enhanced MRIs. *Ann. Rheum. Dis.* **2010**, *69*, 1779–1783. [CrossRef] [PubMed]
32. Mapp, P.; Kidd, B.; Gibson, S.; Terry, J.; Revell, P.; Ibrahim, N.; Blake, D.; Polak, J. Substance P-, calcitonin gene-related peptide-and C-flanking peptide of neuropeptide Y-immunoreactive fibres are present in normal synovium but depleted in patients with rheumatoid arthritis. *Neuroscience* **1990**, *37*, 143–153. [CrossRef]
33. Zou, Y.; Liu, C.; Wang, Z.; Li, G.; Xiao, J. Neural and immune roles in osteoarthritis pain: Mechanisms and intervention strategies. *J. Orthop. Transl.* **2024**, *48*, 123–132. [CrossRef]

34. Nanus, D.E.; Wijesinghe, S.N.; Pearson, M.J.; Hadjicharalambous, M.R.; Rosser, A.; Davis, E.T.; Lindsay, M.A.; Jones, S.W. Regulation of the Inflammatory Synovial Fibroblast Phenotype by Metastasis-Associated Lung Adenocarcinoma Transcript 1 Long Noncoding RNA in Obese Patients With Osteoarthritis. *Arthritis Rheumatol.* **2019**, *72*, 609–619. [CrossRef] [PubMed]
35. Wijesinghe, S.N.; Lindsay, M.A.; Jones, S.W. Oligonucleotide Therapies in the Treatment of Arthritis: A Narrative Review. *Biomedicines* **2021**, *9*, 902. [CrossRef]
36. Wijesinghe, S.N.; Badoume, A.; Nanus, D.E.; Sharma-Oates, A.; Farah, H.; Certo, M.; Alnajjar, F.; Davis, E.T.; Mauro, C.; Lindsay, M.A.; et al. Obesity defined molecular endotypes in the synovium of patients with osteoarthritis provides a rationale for therapeutic targeting of fibroblast subsets. *Clin. Transl. Med.* **2023**, *13*. [CrossRef]
37. Wijesinghe, S.N.; Ditchfield, C.; Flynn, S.; Agrawal, J.; Davis, E.T.; Dajas-Bailador, F.; Chapman, V.; Jones, S.W. Immunomodulation and fibroblast dynamics driving nociceptive joint pain within inflammatory synovium: Unravelling mechanisms for therapeutic advancements in osteoarthritis. *Osteoarthr. Cartil.* **2024**, *32*, 1358–1370. [CrossRef]
38. Farah, H.; Wijesinghe, S.N.; Nicholson, T.; Alnajjar, F.; Certo, M.; Alghamdi, A.; Davis, E.T.; Young, S.P.; Mauro, C.; Jones, S.W. Differential Metabotypes in Synovial Fibroblasts and Synovial Fluid in Hip Osteoarthritis Patients Support Inflammatory Responses. *Int. J. Mol. Sci.* **2022**, *23*, 3266. [CrossRef] [PubMed]
39. Zhou, Y.-Q.; Liu, Z.; Liu, Z.-H.; Chen, S.-P.; Li, M.; Shahveranov, A.; Ye, D.-W.; Tian, Y.-K. Interleukin-6: An emerging regulator of pathological pain. *J. Neuroinflamm.* **2016**, *13*, 1–9. [CrossRef]
40. Lin, Y.; Liu, L.; Jiang, H.; Zhou, J.; Tang, Y. Inhibition of interleukin-6 function attenuates the central sensitization and pain behavior induced by osteoarthritis. *Eur. J. Pharmacol.* **2017**, *811*, 260–267. [CrossRef]
41. Vazquez, E.; Kahlenbach, J.; von Banchet, G.S.; König, C.; Schaible, H.; Ebersberger, A. Spinal interleukin-6 is an amplifier of arthritic pain in the rat. *Arthritis Rheum.* **2012**, *64*, 2233–2242. [CrossRef] [PubMed]
42. Nowell, M.A.; Williams, A.S.; Carty, S.A.; Scheller, J.; Hayes, A.J.; Jones, G.W.; Richards, P.J.; Slinn, S.; Ernst, M.; Jenkins, B.J.; et al. Therapeutic Targeting of IL-6 *Trans* Signaling Counteracts STAT3 Control of Experimental Inflammatory Arthritis. *J. Immunol.* **2009**, *182*, 613–622. [CrossRef] [PubMed]
43. Bjurström, M.F.; Bodelsson, M.; Irwin, M.R.; Orbjörn, C.; Hansson, O.; Mattsson-Carlgren, N. Decreased pain sensitivity and alterations of cerebrospinal fluid and plasma inflammatory mediators after total hip arthroplasty in patients with disabling osteoarthritis. *Pain Pr.* **2021**, *22*, 66–82. [CrossRef] [PubMed]
44. Bastiaansen-Jenniskens, Y.M.; Clockaerts, S.; Feijt, C.; Zuurmond, A.-M.; Stojanovic-Susulic, V.; Bridts, C.; de Clerck, L.; DeGroot, J.; Verhaar, J.A.; Kloppenburg, M.; et al. Infrapatellar fat pad of patients with end-stage osteoarthritis inhibits catabolic mediators in cartilage. *Ann. Rheum. Dis.* **2012**, *71*, 288–294. [CrossRef]
45. Pritzker, K.P.H.; Gay, S.; Jimenez, S.A.; Ostergaard, K.; Pelletier, J.-P.; Revell, P.A.; Salter, D.; van den Berg, W.B. Osteoarthritis cartilage histopathology: Grading and staging. *Osteoarthr. Cartil.* **2006**, *14*, 13–29. [CrossRef]

cells

MDPI

Article

Cilastatin Modulates DPEP1- and IQGAP1-Associated Neuro-Glio-Vascular Inflammation in Oxaliplatin-Induced Peripheral Neurotoxicity

Rita Martín-Ramírez [1,2,†], María Ángeles González-Nicolás [3,4,†], Karen Álvarez-Tosco [1,2,5], Félix Machín [2,5,6], Julio Ávila [1,2], Manuel Morales [2,7,8], Alberto Lázaro [3,4,*,‡] and Pablo Martín-Vasallo [1,2,*,‡]

1 Laboratorio de Biología del Desarrollo, UD de Bioquímica y Biología Molecular, Universidad de La Laguna, 38206 San Cristóbal de La Laguna, Spain; rmartira@ull.edu.es (R.M.-R.); karenalvtos@gmail.com (K.Á.-T.); javila@ull.edu.es (J.Á.)
2 Instituto de Tecnologías Biomédicas, Universidad de La Laguna, 38071 San Cristóbal de La Laguna, Spain; fmachin@fciisc.es (F.M.); mmoraleg@ull.edu.es (M.M.)
3 Laboratorio de Fisiopatología Renal, Departamento de Nefrología, Instituto de Investigación Sanitaria Gregorio Marañón, Hospital General Universitario Gregorio Marañón, 28007 Madrid, Spain; mangeleg@ucm.es
4 Departamento de Fisiología, Facultad de Medicina, Universidad Complutense de Madrid, 28040 Madrid, Spain
5 Unidad de Investigación, Hospital Universitario Nuestra Señora de la Candelaria, Instituto de Investigación Sanitaria de Canarias (IISC), 38010 Santa Cruz de Tenerife, Spain
6 Facultad de Ciencias de la Salud, Universidad Fernando Pessoa Canarias, 35450 Las Palmas de Gran Canaria, Spain
7 Departamento de Medicina, Facultad de Ciencias de la Salud, Universidad de La Laguna, 38200 San Cristóbal de La Laguna, Spain
8 Servicio de Oncología Médica, Hospiten Rambla, Grupo Hospiten, 38001 Santa Cruz de Tenerife, Spain
* Correspondence: alberlaz@ucm.es (A.L.); pmartin@ull.edu.es (P.M.-V.); Tel.: +34-922-318358 (P.M.-V.)
† These authors contributed equally to this work and both are first author.
‡ These authors contributed equally to this work and both are senior authors.

Abstract: Oxaliplatin-induced peripheral neurotoxicity (OIPN) represents a major challenge in cancer therapy, characterized by dorsal root ganglia (DRG) inflammation and disruption of neuro-glio-vascular unit function. In this study, we investigated the involvement of the scaffold protein IQ Motif Containing GTPase Activating Protein 1 (IQGAP1) and dehydropeptidase-1 (DPEP1) in the DRG response to oxaliplatin (OxPt) and the modulatory effect of cilastatin. Behavioral assessment showed a robust nocifensive response to cold stimuli in OxPt-treated rats, attenuated by cilastatin co-treatment. Our confocal study revealed different cellular and subcellular expression patterns of IQGAP1 and DPEP1 in neurons, glia, and endothelial cells, where both signals overlap approximately one-third. OxPt enhanced cytosolic aggregation of IQGAP1 in neurons and upregulation of signal in glia, accompanied by co-expression of TNFα and IL-6, indicating involvement in the inflammatory process. DPEP1 showed altered subcellular distribution in OxPt-treated animals, suggesting a potential role in the inflammatory cascade. Notably, IQGAP1 expression was diminished in endothelial membranes under OxPt, while cilastatin preserved endothelial IQGAP1-CD31 colocalization, suggesting partial restoration of blood-nerve barrier integrity. These findings identify IQGAP1 and DPEP1 as key players in DRG inflammation and position cilastatin as a promising modulator of OIPN through neuro-glio-vascular stabilization.

Keywords: IQGAP1; dorsal root ganglion (DRG); oxaliplatin; neurotoxicity; allodynia; dehydropeptidase-1; DRG-inflammation; cilastatin; peripheral neuropathy

1. Introduction

Oxaliplatin (OxPt)-induced peripheral neuropathy (OIPN) is a common and dose-limiting adverse effect of OxPt-based chemotherapy and part of the broader condition known as chemotherapy-induced peripheral neuropathy (CIPN), a form of neuropathic pain resulting from damage to the somatosensory system [1]. Unlike motor neuropathies, CIPN primarily affects sensory neurons, leading to symptoms such as paresthesia, dysesthesia, and tactile deficits [2–5] and scarcely affects motor neurons [6,7]. OIPN manifests in both acute and chronic forms. The acute phase occurs shortly after OxPt infusion and is transient, often triggered by cold exposure and characterized by jaw tightness, pharyngolaryngeal paresthesia, and cold hypersensitivity. In contrast, the chronic form develops cumulatively and may persist or worsen over time, significantly impairing quality of life and frequently necessitating dose reduction or discontinuation of treatment [8–10]. A key pathological feature of OIPN is the preferential accumulation of OxPt in dorsal root ganglia (DRG), facilitated by the absence of a blood-nerve barrier and the presence of fenestrated capillaries. Uptake is mediated via passive diffusion and metal transporters [11]. Then, inside DRG neurons, OxPt disrupts mitochondrial function by altering ion transport, inducing calcium overload, generating reactive oxygen species (ROS), and impairing ATP production, all of which contribute to neuronal apoptosis and axonal degeneration [12–15]. Despite the clinical impact of OIPN, current treatments for preventing or reversing it remain limited. Understanding the inflammatory mechanisms underlying OIPN is essential to developing effective therapeutic strategies that preserve anticancer efficacy without compromising neurological function.

Recently, we reported the expression and subcellular localization of Dehydropeptidase-1 (DPEP1) in DRG of rats, both in control conditions and six days post-OxPt treatment [16]. The expression of DPEP1 in neurons, glia, and endothelial cells of DRG suggested this protein as a novel target in the prevention of OIPN [16]. Furthermore, in the kidney, DPEP1 inhibition by cilastatin has shown therapeutic potential in renal injury models [17,18]. We demonstrated the usefulness of cilastatin as a nephroprotective agent in an animal model of cisplatin-induced nephrotoxicity, which points to an important relevance of this drug for the preservation of renal function in patients with cancer [17,19].

DPEP1 is a zinc-dependent glycosylated homodimer, originally identified in the brush border of renal proximal tubules, with dipeptidase activity [18,20]. DPEP1 belongs to a family of glycosylphosphatidylinositol (GPI)-anchored membrane-bound enzymes alongside DPEP2 and DPEP3, differing in substrate specificity and tissue distribution [21]. It hydrolyzes diverse dipeptides, including antibiotics like carbapenems, and is involved in glutathione (GSH), leukotriene, and lipid metabolism [20–22]. Structurally, DPEP1 features N-glycosylation, disulfide bridges, and Zn^{2+} coordination critical for enzymatic function [23]. In GSH catabolism, DPEP1 degrades Cys-Gly, regulating antioxidant defenses and ferroptosis susceptibility [24–26]. Its inhibition by cilastatin protects against nephrotoxicity by preserving GSH and preventing cisplatin-induced ferroptosis [19,27]. DPEP1 also metabolizes leukotriene D4, modulating vascular permeability in the lung, liver, and kidneys [28,29]. Beyond enzymatic roles, DPEP1 acts as a neutrophil adhesion receptor, contributing to inflammatory pathology [29]. DPEP1 overexpression promotes tumorigenesis in colorectal and hepatoblastoma cells via c-Myc and PI3K/Akt/mTOR pathways [30,31]. However, its expression may also suppress invasion in pancreatic and breast cancers [32,33]. Expression varies across tissues—including kidney, lung, liver, and testis—and localizes to membranes, cytosol, and nucleus [21,34]. The role of DPEP1 in the nervous system remains largely unexplored.

A differential screening in patients treated with OxPt between those who suffer from OIPN and those who did not develop it showed IQGAP1 (IQ motif-containing GTPase-activating protein 1) as one of the proteins with higher variation in its expression level [35]. IQGAP1 is a multifunctional scaffold protein that regulates diverse signaling pathways critical for cellular architecture and dynamics [36]. As the best-characterized member of the IQGAP family, IQGAP1 integrates signaling through interactions with over 90 protein partners via its six modular domains, including the calponin homology domain (CHD), IQ domains, and the RasGAP-related domain [36–40]. These structural motifs enable binding to cytoskeletal regulators (e.g., F-actin, N-WASP), kinases (e.g., ERK1/2, B-Raf), small GTPases (e.g., Cdc42, Rac1), and adhesion molecules (e.g., E-cadherin, β-catenin), positioning IQGAP1 as a key mediator of actin dynamics, cell polarity, and MAPK signaling [37,40,41].

Subcellularly, IQGAP1 is primarily localized to the plasma membrane but also functions in the nucleus, where it modulates cell cycle progression and DNA replication [26]. In the nervous system, IQGAP1 is essential for neurite outgrowth [39,42], dendritic spine development, and neurogenesis. Its interaction with Lis1 [43] Rho-family GTPases support neural progenitor migration and differentiation, while decreased IQGAP1 expression impairs neuronal plasticity and may contribute to neurodegenerative and psychiatric disorders [39]. Collectively, IQGAP1 plays a pivotal role in orchestrating cellular processes across various tissues, with particular relevance to neural development and function [40,44–47]. The passage of neural stem cells and progenitor cells to mature neurons is promoted by vascular endothelial growth factor (VEGF). Balenci et al. [48] suggest that VEGF released by astrocytes is essential in neuronal progenitors' recruitment to perivascular niches where neuronal differentiation takes place. This function is also related to cell motility activation by IQGAP1 complexes with Rho family GTPases and Lis1 [48]. Finally, IQGAP functions to control dendrite formation, which is related to neuronal plasticity and synaptic input processes [39,49]. Lower levels of IQGAP1 in neurons lead to a decreased number of dendritic tips [50].

Continuing our line of research, we decided to investigate the expression of DPEP1 and IQGAP1 in DRG during OIPN, trying to elucidate their mechanistic interplay and assess the neuroprotective potential of their combined pharmacological modulation by cilastatin in OxPt-treated rats. This study reports behavioral and confocal analyses showing that IQGAP1 and DPEP1 are differentially expressed in neurons, glia, and endothelial cells during OxPt-induced inflammation. OxPts altered their subcellular distribution and promoted co-expression with TNFα and IL-6, implicating them in neuroinflammatory processes. Cilastatin attenuated these effects, suggesting stabilization of the DRG structure-function. These findings present IQGAP1 and DPEP1 as modulators of OIPN and position cilastatin as a potential neuro-glio-vascular protective agent.

2. Materials and Methods

2.1. Subjects

Animal handling was carried out according to the current legal regulations on the protection of animals used for experimental and other scientific purposes: RD 118/2021, of 23 February; Law 32/2007, of 7 November; and ECC/566/2015, of 20 March. Adult male Wistar rats weighing approximately 250 g were used throughout the OxPt-induced neuropathic allodynia model. These animals were supplied by the Instituto de Investigación Sanitaria Gregorio Marañón animal facility, Hospital General Universitario Gregorio Marañón (HGUGM), Madrid, Spain.

Rats were stabled in conventional cages in pairs, without food/water restriction, stable temperature, and humidity conditions (T = 22 ± 2 °C and HR = 45–65%). Based on recent results showing sex dimorphism in rodents in inflammatory pain regulation and in immune cell signaling in neuropathic pain [51,52], we decided to use males, as other models for chemotherapy-induced peripheral neuropathy (PN) did [53–55].

2.2. Animal Models

2.2.1. OxPt-Induced Neuropathic Allodynia Animal Model

Animals were classified in four groups: (1) OxPt, n = 6, (2) OxPt + cilastatin, n = 6, (3) control, n = 4, and (4) cilastatin, n = 4. OxPt was supplied by the HGUGM Pharmacy Service at an initial concentration of 5 mg/mL dissolved in 5% glucose solution (Braun Medical S.A., Barcelona, Spain) and administered at a final concentration of 6 mg/kg, in a single dose intraperitoneal linjection, at the beginning of the study that lasted 6 days. The control group received an injection with the same vehicle, in the same conditions and volumes as the treated groups. Cilastatin, an inhibitor of the renal enzyme DPEP1, was supplied by ACS DOBFAR (s.p.a., Tribiano, Milan, Italy). The concentration used was 150 mg/kg dissolved in a volume of 0.5 mL of 0.9% saline solution. Cilastatin was injected intraperitoneally, immediately after OxPt administration, and every 24 h. The dose and administration regimen of cilastatin were established by literature review and by the group's experience [56]. The rats were weighed daily to adjust the dose of cilastatin and to check their behavior.

On day 6, euthanasia took place. The rats were anesthetized with sevoflurane (Abbvie, Madrid, Spain) at 5% and maintained during surgery at 2%. Once the rats were weighed and placed in the surgical field, they were opened longitudinally, removing only the first layer of skin at the level of the sternum for blood extraction by cardiac puncture. Using scissors, the hair was parted along the spine in a distal direction, and the spine was isolated by cutting on both sides, along the spine beyond the pelvic bone. With the aid of a magnifying lens and microsurgical material, the dorsal (lumbar) root ganglia and the sciatic nerve were located and removed. These ganglia were placed on a dark surface for better manipulation, and a drop of paraformaldehyde (PFA) was added. They were then placed in cassettes with PFA for 24 h. Finally, samples were placed in 70% ethanol (VWR, Radnor, PA, USA). After this, other tissues and organs such as the liver, colon, and kidney were collected and stored at −80 °C for further analysis.

No mortality, diarrhea, or signs of alopecia were observed in any group of animals during the study. Rats were weighed before chemotherapy administration to record baseline weight and on each day of the study. No significant increase or decrease in body weight was observed in the rats during the study.

2.2.2. Allodynia Test

The acetone test was used to evaluate cold allodynia by touching the plantar skin of hind paws with a 200 μL droplet of acetone (PanReac, Barcelona, Spain) [57] from an insulin-type syringe (B. Braun Medical S.A., Madrid, Spain). Dripping acetone was performed on only one hind paw at a time, alternating the left and right sides each time with the intention of preventing habituation. Every day, the times of flicking, biting, or licking the stimulated paw were counted for 2 min. The times registered were the median of testing three times at a 1 h interval.

2.3. Antibodies

Primary and secondary antibodies used for immunohistochemistry are listed in Table 1.

Table 1. Antibodies for immunohistochemistry used in this study.

		Primary Antibodies			
Target		**Host/Class**	**Dilution**	**Source**	**Cat.#**
Anti-DPEP1		Rabbit polyclonal	1:100	Martín-Vasallo/Ávila [16]	DPEP1C
Anti-IQGAP1		Rabbit polyclonal	1:500	Millipore-Sigma Darmstadt, Germany	ABT186
Anti-IQGAP1		Mouse monoclonal	1:100	Santa Cruz Biotechnology Dallas, TX, USA	sc-376021
Anti-MAP2		Mouse monoclonal	1:500	Merck-Millipore	MAB378
Anti-GFAP		Mouse monoclonal	1:100	Santa Cruz Biotechnology Dallas, TX, USA	sc-33673
Anti-CD31		Mouse monoclonal	1:150	Santa Cruz Biotechnology Dallas, TX, USA	sc-376764
Anti-TNFα		Mouse monoclonal	1:150	Santa Cruz Biotechnology Dallas, TX, USA	sc-52B83
Anti-IL-6		Mouse monoclonal	1:200	Santa Cruz Biotechnology Dallas, TX, USA	sc-28343
Anti-Laminin		Mouse monoclonal	1:200	Santa Cruz Biotechnology Dallas, TX, USA	sc-365962
Anti-α-Actin		Mouse monoclonal	1:60,000	Sigma Aldrich/Merck Millipore Saint Louis, MO, USA/Darmstadt, Germany	A5441
		Secondary antibodies			
Target	**Conjugation**	**Host/Class**	**Dilution**	**Source**	**Cat.#**
Rabbit IgG	FITC	Goat polyclonal	1:200	Sigma-Aldrich Saint Louis, MO, USA	F9887
Mouse IgG	DyLight®650	Goat polyclonal	1:100	Abcam Cambridge, UK	ab97018

2.4. Western Blot

After homogenizing, tissue samples (kidney, liver, and colon) were boiled in LSB buffer at 95 °C for 10 min. The resulting whole cell lysates were separated on a 4–20% denaturing polyacrylamide gel and transferred onto Immobilon™-158 membranes (Millipore) via electroblotting. Membranes were then blocked for 1 h in PBS containing 5% BSA. Proteins were detected using primary antibodies against DPEP1 or IQGAP1 (refer to Table 1), followed by a horseradish peroxidase-conjugated anti-rabbit IgG secondary antibody (Cat. # NA9340, GE Healthcare, Madrid, Spain). Visualization was carried out using the ChemiDoc XRS imaging system (Bio-Rad Laboratories, Hercules, CA, USA) with the Immobilon Western Chemiluminescent HRP substrate (Merck Millipore, Darmstadt, Germany), according to the manufacturer's protocol. Laminin and α-Actin were used as loading controls.

2.5. Immunohistochemistry

DRG tissue samples were embedded in paraffin and cut into five-micron-thick sections. Tissue sections were deparaffinized in xylene and rehydrated in a 100%, 96%, and 70% alcohol bath, sequentially. Epitope retrieval was performed by heating samples in sodium citrate buffer (pH 6.0) at 120 °C for 10 min in an autoclave. Then, non-specific sites were blocked with 5% BSA or serum in Tris-buffered saline (TBS) for 1 h at room temperature.

Finally, for the immunofluorescence staining, tissue sections were incubated with primary antibodies (see Table 1) overnight at 4 °C. Samples incubated without primary antibodies were used as a negative control. Slides were incubated for 1 h at room temperature in the dark with secondary antibodies raised in different species and conjugated to different fluorochromes. Slides were mounted with ProLong® Diamond Anti-fade Mountant with DAPI to visualize cell nuclei.

2.6. Confocal Analysis

Slides were analyzed, and digital images were captured using Zeiss LSM980 Airyscan-2 (Zeiss, Oberkochen, Germany) and Leica SP8 (Leica Microsystems, Wetzlar, Germany) confocal microscopes. Raw images in Carl Zeiss Image Data File (CZI) or Leica Image Format (LIF) were exported as Joint Photographic Experts Group (JPEG) at 300 dots per inch (dpi). Figures were assembled using Adobe Photoshop CC 2018 and exported at 300 ppi.

2.7. Image Quantitative Analysis

Image analysis was carried out using ImageJ software, version 1.54p (National Institutes of Health; Bethesda, MD, USA) with the EzColocalization plugin (10.1038/s41598-018-33592-8). Confocal images used for analysis were taken using the same parameters. Laser power and detector gain settings were optimized to cover fluorescence signals in a 16-bit depth range without saturation. Changes in fluorescence from baseline were measured as the mean intensity of selected regions of interest; 50 measurements (at least) were made per photograph in four photographs (at least) of different fields of each different immunostaining. Complementarily, the "Cell Counter" plug-in was used to ensure that neurons or glia were counted only once. The Pearson correlation coefficient (PCC) was used to quantify the degree of colocalization between fluorophores [58,59].

2.8. Image Qualitative and Semi-Quantitative Analysis

Although in this report we will only present the results of the quantitative analysis, being aware of the possible errors in the use of plugins [58], we performed a blind qualitative and semi-quantitative analysis, which we then compared with the quantitative analysis specified below in the quantitative analysis section. The analysis was conducted as follows: Two independent observers evaluated preparations and photographs blindly, grading the staining intensities as absent (–), faint (+), moderate (++), or strong (+++). Cutoffs were established by consensus among observers after an initial review of several samples of varying appearance and blind coding. In cases where the score data differed by more than one unit, the means of the score data were calculated. A table with the results of this analysis is included in the Supplementary Materials section.

2.9. Statistical Analysis

SPSS version 29.0 for Windows (IBM Corp., Armonk, NY, USA) was used for statistical analysis. Values were subjected to two-way ANOVA using means per day, followed by the Newman–Keuls test; acetone test data for allodynia are presented as mean $\pm$ standard error of the mean (SEM). Values were subjected to analyses. A probability value (p) < 0.05 was considered statistically significant. Quantification of immunofluorescence intensity was performed for each specific immunohistochemical marker corresponding to the analyzed proteins. The non-normal distribution of the data was confirmed using the Kolmogorov–Smirnov test. Differences in staining intensity between the control and specific treatments were assessed using the non-parametric Kruskal–Wallis test [60]. A p-value of < 0.05 was considered statistically significant. To determine the significance, quantitative variables were summarized as the mean $\pm$ SEM. Equality of variances was tested with Levene's test. Normally distributed continuous variables with equal variances were analyzed with analysis of variance, and Student's t-test was used to test the difference between the responses of two groups (IQGAP1-DPEP1, -MAP2, -CD31, -GFAP, TNFα, IL-6, and any other reported in this study) as a means test of the points obtained; if variances were not equal, the Kruskal–Wallis test was performed. $p < 0.05$ was considered statistically significant.

3. Results

3.1. Differential Nocifensive Response to Allodynia Among Controls, OxPt-, and Cilastatin-Treated Rats

Nocifensive response to drops of cold acetone stimulation was evaluated every 24 h, at the same time during the day, over the six days of the study, by an observer unaware of the experimental conditions of the rat group. The number of times the animal bit, flicked, or licked the examined paw increased by a factor of two to fifteen in rats treated with OxPt compared to those in the control and control + cilastatin groups (Figure 1B). Control and cilastatin-treated rats did not show any considerable movement; this number varied from 8 to 18 in rats undergoing OxPt and only 3–6 times/2 min in the animals pretreated with cilastatin and undergoing OxPt. Significant differences ($p < 0.05$) were found from the second day on between the OxPt- and OxPt + cilastatin-treated groups, showing a protective effect of cilastatin from OxPt-elicited allodynia.

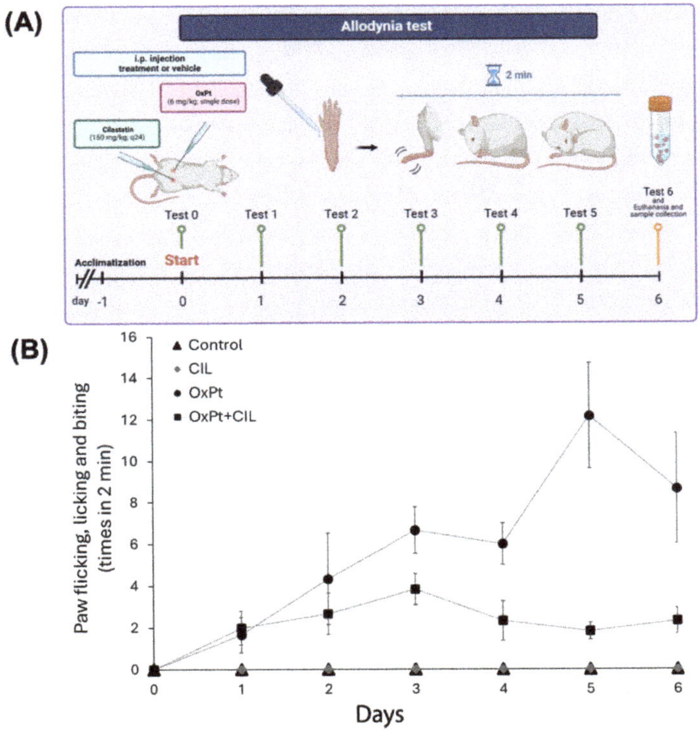

Figure 1. Workflow of the model generation. (**A**) Cold allodynia was tested with the acetone test performed on all six days by touching the plantar skin of both hind paws with a 200 μL droplet of acetone. The times of flicking, licking, or biting the stimulated paw were counted for 2 min. The times were considered as the average of testing three times at a 1 h interval. (**B**) The allodynia test was run on control, OxPt, cilastatin (CIL), and OxPt + CIL rats every 24 h. The data are expressed as the mean ±SEM. Blue asterisks indicate statistical significance of the OxPt group compared to all other groups ($p < 0.01$). The red asterisk indicates statistical significance compared to the Control ± CIL group, but not compared to the OxPt + CIL group. There is a significant difference ($p < 0.05$) between controls and OxPt- and OxPt + cilastatin-treated groups from the second day.

3.2. Tissue-Specific Expression Patterns of DPEP1 and IQGAP1 by Western Blot Reveal the Need for Cellular-Level Analysis in Peripheral Neuropathy

Figure 2 shows the expression of DPEP1 and IQGAP1 in two or three different samples of homogenates of colon and kidney from rats of the different studied groups (control, cilastatin, OxPt, and OxPt plus cilastatin). The quantification showed slight variation or no variation in intensity. However, when these tissues were studied by immunohistochemistry, remarkable differences were found among cells and tissue structures (renal glomerulus, tubules) [61]. Due to this fact, we decided to perform a confocal study using specific and well-characterized antibodies in order to find specific effects in subpopulations of cells among those of DRG, the most affected organ in PN.

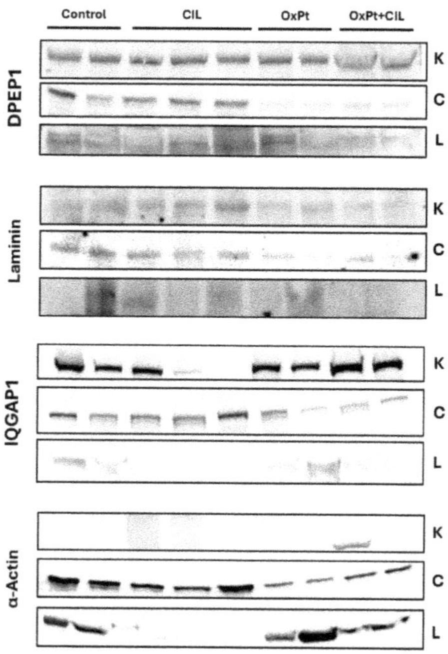

Figure 2. Western blot analysis probed with DPEP1C and IQGAP1 antibodies in rat kidney (K), colon (C), and liver (L) from control, cilastatin (CIL)-, OxPt-, and OxPt + CIL-treated groups. Samples are from two (Control, OxPt, and OxPt + CIL) or three (CIL) different animals. Data were analyzed using a mixed-effects model. For DPEP1 in K, a significant difference was observed between the OxPt + CIL and Control groups, $p < 0.0378$. No statistically significant differences were found in C or L. IQGAP1: No statistically significant differences were detected in K, C, or L.

3.3. Immunohistochemical-Confocal Localization of DPEP1 and IQGAP1 in DRG from OxPt- and Cilastatin-Treated Rats

In samples from rats of the control group, immunofluorescence signal for DPEP1 was found in the cytosol of neurons, ranging from medium to high intensity and following a reticulated pattern, and in satellite glial cells at a slightly higher intensity level, Figure 3, upper lane. As for IQGAP1, it exhibited specific immunofluorescence at a high intensity level localized in the plasma membrane of most neurons, and in cytosol at a much lower level in a homogeneously granulated fashion. In satellite glial cells, IQGAP1 immunofluorescence was present in the plasma membrane at high levels.

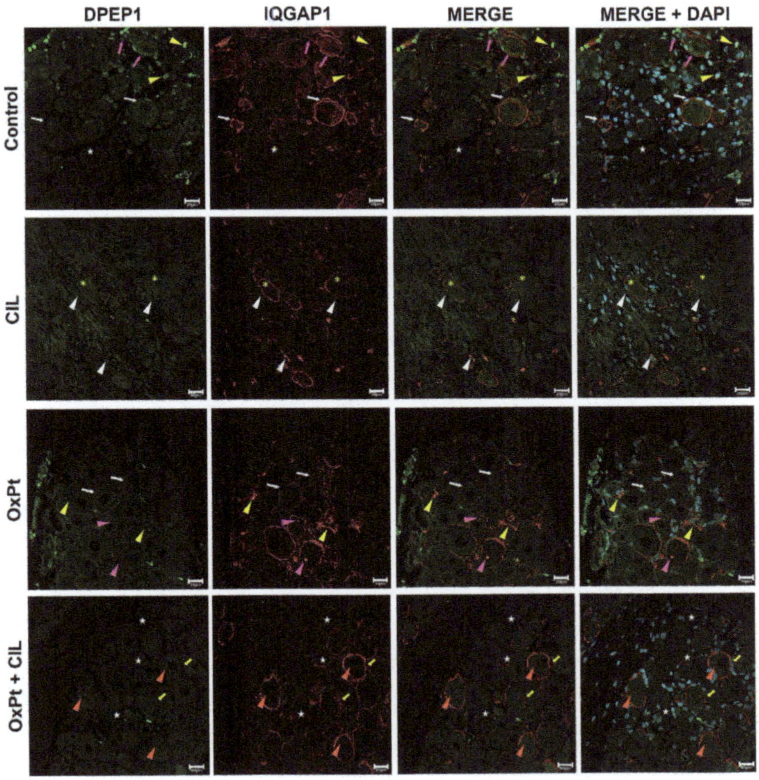

Figure 3. Immunostaining for DPEP1 (green) and IQGAP1 (red). Control DPEP1 neurons cytosol signal (white stars), reticulated pattern, in glia (yellow arrowheads). IQGAP1 neuronal plasma membranes (white arrows), faint and granular in cytosol (yellow asterisks), and glial plasma membranes (magenta arrows). Cilastatin: DPEP1 in neurons and glia (magenta stars). IQGAP1 granular cytosolic signal (yellow asterisks), plasma membrane labeling (white arrowheads). OxPt: DPEP1 reticulated cytosolic pattern (yellow asterisks), brighter in neuronal membranes (white arrows). IQGAP1 increased in satellite glia cells (yellow arrowheads), near the plasma membrane, as cytosolic dots in some neurons (magenta arrowheads). OxPt + CIL: DPEP1 reticulated, increased intensity in some neurons (red arrowheads), and homogeneous distribution (white stars). IQGAP1, a strong membrane signal in neurons and glia (yellow arrows), is absent in axons and some neurons (yellow stars). Scale bar 20 μm. (*n*-Control = 4, 178 photographs, 7 DRGs; *n*-CIL = 4, 184 photographs, 8 DRGs; n-OxPt = 6, 215 photographs, 10 DRGs; *n*-Ox-Pt + CIL = 6, 185 photographs, 7 DRGs). OxPt increased IQGAP1 expression in neurons and glia 2–3-fold, with respect to control. $p < 0.05$, within a heterogeneous population. Magnified views of highlighted regions are available in Supplementary Figure S1.

DPEP1-specific immunofluorescence signals in cilastatin samples were homogeneously distributed at medium intensity in neurons and glial cells, resulting in a reticulated pattern (magenta stars in the second lane of Figure 3). The IQGAP1 immunofluorescence signal was ubiquitously localized and in a granulated pattern (yellow asterisks) as in the control panels. Furthermore, a higher-intensity red signal was present in the plasma membrane of neurons and satellite cells (white arrowheads).

OxPt panels in Figure 3 display representative images of DPEP1-specific immunofluorescence signal within neurons, in a similar reticulated pattern to that of images above (yellow asterisks) and slightly brighter in the membrane of neurons (white arrows). IQGAP1-specific immunostaining signal panels depicted higher intensity compared to that of the control in satellite cells (yellow arrowheads). The signal was mainly present near the plasma membrane and at a lesser intensity in body cells. High-intensity fluorescence specific for IQGAP1 displayed a spot pattern in the cytosol of some neurons (magenta arrowheads).

OxPt + cilastatin panels, Figure 3, showed a reticulated pattern of DPEP1 signal, presented with higher intensity in a few neurons (red arrowheads); white stars in images mark a more homogeneous distribution of DPEP1 immunofluorescence signal within neurons. A high level of IQGAP1 immunostaining signal was evident in the plasma membrane of glia cells and neurons (yellow arrows). There was no specific signal of IQGAP1 inside axons from the root and in certain neurons (yellow stars).

3.4. IQGAP1 Expression in Glial Cells in DRGs from OxPt- and Cilastatin-Treated Rats

In order to learn the effects of OxPt and cilastatin on IQGAP1 of DRG glial cells and to discriminate them from other cells similar in localization and size, such as endothelial cells, co-labeling for this scaffoldine and GFAP was performed. Figure 4 shows exemplary confocal microscopy images of immunostaining for IQGAP1 and GFAP. In control samples, the IQGAP1 signal showed the typical pattern in neurons, as described in Figure 3. Merge image in control shows a high IQGAP1-GFAP co-labeling level in presumably activated GFAP-specific fluorescence in samples from cilastatin- or OxPt + cilastatin-treated rats, apparently and quantitatively lower than in samples from animals that did not undergo cilastatin.

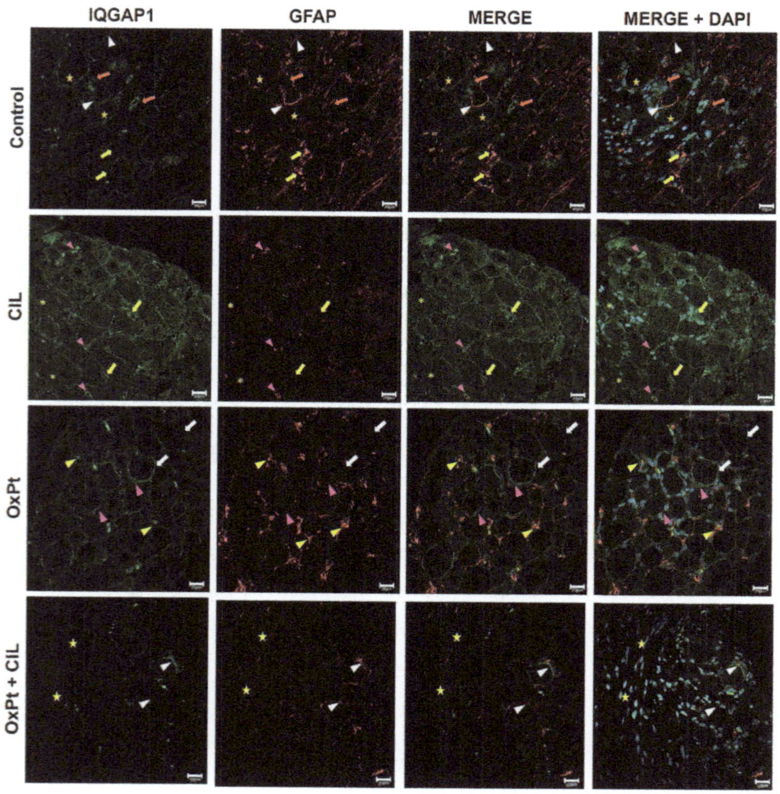

Figure 4. Immunostaining for IQGAP1 (green) and GFAP (red). IQGAP1 is intense in membranes and lower in neuronal cytosol (orange stars). GFAP in glia (white arrowheads), absent in axons (magenta

arrowheads), and higher levels in activated glia (yellow arrows). Glial IQGAP1 without GFAP (red arrows). Cilastatin: IQGAP1 increased in neurons (white stars) and colocalized with GFAP in glia (magenta arrowheads, yellow arrows). GFAP at low–medium intensity, dot-like, and present in root glia (yellow asterisks). OxPt: IQGAP1 has strong fluorescence in neuronal membranes (white arrows), lower in cytosol (yellow stars), and increased in glial membranes (magenta arrowheads). GFAP is strong in activated glia lacking IQGAP1 (yellow arrowheads). OxPt + CIL: IQGAP1 is intense in neuron/glia membranes and low in axons (yellow stars), with a medium/low cytosolic pattern (yellow stars). IQGAP1/GFAP colocalization is evident (white arrowheads). Root glial GFAP is low/medium (magenta asterisks). Scale bar 20 μm. (*n*-Control = 3, 163 photographs, 5 DRGs; *n*-CIL = 3, 204 photographs, 7 DRGs; *n*-OxPt = 3, 200 photographs, 8 DRGs; *n*-Ox-Pt + CIL = 2, 178 photographs, 5 DRGs). OxPt increased GFAP expression in glia with respect to CIL and OxPt + CIL, $p < 0.05$. See Supplementary Figure S2 for zoomed-in images of selected areas.

3.5. IQGAP1 Expression in Endothelial Cells of DRG

Co-immunostaining for IQGAP1 (green) and CD31 (red) in Figure 5, control panels display IQGAP1 immunofluorescence representative images in neurons following the distribution and expression patterns described in previous figures and in this Section 3. The IQGAP1 signal of brighter intensity around neurons corresponded mainly to satellite glial cells (yellow arrows). The CD31 immunostaining signal identified endothelial cells near neurons (magenta arrowheads) and surrounding axons of the root, where a specific signal for IQGAP1 at a medium intensity level was evident in CD31-positive cells.

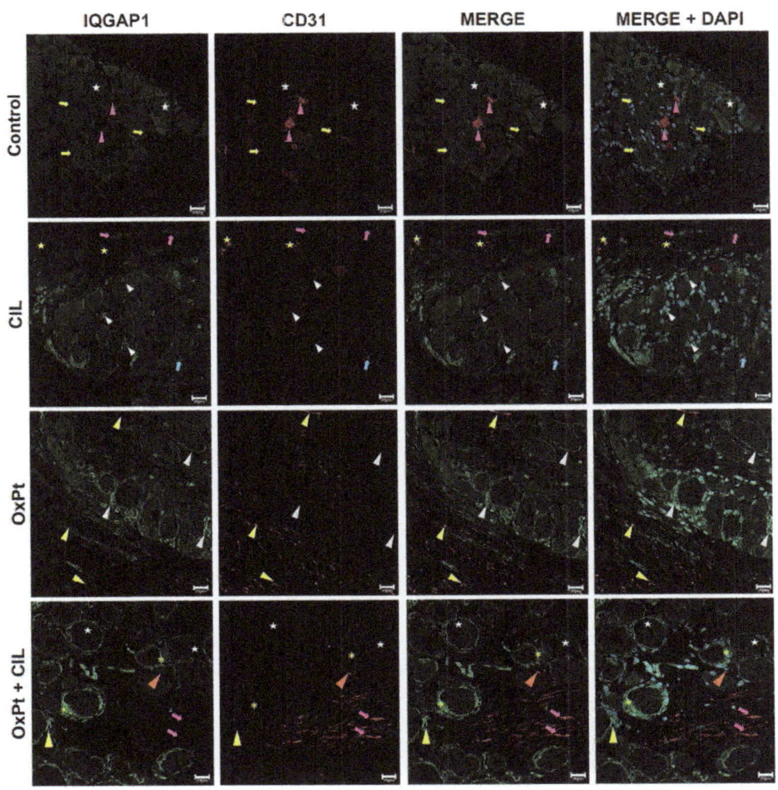

Figure 5. Co-immunostaining of IQGAP1 (green) and CD31 (red) in DRG endothelial cells. Control: IQGAP1 neurons (white stars), glia (yellow arrows), and endothelial cells near neurons (magenta arrowheads) with CD31 signal. Cilastatin (CIL): IQGAP1 is higher in glial membranes (white arrowheads)

and root (magenta arrows). CD31 is faint except near axons (blue arrow, yellow stars). OxPt: IQGAP1 is bright at neuronal periphery and dot-like in glia (white arrowheads). CD31 at medium/high intensity in the root, no colocalization with IQGAP1 (yellow arrowheads). OxPt + CIL: IQGAP1 has a reticulated pattern in glia (yellow asterisks) and in neurons (white stars). Endothelial cells CD31+ near axons (magenta arrows) and neurons (orange arrowheads), some co-labeling with IQGAP1 (yellow arrows, yellow asterisk). Scale bar 20 μm. (*n*-Control = 4, 155 photographs, 4 DRGs; *n*-CIL = 4, 190 photographs, 4 DRGs; *n*-OxPt = 6, 175 photographs, 9 DRGs; *n*-Ox-Pt + CIL = 4, 185 photographs, 6 DRGs). OxPt increased IQGAP1 expression in endothelial cells of roots with respect to control and CIL ($p < 0.05$). No significant OxPt vs. OxPt + CIL. Detailed close-ups of specific regions are provided in Supplementary Figure S3.

In cilastatin-treated rats, a higher intensity of the immunostaining signal for IQGAP1 was present in the membranes of glia cells and neurons, and a few of the highest intensities were present in plasma membranes, even in axons of the root. A faint CD31 signal was shown in the ganglion, with the exception of some points. The immunofluorescence signal of CD31 was brighter around axons in the root (yellow stars).

In samples from OxPt-treated rats, neurons presented an IQGAP1-specific bright signal in the cell periphery, touching the plasma membrane, and a lower intensity in the cytosol, exhibiting a reticular pattern. A higher level of IQGAP1 signal was present in some satellite glial cells in a dot pattern (white arrowheads). The CD31 immunostaining signal was found at medium/high intensity in the root, at a lower intensity level. Co-immunolabeling between CD31 and IQGAP1 was not present in endothelial cells at the root of axons in merged images.

When rats were treated with cilastatin and OxPt at the same time, a high-intensity immunofluorescence signal for IQGAP1 was in the cytosol of glia cells, depicting a reticular pattern, and at a lower level was present in the cytosol of neurons in a reticulated fashion. In comparison with the other groups of images, the CD31 immunostaining signal evidenced endothelial cells around axons of the root (magenta arrows), and some endothelial cells in the ganglion body, near neurons, showed medium/low intensity of red immunofluorescence signal. Some evidence of co-labeling between IQGAP1 and CD31 signals was found.

3.6. *IQGAP1 in the Short-Time DRG Inflammation Response to OxPt Chemotherapy*

Short-term involvement of IQGAP1 in the inflammation response was checked by co-labeling with TNFα (Figure 6). In control samples, a faint or no signal for TNFα was found in neurons. IQGAP1-specific fluorescence presented the typical picture of heterogeneous intensity with higher intensity in the plasma membrane of most neurons and, at a much lower level, in the cytosol, in a reticular and grainy background. A similar pattern was observed in satellite glial cells surrounding neurons. In cilastatin, IQGAP1 and TNFα co-staining displayed distribution and intensity patterns of fluorescence signals similar to those of controls. However, in OxPt panels, a higher intensity of immunostaining signal for IQGAP1 in the cytoplasmic membrane and in the cytosol than in control panels was seen, along with a high-intensity immunofluorescence-specific signal for TNFα following a dot pattern inside neurons (magenta arrowheads). Merges portray a heterogeneous fashion of co-expression, mainly in the cytosol. Interestingly, OxPt + cilastatin panels show specific fluorescence signals for either IQGAP1 or TNFα, similar to those of control and control + cilastatin.

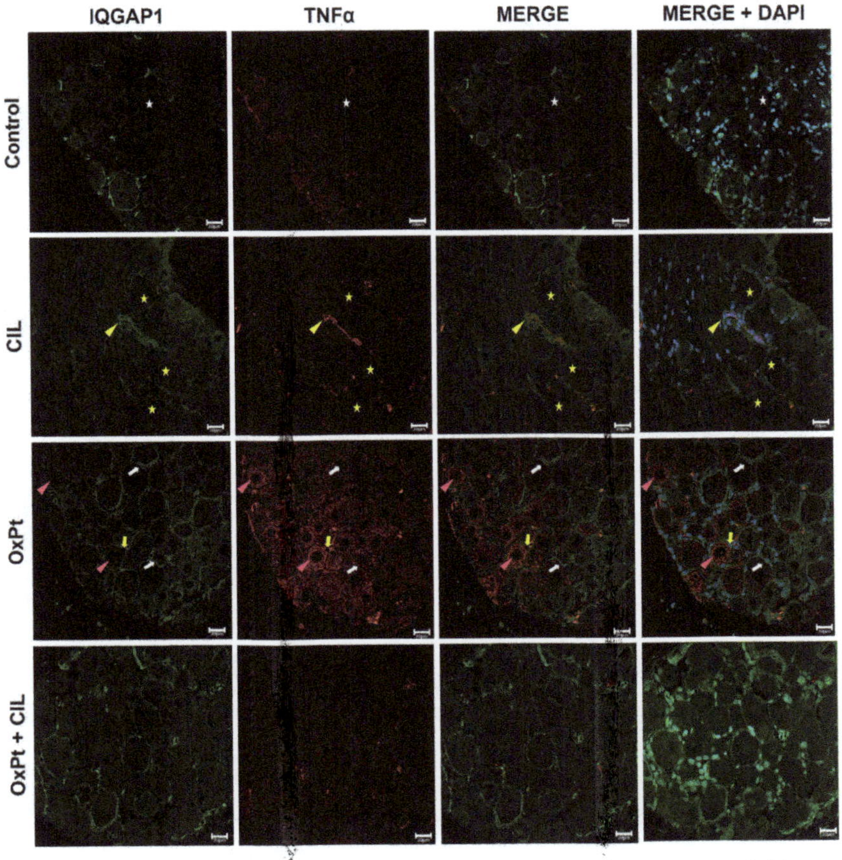

Figure 6. Immunostaining for IQGAP1 and TNFα. TNFα is absent or faint in neurons (white stars). IQGAP1 shows a heterogeneous distribution (as in Figure 3). Cilastatin (CIL): TNFα (yellow stars), in neurons and other cells (yellow arrowhead). OxPt: IQGAP1 in membrane and cytosol (white arrows). TNFα is strong in neuronal cytoplasm in a dot-like pattern (magenta arrowheads); the merge is heterogeneous. TNFα is expressed at medium-high levels in glia cells (yellow arrows). Scale bar 20 μm. OxPt + CIL: IQGAP1 and TNFα show distributions similar to controls. (*n*-Control = 4, 160 photographs, 7 DRGs; *n*-CIL = 4, 153 photographs, 7 DRGs; n-OxPt = 6, 297 photographs, 12 DRGs; *n*-Ox-Pt + CIL = 6, 217 photographs, 11 DRGs). OxPt increased TNFa expression in neurons and in glia 2–3-fold with respect to control and OxPt + CIL, $p < 0.05$. Supplementary Figure S4 includes higher-magnification images of representative regions from this figure.

Representative images of the immunostaining for IQGAP1 and IL-6 are shown in Figure 7. In control panels, immunofluorescence signal for IQGAP1 was found at heterogeneous intensity levels varying from medium to high, brighter in the cytoplasmic membrane and peripheral areas of cytosol, showing the cell's shape in neurons, satellite glial, and capillaries, and leaving a separation interface between the cells. A low-intensity signal for IL-6 immunostaining was shown in controls. Medium- to low-intensity signal was found in satellite cells where IQGAP1 and IL-6 signals colocalize in endothelial capillary cells (erythrocytes inside, yellow stars). IQGAP1 and IL-6 specific signals in DRG samples from cilastatin-treated rats were similar in intensity and localization to those shown in control panels. Cells in OxPt panels portray photographs with heterogeneous signals for IQGAP1, ranging within the medium intensity level, not as bright as in control or cilastatin samples.

Quantitation demonstrated a higher-intensity signal for IQGAP1 inside glia cells in this group compared to the control. IL-6 and IQGAP1 signals colocalized to a great extent. As in the case of IQGAP1-TNFα OxPt + cilastatin samples, IQGAP1-specific immunofluorescence was present in a heterogeneous intensity pattern, mainly in neurons displaying a high-intensity signal in the cytoplasmic membrane and neighboring areas of cytosol, very similar to that in control panels. The signal for IL-6 was at a moderate intensity level.

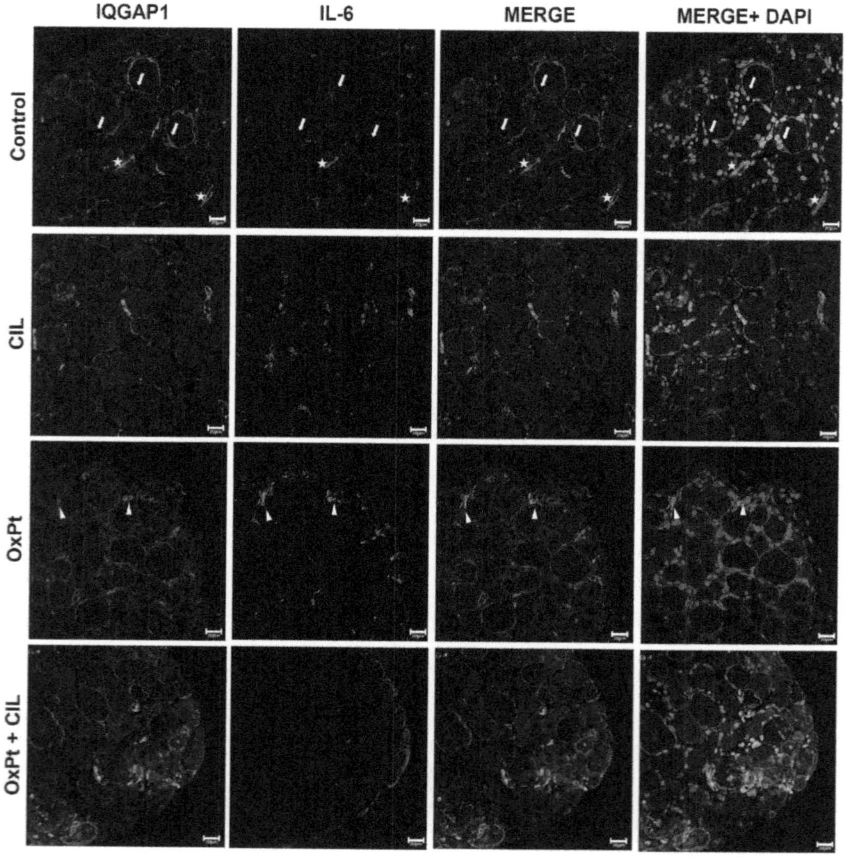

Figure 7. Immunostaining for IQGAP1 and IL-6. Control: IQGAP1 shows medium–high heterogeneous signal in neurons, glia, and capillaries (white arrows), outlining cell contours. IL-6 is low, with colocalization in capillary endothelial cells (yellow stars). Cilastatin (CIL): IQGAP1 and IL-6 signals remain comparable to controls. OxPt: IQGAP1 is medium and heterogeneous, lower than controls; increased in glia (white arrowheads). IL-6 colocalizes with IQGAP1 at high levels. OxPt + CIL: IQGAP1 remains high in membranes and perisomatic cytosol. IL-6 does not display a high-intensity signal. Scale bar 20 µm. (*n*-Control = 4, 183 photographs, 3 DRGs; *n*-CIL = 4, 200 photographs, 8 DRGs; n-OxPt = 6, 260 photographs, 10 DRGs; *n*-Ox-Pt + CIL = 6, 230 photographs, 8 DRGs). Enlarged views of key areas are shown in Supplementary Figure S4.

Figure 8 displays the compilation of results of the expression and localization of TNF-α, DPEP1, and IQGAP1 in DRG neurons, glia, and endothelial cells in control, cilastatin-, OxPt-, and OxPt–cilastatin-treated groups of rats and statistical significance.

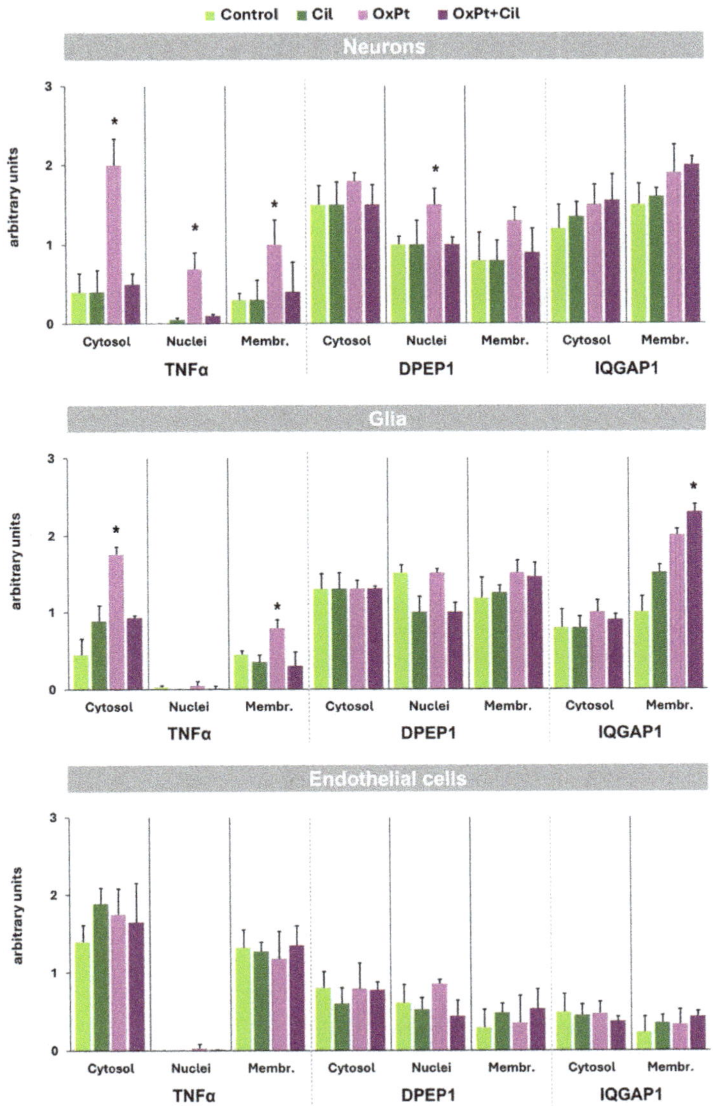

Figure 8. Intensity of specific fluorescence signal for TNF-α, DPEP1, and IQGAP1 in DRG neurons, glia, and endothelial cells in control, cilastatin (CIL)-, OxPt-, and OxPt + cilastatin-treated groups of rats. Neurons: a significant increase in TNF-α was observed in cytosol and nuclei of neurons in OxPt-treated animals, in contrast with basal levels of this inflammatory cell marker in the OxPt + CIL-treated group. The DPEP1 immunostaining signal was heterogeneous among the population of neurons. Heterogeneity in fluorescence intensities between cells reveals different response populations, which, however, when measured in all cells, hides possible statistical differences. There was a significant increase in DPEP1 signal in the nucleus of neurons of the OxPt-treated group versus basal levels in the rest of the groups, and a slight increase in IQGAP1 intensity signal in the cilastatin- and OxPt-treated groups compared with control. Glia: a significant increase in TNF-α was observed in the cytosol of satellite glia cells in OxPt-treated animals and the plasma membrane in the OxPt + cilastatin-treated group. The DPEP1 immunostaining signal was heterogeneous in the glial cell population. Endothelial cells: No statistical significance was found among the cells of the groups. Data are expressed as the mean ± SEM; * indicates statistical significance, $p < 0.05$.

4. Discussions

This study identifies IQGAP1 and DPEP1 as participants in DRG inflammation in OIPN; concurrently, OxPt altered their subcellular distribution in neurons, glia, and endothelial cells, promoting inflammatory signaling and neuro-glio-vascular disruption. Notably, cilastatin attenuated nocifensive responses and prevented inflammation and allodynia, suggesting partial morpho-functional integrity. These findings highlight the mechanistic involvement of IQGAP1 and DPEP1 in OIPN and propose cilastatin as a potential therapeutic modulator targeting neuro-glio-vascular dysfunction.

The graph in Figure 1B shows how the administration of cilastatin together with OxPt effectively prevents allodynia in the same way that it protects renal function [17,61] and, at the same time, confirms the involvement of DPEP1 in the neuropathic and inflammatory process.

Differential immunofluorescence patterns of DPEP1, IQGAP1, TNFα, and IL-6 across treatment groups highlight the pleiotropic role of IQGAP1 in the context of OxPt-induced DRG inflammation, as well as the modulatory capacity of cilastatin. IQGAP1 is involved in signal transduction and cytoskeletal and membrane dynamics (specified in the Section 1) performing variable functions depending on cell type (neurons, GFAP+ glia, and CD31+ endothelial cells) and modulating the effects according to the treatment conditions, suggesting a multifaceted role in the neuroinflammatory response and confirming, this way, the qualifying of the puppeteer [21].

IQGAP1 was predominantly located in the plasma membrane and pericytosolic areas in DRG neurons from control and cilastatin-treated animals, displaying a granular and reticulated pattern, in accordance with its function in organizing cortical actin and stabilizing cell polarity. Upon OxPt treatment, IQGAP1 distribution was altered significantly, showing enhanced cytosolic dot-like aggregates and intensified membrane labeling, indicating a stress-induced response of IQGAP1 associated with neuronal cytoskeletal remodeling and impaired trafficking or reactive plasticity. Interestingly, the co-localization with TNFα and IL-6 in the cytosol of neurons supports a role of IQGAP1 in inflammation, probably linked to cytoskeletal perturbations, a glimpse into neuroinflammatory processes. Ghosh et al. [62] showed the link between the rise in pro-inflammatory molecules, such as TNFα, and the rearrangement in cytoskeleton structure under oxidative stress in glioma cells. There were no changes in quantitative expression of IQGAP1, though intensity of interaction with other molecules was described, as, e.g., Cdc42 [62], a protein from the Rho family, activated via IQGAP1, which contributes to its translocation to the plasma membrane, where it regulates actin polymerization [63].

On the other hand, changes in subcellular localization of this scaffold protein have been previously described in tumoral cells; Rotoli et al. [64] showed an altered expression of IQGAP1 in colorectal carcinoma cells in comparison to healthy samples in which increased levels of IQGAP1 were found around plasma membranes, probably related to modifications in cell polarity in line with tumoral progression [65,66].

IQGAP1 has a key role in cytoskeleton structure and rearrangements in cells under stress conditions, intimately related to different signal pathways activation as a result of inflammatory responses [21]. Our results showed that this function of IQGAP1 is also activated in neurons and could contribute to OIPN.

In satellite glial cells (GFAP+), OxPt induced a clear upregulation of IQGAP1, particularly in perinuclear regions and membrane domains, suggesting reactive gliosis and potential involvement of IQGAP1 in glial activation and neuro-glial crosstalk. The strong co-expression with TNFα and IL-6 in this context further supports IQGAP1 participation in amplifying the inflammatory microenvironment through glial cells. The two subpopulations of TNFα, in the cytosol and at the plasma membrane, could indicate an increase in biosynthesis and an increase in extracellular binding to its membrane receptor. Interestingly, cilastatin significantly attenuated these alterations, restoring IQGAP1 patterns in glia to those seen in controls and reducing cytokine co-expression, which may reflect a protective effect of cilastatin by preserving glial homeostasis and limiting neuro-glial pro-inflammatory signaling pathways. Effects of oxidative stress in glial cells have been previously described [67] and were related to GFAP expression and cytoskeleton reorganization leading to cell death, but this is the first time that IQGAP1 expression is studied in these cells.

Endothelial cells (CD31+), although less explored in neuropathic models, also showed intriguing behavior. In OxPt-treated DRGs, while CD31 immunostaining was preserved around axons, IQGAP1 was absent in endothelial membranes, contrasting with its presence in controls, suggesting a vascular disassembly or dysfunction possibly contributing to the neuroinflammatory milieu. Importantly, cilastatin partially restored IQGAP1 expression in endothelial cells, indicating a potential endothelial-protective effect of cilastatin that could contribute to maintaining blood-nerve barrier integrity. Two studies from Yamaoka-Tojo et al. addressed this topic and found a relationship between reduction in cell adhesion through endothelial cells and promotion of angiogenesis mediated by IQGAP1 and vascular endothelial (VE)-cadherin [68,69]. IQGAP1 demonstrated a dual function in endothelial cells; in basal conditions, it has a role in adherence junctions and maintaining cell-to-cell adhesion; however, under stress circumstances and increased levels of ROS, IQGAP1 promotes phosphorylation of VE-cadherin, loss of cell-to-cell contact, and VEGF2 recruitment [68–70].

The heterogeneity in the expression of markers, as well as of IQGAP1 and DPEP1 in neurons and glia, leads to the possibility of polymorphisms; however, to our knowledge, no association studies exist between IQGAP1 expression or protein variants and OIPN. However, different basal levels of IQGAP1 (often regulated through polymorphisms either in the gene itself or regulatory microRNAs) have been linked to human cognitive performance and multiple sclerosis [71,72]. As for DPEP1, only a clear link between gene variants and osteoarthritis has been established [73]. In addition, DPEP SNPs have been associated with changes in plasma homocysteine levels in healthy women, which epidemiologically is linked to later development of cardiovascular diseases and loss of cognitive function [74]. Finally, other Genome-Wide Association Study (GWAS) and meta-analysis studies have found associations between DPEP1 variants and other traits (plasma and urine metabolites, blood pressure, etc.) [75]. Thus, it is indeed plausible that IQGAP1 and/or DPEP1 genetic variants regulate their expression, location, and/or responsiveness upon oxaliplatin in DRG.

Collectively, these data support a pleiotropic role of IQGAP1, acting not only as a cytoskeletal organizer but also as an integrator of inflammatory signals within the DRG microenvironment, modulating neuronal, glial, and endothelial compartments in a cell type-specific and stimulus-dependent manner. Cilastatin appears to modulate these responses by preserving physiological patterns of IQGAP1 localization and attenuating the aberrant inflammatory cascade induced by OxPt. This positions IQGAP1 as a critical node in the neuro-glio-vascular unit dysfunction during chemotherapy-induced peripheral neuropathy and cilastatin as a potential modulator of these pleiotropic effects.

Supplementary Materials: The following supporting information can be downloaded at https://www.mdpi.com/article/10.3390/cells14161294/s1: Table S1: Intensity of specific fluorescence signal. Figure S1: DPEP1 (green) and IQGAP1 (red) magnification panels. Magnified views of selected regions from the main figure (Figure 3) panels presented in the manuscript. Figure S2:

IQGAP1 (green) and GFAP (red) magnification panels. Magnified views of selected regions from the main figure (Figure 4) panels presented in the manuscript. Figure S3: IQGAP1 (green) and CD31 (red) magnification panels. Magnified views of selected regions from the main figure (Figure 5) panels presented in the manuscript. Figure S4: IQGAP1 (green) and TNFα (red) magnification panels. Magnified views of selected regions from the main figure (Figure 6) panels presented in the manuscript. Figure S5: IQGAP1 (green) and IL-6 (red) magnification panels. Magnified views of selected regions from the main figure (Figure 7) panels presented in the manuscript.

Author Contributions: Conceptualization, K.Á.-T., M.M., A.L. and P.M.-V.; methodology and investigation, R.M.-R.; M.Á.G.-N., K.Á.-T., F.M., J.Á. and A.L.; validation and formal analysis, M.M., J.Á., A.L. and P.M.-V.; data curation, writing and original draft preparation R.M.-R., M.Á.G.-N., K.Á.-T., M.M., A.L. and P.M.-V.; writing—review and editing, R.M.-R., M.Á.G.-N., K.Á.-T., F.M., M.M., A.L. and P.M.-V.; project administration, R.M.-R.; M.Á.G.-N. and A.L.; funding acquisition, F.M., M.M., J.Á. and A.L. All authors have read and agreed to the published version of the manuscript.

Funding: This research was funded by grants numbers OA21/110 from Fundación MAPFRE-Guanarteme and PIFIISC22/34 from Fundación Canaria Instituto de Investigación Sanitaria de Canarias (FUNCANIS) to M.M. F.M. was supported by the Spanish Ministry of Science and Innovation (MCIN/AEI/10.13039/501100011033) through the grant PID2021-123716OB-I100, cofinanced by the EU-ERDF "A way of making Europe". A.L.'s funding came from Spanish Ministry of Economy and Competitiveness Instituto de Salud Carlos III-Fondo de Investigación en Salud, grant numbers PI20/01577, and ICI21/00111, cofinanced by Fondo Europeo de Desarrollo Regional (FEDER) Funds from the European Commission, "A way of making Europe", ISCIII-RICORS program to RICORS2040 (Kidney Disease) and RICORS2040-Renal, grant numbers RD21/0005/0029 and RD24/0004/0023, respectively, co-funded by the European Union, the Spanish Ministry of Science and Innovation, grant number CPP2022-009633 (financed by the European Union, NextGenerationUE, Plan de Recuperación, Transformación y Resilencia), Comunidad de Madrid, grant number 2022/BMD-7223 (CIFRA_COR-CM) and Fundación Mutua Madrileña (XX Convocatoria de Ayudas a la Investigación en Salud). J.Á. was funded by the Gobierno de Canarias (cofounded by FEDER), grant number ProID2020010073.

Institutional Review Board Statement: The animal study protocol was approved by the Institutional Review Board of Gregorio Marañón Hospital (Registration code 07-2008), and the animals were handled according to legal regulations stipulated by RD 118/2021, of 23rd February.

Informed Consent Statement: Not applicable.

Data Availability Statement: The data presented in this study are available on request from corresponding authors.

Acknowledgments: We thank Felipe Rosa, Department of Mathematics, Statistics and Operational Research, University of La Laguna, for statistical analysis and advice.

Conflicts of Interest: The authors declare no conflicts of interest.

Abbreviations

The following abbreviations are used in this manuscript:

CD31	Cluster of differentiation 31 (also known as platelet endothelial cell adhesion molecule)
CIL	Cilastatin
CIPN	Chemotherapy-induced peripheral neuropathy
CZI	Carl Zeiss Image Data File
DPEP1	Dehydropeptidase-1
DRG	Dorsal root ganglion
GFAP	Glial fibrillary acidic protein
GPI	Glycosylphosphatidylinositol

GSH	Glutathione
IL-6	Interlukin 6
IQGAP1	IQ motif-containing GTPase-activating protein 1
LIF	Leica Image Format
MAP2	Microtubule-associated protein 2
OIPN	Oxaliplatin-induced peripheral neuropathy
OxPt	Oxaliplatin
PCC	Pearson correlation coefficient
PFA	Paraformaldehyde
ROS	Reactive oxygen species
SEM	Standard error of the mean
TBS	Tris-buffered saline
TNFα	Tumor Necrosis Factor-alpha

References

1. Hu, L.-Y.; Mi, W.-L.; Wu, G.-C.; Wang, Y.-Q.; Mao-Ying, Q.-L. Prevention and Treatment for Chemotherapy-Induced Peripheral Neuropathy: Therapies Based on CIPN Mechanisms. *Curr. Neuropharmacol.* **2019**, *17*, 184–196. [CrossRef]
2. Bouhassira, D. Neuropathic Pain: Definition, Assessment and Epidemiology. *Rev. Neurol.* **2019**, *175*, 16–25. [CrossRef]
3. Sałat, K. Chemotherapy-Induced Peripheral Neuropathy: Part 1—Current State of Knowledge and Perspectives for Pharmacotherapy. *Pharmacol. Rep.* **2020**, *72*, 486–507. [CrossRef] [PubMed]
4. von Hehn, C.A.; Baron, R.; Woolf, C.J. Deconstructing the Neuropathic Pain Phenotype to Reveal Neural Mechanisms. *Neuron* **2012**, *73*, 638–652. [CrossRef] [PubMed]
5. Seretny, M.; Currie, G.L.; Sena, E.S.; Ramnarine, S.; Grant, R.; MacLeod, M.R.; Colvin, L.A.; Fallon, M. Incidence, Prevalence, and Predictors of Chemotherapy-Induced Peripheral Neuropathy: A Systematic Review and Meta-Analysis. *Pain* **2014**, *155*, 2461–2470. [CrossRef] [PubMed]
6. Xiao, W.H.; Zheng, H.; Zheng, F.Y.; Nuydens, R.; Meert, T.F.; Bennett, G.J. Mitochondrial Abnormality in Sensory, but Not Motor, Axons in Paclitaxel-Evoked Painful Peripheral Neuropathy in the Rat. *Neuroscience* **2011**, *199*, 461–469. [CrossRef]
7. Woolf, C.J.; Mannion, R.J. Neuropathic Pain: Aetiology, Symptoms, Mechanisms, and Management. *Lancet* **1999**, *353*, 1959–1964. [CrossRef]
8. Argyriou, A.A.; Polychronopoulos, P.; Iconomou, G.; Chroni, E.; Kalofonos, H.P. A Review on Oxaliplatin-Induced Peripheral Nerve Damage. *Cancer Treat. Rev.* **2008**, *34*, 368–377. [CrossRef]
9. Alcindor, T.; Beauger, N. Oxaliplatin: A Review in the Era of Molecularly Targeted Therapy. *Curr. Oncol.* **2011**, *18*, 18–25. [CrossRef]
10. Pachman, D.R.; Qin, R.; Seisler, D.K.; Smith, E.M.L.; Beutler, A.S.; Ta, L.E.; Lafky, J.M.; Wagner-Johnston, N.D.; Ruddy, K.J.; Dakhil, S.; et al. Clinical Course of Oxaliplatin-Induced Neuropathy: Results From the Randomized Phase III Trial N08CB (Alliance). *J. Clin. Oncol.* **2015**, *33*, 3416–3422. [CrossRef]
11. Fujita, S.; Hirota, T.; Sakiyama, R.; Baba, M.; Ieiri, I. Identification of Drug Transporters Contributing to Oxaliplatin-induced Peripheral Neuropathy. *J. Neurochem.* **2019**, *148*, 373–385. [CrossRef]
12. McWhinney, S.R.; Goldberg, R.M.; McLeod, H.L. Platinum Neurotoxicity Pharmacogenetics. *Mol. Cancer Ther.* **2009**, *8*, 10–16. [CrossRef] [PubMed]
13. Park, S.; Krishnan, A.; Lin, C.; Goldstein, D.; Friedlander, M.; Kiernan, M. Mechanisms Underlying Chemotherapy-Induced Neurotoxicity and the Potential for Neuroprotective Strategies. *Curr. Med. Chem.* **2008**, *15*, 3081–3094. [CrossRef] [PubMed]
14. Argyriou, A. Updates on Oxaliplatin-Induced Peripheral Neurotoxicity (OXAIPN). *Toxics* **2015**, *3*, 187–197. [CrossRef]
15. Kang, L.; Tian, Y.; Xu, S.; Chen, H. Oxaliplatin-Induced Peripheral Neuropathy: Clinical Features, Mechanisms, Prevention and Treatment. *J. Neurol.* **2021**, *268*, 3269–3282. [CrossRef]
16. Álvarez-Tosco, K.; González-Fernández, R.; González-Nicolás, M.Á.; Martín-Ramírez, R.; Morales, M.; Gutiérrez, R.; Díaz-Flores, L.; Arnau, M.R.; Machín, F.; Ávila, J.; et al. Dorsal Root Ganglion Inflammation by Oxaliplatin Toxicity: DPEP1 as Possible Target for Peripheral Neuropathy Prevention. *BMC Neurosci.* **2024**, *25*, 44. [CrossRef]
17. Lau, A.; Rahn, J.J.; Chappellaz, M.; Chung, H.; Benediktsson, H.; Bihan, D.; von Mässenhausen, A.; Linkermann, A.; Jenne, C.N.; Robbins, S.M.; et al. Dipeptidase-1 Governs Renal Inflammation during Ischemia Reperfusion Injury. *Sci. Adv.* **2022**, *8*, eabm0142. [CrossRef]

18. Humanes, B.; Lazaro, A.; Camano, S.; Moreno-Gordaliza, E.; Lazaro, J.A.; Blanco-Codesido, M.; Lara, J.M.; Ortiz, A.; Gomez-Gomez, M.M.; Martín-Vasallo, P.; et al. Cilastatin Protects against Cisplatin-Induced Nephrotoxicity without Compromising Its Anticancer Efficiency in Rats. *Kidney Int.* **2012**, *82*, 652–663. [CrossRef]
19. Shayan, M.; Elyasi, S. Cilastatin as a Protective Agent against Drug-Induced Nephrotoxicity: A Literature Review. *Expert. Opin. Drug Saf.* **2020**, *19*, 999–1010. [CrossRef]
20. Hooper, N.M.; Low, M.G.; Turner, A.J. Renal Dipeptidase Is One of the Membrane Proteins Released by Phosphatidylinositol-Specific Phospholipase C. *Biochem. J.* **1987**, *244*, 465–469. [CrossRef]
21. Wang, L.; Tian, G. Insight into Dipeptidase 1: Structure, Function, and Mechanism in Gastrointestinal Cancer Diseases. *Transl. Cancer Res.* **2024**, *13*, 7015–7025. [CrossRef] [PubMed]
22. Eisenach, P.A.; Soeth, E.; Röder, C.; Klöppel, G.; Tepel, J.; Kalthoff, H.; Sipos, B. Dipeptidase 1 (DPEP1) Is a Marker for the Transition from Low-Grade to High-Grade Intraepithelial Neoplasia and an Adverse Prognostic Factor in Colorectal Cancer. *Br. J. Cancer* **2013**, *109*, 694–703. [CrossRef] [PubMed]
23. Nitanai, Y.; Satow, Y.; Adachi, H.; Tsujimoto, M. Crystal Structure of Human Renal Dipeptidase Involved in β-Lactam Hydrolysis. *J. Mol. Biol.* **2002**, *321*, 177–184. [CrossRef] [PubMed]
24. Ballatori, N.; Krance, S.M.; Notenboom, S.; Shi, S.; Tieu, K.; Hammond, C.L. Glutathione Dysregulation and the Etiology and Progression of Human Diseases. *Biol. Chem.* **2009**, *390*, 191–214. [CrossRef]
25. Baudouin-Cornu, P.; Lagniel, G.; Kumar, C.; Huang, M.-E.; Labarre, J. Glutathione Degradation Is a Key Determinant of Glutathione Homeostasis. *J. Biol. Chem.* **2012**, *287*, 4552–4561. [CrossRef]
26. Tang, D.; Chen, X.; Kang, R.; Kroemer, G. Ferroptosis: Molecular Mechanisms and Health Implications. *Cell Res.* **2021**, *31*, 107–125. [CrossRef]
27. Guan, Y.; Liang, X.; Ma, Z.; Hu, H.; Liu, H.; Miao, Z.; Linkermann, A.; Hellwege, J.N.; Voight, B.F.; Susztak, K. A Single Genetic Locus Controls Both Expression of DPEP1/CHMP1A and Kidney Disease Development via Ferroptosis. *Nat. Commun.* **2021**, *12*, 5078. [CrossRef]
28. Kozak, E.M.; Tate, S.S. Glutathione-Degrading Enzymes of Microvillus Membranes. *J. Biol. Chem.* **1982**, *257*, 6322–6327. [CrossRef]
29. Inamura, T.; Pardridge, W.M.; Kumagai, Y.; Black, K.L. Differential Tissue Expression of Immunoreactive Dehydropeptidase I, a Peptidyl Leukotriene Metabolizing Enzyme. *Prostaglandins Leukot. Essent. Fat. Acids* **1994**, *50*, 85–92. [CrossRef]
30. Liu, Q.; Deng, J.; Yang, C.; Wang, Y.; Shen, Y.; Zhang, H.; Ding, Z.; Zeng, C.; Hou, Y.; Lu, W.; et al. DPEP1 Promotes the Proliferation of Colon Cancer Cells via the DPEP1/MYC Feedback Loop Regulation. *Biochem. Biophys. Res. Commun.* **2020**, *532*, 520–527. [CrossRef]
31. Cui, X.; Liu, X.; Han, Q.; Zhu, J.; Li, J.; Ren, Z.; Liu, L.; Luo, Y.; Wang, Z.; Zhang, D.; et al. DPEP1 Is a Direct Target of MiR-193a-5p and Promotes Hepatoblastoma Progression by PI3K/Akt/MTOR Pathway. *Cell Death Dis.* **2019**, *10*, 701. [CrossRef]
32. Zhang, G.; Schetter, A.; He, P.; Funamizu, N.; Gaedcke, J.; Ghadimi, B.M.; Ried, T.; Hassan, R.; Yfantis, H.G.; Lee, D.H.; et al. DPEP1 Inhibits Tumor Cell Invasiveness, Enhances Chemosensitivity and Predicts Clinical Outcome in Pancreatic Ductal Adenocarcinoma. *PLoS ONE* **2012**, *7*, e31507. [CrossRef]
33. Green, A.R.; Krivinskas, S.; Young, P.; Rakha, E.A.; Paish, E.C.; Powe, D.G.; Ellis, I.O. Loss of Expression of Chromosome 16q Genes DPEP1 and CTCF in Lobular Carcinoma in Situ of the Breast. *Breast Cancer Res. Treat.* **2009**, *113*, 59–66. [CrossRef]
34. Kinoshita, T. Biosynthesis and Biology of Mammalian GPI-Anchored Proteins. *Open Biol.* **2020**, *10*, 190290. [CrossRef]
35. Morales, M.; Ávila, J.; González-Fernández, R.; Boronat, L.; Soriano, M.; Martín-Vasallo, P. Differential Transcriptome Profile of Peripheral White Cells to Identify Biomarkers Involved in Oxaliplatin Induced Neuropathy. *J. Pers. Med.* **2014**, *4*, 282–296. [CrossRef] [PubMed]
36. Weissbach, L.; Settleman, J.; Kalady, M.F.; Snijders, A.J.; Murthy, A.E.; Yan, Y.X.; Bernards, A. Identification of a Human RasGAP-Related Protein Containing Calmodulin-Binding Motifs. *J. Biol. Chem.* **1994**, *269*, 20517–20521. [CrossRef] [PubMed]
37. Abel, A.M.; Schuldt, K.M.; Rajasekaran, K.; Hwang, D.; Riese, M.J.; Rao, S.; Thakar, M.S.; Malarkannan, S. IQGAP1: Insights into the Function of a Molecular Puppeteer. *Mol. Immunol.* **2015**, *65*, 336–349. [CrossRef] [PubMed]
38. Weissbach, L.; Bernards, A.; Herion, D.W. Binding of Myosin Essential Light Chain to the Cytoskeleton-Associated Protein IQGAP1. *Biochem. Biophys. Res. Commun.* **1998**, *251*, 269–276. [CrossRef]
39. Hedman, A.C.; Smith, J.M.; Sacks, D.B. The Biology of IQGAP Proteins: Beyond the Cytoskeleton. *EMBO Rep.* **2015**, *16*, 427–446. [CrossRef]
40. White, C.D.; Erdemir, H.H.; Sacks, D.B. IQGAP1 and Its Binding Proteins Control Diverse Biological Functions. *Cell. Signal.* **2012**, *24*, 826–834. [CrossRef]
41. Mateer, S.C.; McDaniel, A.E.; Nicolas, V.; Habermacher, G.M.; Lin, M.-J.S.; Cromer, D.A.; King, M.E.; Bloom, G.S. The Mechanism for Regulation of the F-Actin Binding Activity of IQGAP1 by Calcium/Calmodulin. *J. Biol. Chem.* **2002**, *277*, 12324–12333. [CrossRef]

42. Li, Z.; McNulty, D.E.; Marler, K.J.M.; Lim, L.; Hall, C.; Annan, R.S.; Sacks, D.B. IQGAP1 Promotes Neurite Outgrowth in a Phosphorylation-Dependent Manner. *J. Biol. Chem.* **2005**, *280*, 13871–13878. [CrossRef] [PubMed]
43. Kholmanskikh, S.S.; Koeller, H.B.; Wynshaw-Boris, A.; Gomez, T.; Letourneau, P.C.; Ross, M.E. Calcium-Dependent Interaction of Lis1 with IQGAP1 and Cdc42 Promotes Neuronal Motility. *Nat. Neurosci.* **2006**, *9*, 50–57. [CrossRef] [PubMed]
44. Johnson, M.; Sharma, M.; Brocardo, M.G.; Henderson, B.R. IQGAP1 Translocates to the Nucleus in Early S-Phase and Contributes to Cell Cycle Progression after DNA Replication Arrest. *Int. J. Biochem. Cell Biol.* **2011**, *43*, 65–73. [CrossRef] [PubMed]
45. Rittmeyer, E.N.; Daniel, S.; Hsu, S.-C.; Osman, M.A. A Dual Role for IQGAP1 in Regulating Exocytosis. *J. Cell Sci.* **2008**, *121*, 391–403. [CrossRef]
46. Sakurai-Yageta, M.; Recchi, C.; Le Dez, G.; Sibarita, J.-B.; Daviet, L.; Camonis, J.; D'Souza-Schorey, C.; Chavrier, P. The Interaction of IQGAP1 with the Exocyst Complex Is Required for Tumor Cell Invasion Downstream of Cdc42 and RhoA. *J. Cell Biol.* **2008**, *181*, 985–998. [CrossRef]
47. Sharma, S.; Findlay, G.M.; Bandukwala, H.S.; Oberdoerffer, S.; Baust, B.; Li, Z.; Schmidt, V.; Hogan, P.G.; Sacks, D.B.; Rao, A. Dephosphorylation of the Nuclear Factor of Activated T Cells (NFAT) Transcription Factor Is Regulated by an RNA-Protein Scaffold Complex. *Proc. Natl. Acad. Sci. USA* **2011**, *108*, 11381–11386. [CrossRef]
48. Balenci, L.; Saoudi, Y.; Grunwald, D.; Deloulme, J.C.; Bouron, A.; Bernards, A.; Baudier, J. IQGAP1 Regulates Adult Neural Progenitors In Vivo and Vascular Endothelial Growth Factor-Triggered Neural Progenitor Migration In Vitro. *J. Neurosci.* **2007**, *27*, 4716–4724. [CrossRef]
49. Jausoro, I.; Mestres, I.; Remedi, M.; Sanchez, M.; Cáceres, A. IQGAP1: A Microtubule-Microfilament Scaffolding Protein with Multiple Roles in Nerve Cell Development and Synaptic Plasticity. *Histol. Histopathol.* **2012**, *27*, 1385–1394. [CrossRef]
50. Swiech, L.; Blazejczyk, M.; Urbanska, M.; Pietruszka, P.; Dortland, B.R.; Malik, A.R.; Wulf, P.S.; Hoogenraad, C.C.; Jaworski, J. CLIP-170 and IQGAP1 Cooperatively Regulate Dendrite Morphology. *J. Neurosci.* **2011**, *31*, 4555–4568. [CrossRef]
51. Akhilesh; Uniyal, A.; Gadepalli, A.; Tiwari, V.; Allani, M.; Chouhan, D.; Ummadisetty, O.; Verma, N.; Tiwari, V. Unlocking the Potential of TRPV1 Based SiRNA Therapeutics for the Treatment of Chemotherapy-Induced Neuropathic Pain. *Life Sci.* **2022**, *288*, 120187. [CrossRef]
52. Luo, X.; Gu, Y.; Tao, X.; Serhan, C.N.; Ji, R.-R. Resolvin D5 Inhibits Neuropathic and Inflammatory Pain in Male But Not Female Mice: Distinct Actions of D-Series Resolvins in Chemotherapy-Induced Peripheral Neuropathy. *Front. Pharmacol.* **2019**, *10*, 745. [CrossRef]
53. Gadepalli, A.; Ummadisetty, O.; Akhilesh; Chouhan, D.; Anmol; Tiwari, V. Loperamide, a Peripheral Mu-Opioid Receptor Agonist, Attenuates Chemotherapy-Induced Neuropathic Pain in Rats. *Int. Immunopharmacol.* **2023**, *124*, 110944. [CrossRef]
54. Gadepalli, A.; Ummadisetty, O.; Akhilesh; Chouhan, D.; Yadav, K.E.; Tiwari, V. Peripheral Mu-Opioid Receptor Activation by Dermorphin [D-Arg2, Lys4] (1–4) Amide Alleviates Behavioral and Neurobiological Aberrations in Rat Model of Chemotherapy-Induced Neuropathic Pain. *Neurotherapeutics* **2024**, *21*, e00302. [CrossRef]
55. Akhilesh; Chouhan, D.; Ummadisetty, O.; Verma, N.; Tiwari, V. Bergenin Ameliorates Chemotherapy-Induced Neuropathic Pain in Rats by Modulating TRPA1/TRPV1/NR2B Signalling. *Int. Immunopharmacol.* **2023**, *125*, 111100. [CrossRef]
56. Moreno-Gordaliza, E.; González-Nicolás, M.Á.; Lázaro, A.; Barbas, C.; Gómez-Gómez, M.M.; López-Gonzálvez, Á. Untargeted Metabolomics Analysis of Serum and Urine Unveils the Protective Effect of Cilastatin on Altered Metabolic Pathways during Cisplatin-Induced Acute Kidney Injury. *Biochem. Pharmacol.* **2024**, *227*, 116435. [CrossRef]
57. Mizuno, K.; Kono, T.; Suzuki, Y.; Miyagi, C.; Omiya, Y.; Miyano, K.; Kase, Y.; Uezono, Y. Goshajinkigan, a Traditional Japanese Medicine, Prevents Oxaliplatin-Induced Acute Peripheral Neuropathy by Suppressing Functional Alteration of TRP Channels in Rat. *J. Pharmacol. Sci.* **2014**, *125*, 91–98. [CrossRef]
58. McDonald, J.H.; Dunn, K.W. Statistical Tests for Measures of Colocalization in Biological Microscopy. *J. Microsc.* **2013**, *252*, 295–302. [CrossRef]
59. Adler, J.; Parmryd, I. Quantifying Colocalization by Correlation: The Pearson Correlation Coefficient Is Superior to the Mander's Overlap Coefficient. *Cytom. Part A* **2010**, *77A*, 733–742. [CrossRef]
60. Armitage, P.; Berry, G.; Matthews, J.N.S. *Statistical Methods in Medical Research*, 4th ed.; Wiley-Blackwell: Hoboken, NJ, USA, 2001.
61. González-Fernández, R.; González-Nicolás, M.Á.; Morales, M.; Ávila, J.; Lázaro, A.; Martín-Vasallo, P. FKBP51, AmotL2 and IQGAP1 Involvement in Cilastatin Prevention of Cisplatin-Induced Tubular Nephrotoxicity in Rats. *Cells* **2022**, *11*, 1585. [CrossRef]
62. Ghosh, S.; Tewari, R.; Dixit, D.; Sen, E. TNFα Induced Oxidative Stress Dependent Akt Signaling Affects Actin Cytoskeletal Organization in Glioma Cells. *Neurochem. Int.* **2010**, *56*, 194–201. [CrossRef]
63. Swart-Mataraza, J.M.; Li, Z.; Sacks, D.B. IQGAP1 Is a Component of Cdc42 Signaling to the Cytoskeleton. *J. Biol. Chem.* **2002**, *277*, 24753–24763. [CrossRef]

64. Rotoli, D.; Morales, M.; Maeso, M.D.C.; García, M.D.P.; Gutierrez, R.; Valladares, F.; Ávila, J.; Díaz-Flores, L.; Mobasheri, A.; Martín-Vasallo, P. Alterations in IQGAP1 Expression and Localization in Colorectal Carcinoma and Liver Metastases Following Oxaliplatin-Based Chemotherapy. *Oncol. Lett.* **2017**, *14*, 2621–2628. [CrossRef]
65. Noritake, J.; Watanabe, T.; Sato, K.; Wang, S.; Kaibuchi, K. IQGAP1: A Key Regulator of Adhesion and Migration. *J. Cell Sci.* **2005**, *118*, 2085–2092. [CrossRef]
66. Razidlo, G.L.; Burton, K.M.; McNiven, M.A. Interleukin-6 Promotes Pancreatic Cancer Cell Migration by Rapidly Activating the Small GTPase CDC42. *J. Biol. Chem.* **2018**, *293*, 11143–11153. [CrossRef] [PubMed]
67. Funchal, C.; Latini, A.; Jacques-Silva, M.C.; dos Santos, A.Q.; Buzin, L.; Gottfried, C.; Wajner, M.; Pessoa-Pureur, R. Morphological Alterations and Induction of Oxidative Stress in Glial Cells Caused by the Branched-Chain α-Keto Acids Accumulating in Maple Syrup Urine Disease. *Neurochem. Int.* **2006**, *49*, 640–650. [CrossRef] [PubMed]
68. Yamaoka-Tojo, M.; Tojo, T.; Kim, H.W.; Hilenski, L.; Patrushev, N.A.; Zhang, L.; Fukai, T.; Ushio-Fukai, M. IQGAP1 Mediates VE-Cadherin–Based Cell–Cell Contacts and VEGF Signaling at Adherence Junctions Linked to Angiogenesis. *Arter. Thromb. Vasc. Biol.* **2006**, *26*, 1991–1997. [CrossRef] [PubMed]
69. Yamaoka-Tojo, M.; Ushio-Fukai, M.; Hilenski, L.; Dikalov, S.I.; Chen, Y.E.; Tojo, T.; Fukai, T.; Fujimoto, M.; Patrushev, N.A.; Wang, N.; et al. IQGAP1, a Novel Vascular Endothelial Growth Factor Receptor Binding Protein, Is Involved in Reactive Oxygen Species—Dependent Endothelial Migration and Proliferation. *Circ. Res.* **2004**, *95*, 276–283. [CrossRef]
70. Meyer, R.D.; Sacks, D.B.; Rahimi, N. IQGAP1-Dependent Signaling Pathway Regulates Endothelial Cell Proliferation and Angiogenesis. *PLoS ONE* **2008**, *3*, e3848. [CrossRef]
71. Yang, L.; Zhang, R.; Li, M.; Wu, X.; Wang, J.; Huang, L.; Shi, X.; Li, Q.; Su, B. A Functional MiR-124 Binding-Site Polymorphism in IQGAP1 Affects Human Cognitive Performance. *PLoS ONE* **2014**, *9*, e107065. [CrossRef]
72. Keshari, P.K.; Harbo, H.F.; Myhr, K.-M.; Aarseth, J.H.; Bos, S.D.; Berge, T. Allelic Imbalance of Multiple Sclerosis Susceptibility Genes IKZF3 and IQGAP1 in Human Peripheral Blood. *BMC Genet.* **2016**, *17*, 59. [CrossRef]
73. Zhang, Z.; Mei, Y.; Feng, M.; Wang, C.; Yang, P.; Tian, R. The Relationship between Common Variants in the DPEP1 Gene and the Susceptibility and Clinical Severity of Osteoarthritis. *Int. J. Rheum. Dis.* **2021**, *24*, 1192–1199. [CrossRef]
74. Paré, G.; Chasman, D.I.; Parker, A.N.; Zee, R.R.Y.; Mälarstig, A.; Seedorf, U.; Collins, R.; Watkins, H.; Hamsten, A.; Miletich, J.P.; et al. Novel Associations of CPS1, MUT, NOX4, and DPEP1 With Plasma Homocysteine in a Healthy Population. *Circ. Cardiovasc. Genet.* **2009**, *2*, 142–150. [CrossRef]
75. Xcode Life. DPEP1 Gene; Xcode Life. Available online: https://www.xcode.life/gene/DPEP1.html (accessed on 2 August 2025).

cells

MDPI

Review

Protein Kinases as Mediators for miRNA Modulation of Neuropathic Pain

Leah Chang †, Zala Čok † and Lei Yu *

Department of Genetics, Center of Alcohol & Substance Use Studies, Rutgers University, Piscataway, NJ 08854, USA; lc1100@scarletmail.rutgers.edu (L.C.)
* Correspondence: yu@biology.rutgers.edu
† These authors contributed equally to this work.

Abstract: Neuropathic pain is a chronic condition resulting from injury or dysfunction in the somatosensory nervous system, which leads to persistent pain and a significant impairment of quality of life. Research has highlighted the complex molecular mechanisms that underlie neuropathic pain and has begun to delineate the roles of microRNAs (miRNAs) in modulating pain pathways. miRNAs, which are small non-coding RNAs that regulate gene expression post-transcriptionally, have been shown to influence key cellular processes, including neuroinflammation, neuronal excitability, and synaptic plasticity. These processes contribute to the persistence of neuropathic pain, and miRNAs have emerged as critical regulators of pain behaviors by modulating signaling pathways that control pain sensitivity. miRNAs can influence neuropathic pain by targeting genes that encode protein kinases involved in pain signaling. This review focuses on miRNAs that have been demonstrated to modulate neuropathic pain behavior through their effects on protein kinases or their immediate upstream regulators. The relationship between miRNAs and neuropathic pain behaviors is characterized as either an upregulation or a downregulation of miRNA levels that leads to a reduction in neuropathic pain. In the case of miRNA upregulation resulting in an alleviation of neuropathic pain behaviors, protein kinases exhibit a positive correlation with neuropathic pain, whereas decreased protein kinase levels correlate with diminished neuropathic pain behaviors. The only exception is GRK2, which shows an inverse correlation with neuropathic pain. In the case of miRNA downregulation resulting in a reduction in neuropathic pain behaviors, protein kinases display mixed relationships to neuropathic pain, with some kinases exhibiting positive correlation, while others exhibit negative correlation. By exploring how protein kinases mediate miRNA modulation of neuropathic pain, valuable insight may be gained into the pathophysiology of neuropathic pain, offering potential therapeutic targets for developing more effective strategies for pain management.

Keywords: miRNAs; kinases; neuropathic pain

1. Introduction

Neuropathic pain is a chronic, debilitating condition caused by injury or dysfunction of the somatosensory nervous system [1–5]. Unlike nociceptive pain, which stems from acute tissue damage and typically resolves with healing, neuropathic pain persists beyond the initial injury, often becoming long-term or even lifelong. Affecting an estimated 3% to 17% of the global population [6,7], neuropathic pain significantly diminishes quality of life and remains a major challenge within the field of pain management [8–11].

Despite decades of research, effective treatment of neuropathic pain remains a significant clinical challenge [11–15]. Current pharmaceutical approaches, including antiepileptic drugs, antidepressants, and opioids, often fall short, with nearly 50% of patients failing to achieve adequate pain relief [12–14,16,17]. This underscores the urgent need for novel therapeutic strategies. A major obstacle lies within the complex molecular and cellular mechanisms underlying neuropathic pain, which involve neuroinflammation, altered synaptic plasticity, and dysregulated signaling pathways [4,18,19]. A deeper understanding of these processes at the molecular level is essential for identifying new and more effective therapeutic targets.

MicroRNAs (miRNAs) are a class of short, non-coding RNA molecules that regulate gene expression post-transcriptionally by binding to target messenger RNAs (mRNAs), leading to their degradation or translational repression [20–23]. These molecules play essential roles in a variety of biological and pathological processes, including neuroinflammation, neuronal excitability, and synaptic plasticity—factors that contribute to the persistence of neuropathic pain [24,25]. Recent evidence suggests that miRNAs are key modulators of neuropathic pain behaviors, either exacerbating or alleviating pain states by targeting specific signaling pathways.

Among these pathways, protein kinases are critical cellular regulators that mediate diverse biological functions through the phosphorylation of proteins [26–28]. Many protein kinases have been implicated in neuropathic pain modulation, either promoting or suppressing pain signaling depending on their specific roles in various intracellular cascades [29,30]. Given the emerging evidence linking miRNA activity to modified protein kinase signaling in neuropathic pain models, this review focuses on miRNAs that have been demonstrated to modulate neuropathic pain behavior through direct or indirect effects on the mRNAs encoding protein kinases or their immediate upstream regulators. By elucidating these interactions, we aim to provide insight into the regulatory mechanisms of neuropathic pain and potential therapeutic targets. For a more comprehensive list of miRNAs implicated in neuropathic pain, readers are directed to the following review articles [31–37].

2. miRNA Modulation: Upregulation or Downregulation Can Alleviate Neuropathic Pain

A growing body of literature has shown that miRNAs can causally influence neuropathic pain behaviors. We identified two distinct motifs regarding the relationships that characterize the interconnection between changes in miRNA levels and the consequences on neuropathic pain states.

For the first motif, which is more abundantly reported in the literature, the <u>upregulation</u> of specific miRNAs results in the <u>attenuation</u> of neuropathic pain behaviors; in other words, the increased expression of these miRNAs leads to reduced pain. As summarized in Table 1 and illustrated in Figure 1, the miRNAs belonging to this relationship pattern predominantly target protein kinases that exhibit a positive correlation with neuropathic pain; specifically, in this model, when the levels of these protein kinases decrease, neuropathic pain behaviors also diminish. This suggests that these kinases play a pro-nociceptive role in neuropathic pain at the cellular level, and their suppression, whether by miRNAs or other means, may provide therapeutic benefits. However, there is one exception to this pattern of parallel change between the level of kinase activity and the severity of neuropathic pain, G-protein receptor kinase 2 (GRK2). This kinase displays a negative correlation with neuropathic pain [38], with GRK2 levels increasing with miRNA expression as neuropathic pain subsides.

Table 1. Upregulation of miRNA, via kinase mediation, alleviates neuropathic pain.

↑ miRNA ⇒ Kinase ⇒ ↓ Neuropathic Pain						
lncRNA	**miRNA**	**Intermediary**	**Protein Kinase**	**Neuropathic Pain**	**Model**	**Reference**
-	miR-183 ↑	-	MAP3K4 ↓	down ↓	Rat, CCI	[39]
-	miR-15a ↑	-	AKT3 ↓	down ↓	Rat, CCI	[40]
-	miR-150 ↑	-	AKT3 ↓	down ↓	Rat, CCI	[41]
-	miR-20b-5p ↑	-	AKT3 ↓	down ↓	Rat, CCI	[42]
-	miR-101 ↑	-	mTOR ↓	down ↓	Rat, CCI	[43]
-	miR-183 ↑	-	mTOR ↓	down ↓	Rat, CCI	[44]
-	miR-362-3p ↑	PAX2 ↓	MEK1/2 ↓	down ↓	Rat, SCI	[45]
-	miR-206 ↑	-	MEK ↓	down ↓	Rat, CCI DRG	[46]
-	miR-186-5p ↑	CXCL13 ↓ CXCR5 ↓	ERK ↓	down ↓	Mouse, SNL	[47]
-	miR-143 ↑	-	ERK1/2 ↓	down ↓	Rat, SNL-induced DRG	[48]
-	miR-26a-5p ↑	-	MAPK6 ↓	down ↓	Rat, CCI	[49]
-	miR-146a-5p ↑	-	IRAK1 ↓	down ↓	Rat, CCI	[50]
LINC00052 ↓	miR-448 ↑	-	JAK1 ↓	down ↓	Rat, SNL	[51]
-	miR-124 ↑	-	GRK2 ↑	down ↓	Mouse, de novo GRK2 knockout	[38]

Abbreviations of nerve injury models: CCI, chronic constriction injury; DRG, dorsal root ganglia; SCI, spinal cord injury; SNL, spinal nerve ligation. Upward arrows (↑) indicate upregulation; downward arrows (↓) indicate downregulation.

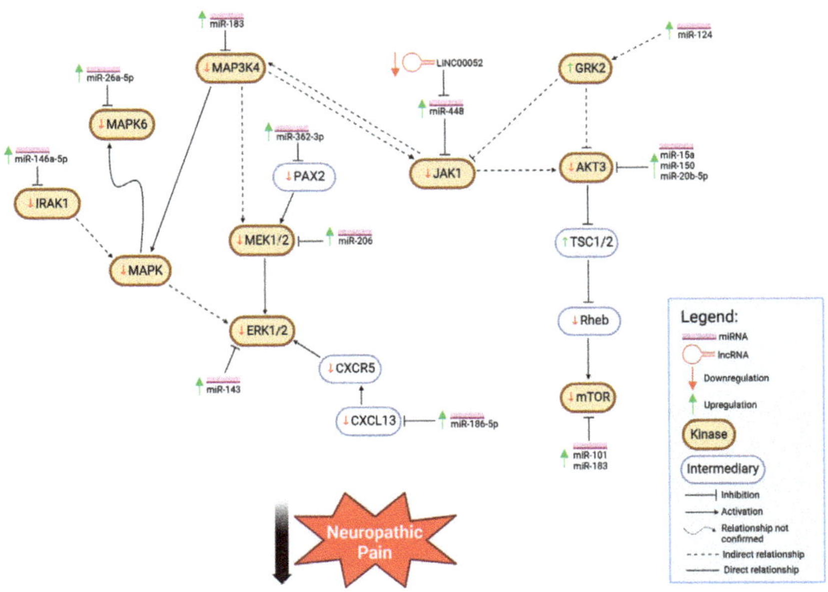

Figure 1. Upregulation of miRNA, via kinase mediation, alleviates neuropathic pain. The upregulation of miRNAs is indicated by an upward green arrow. Reduction in kinase activities or expression levels, which is the case for all kinases directly or indirectly impacted by miRNA, is marked by a downward red arrow, except for GRK2, which shows increased levels with miRNA upregulation with an upward green arrow. Additionally, TSC1/2, an intermediary protein, is also upregulated in this motif. The outcome of all indicated changes in miRNA or other factors' levels is the alleviation of neuropathic pain.

For the second motif, the downregulation of certain miRNAs results in a reduction in neuropathic pain behaviors. As summarized in Table 2 and illustrated in Figure 2, the kinases targeted by the miRNAs within this pattern display mixed relationships to neuropathic pain, with some kinases exhibiting a positive correlation in which kinase downregulation relieves pain, and other kinases exhibiting a negative correlation in which kinase downregulation exacerbates pain. This suggests a more complex regulatory landscape in which the miRNA-mediated modulation of protein kinases has varying effects depending on the specific molecular context.

Table 2. Downregulation of miRNA, via kinase mediation, alleviates neuropathic pain.

↓ miRNA ⇒ Kinase ⇒ ↓ Neuropathic Pain						
lncRNA	**miRNA**	**Intermediary**	**Protein Kinase**	**Neuropathic Pain**	**Model**	**Reference**
-	miR-155 ↓	SOCS1 ↑	p38 ↓	down ↓	Rat, CCI	[52]
-	miR-221 ↓	SOCS1 ↑	p38 ↓	down ↓	Rat, CCI	[53]
-	miR-133a-3p ↓	-	p38 ↓	down ↓	Rat (diabetic), sciatic nerve	[54]
-	miR-15a/16 ↓	-	p38 ↓	down ↓	Mouse, CCI	[55]
-	miR-21 ↓	TLR8 ↓	ERK ↓	down ↓	Mouse, DRG, Tlr8 knockout	[56]
-	miR-142-3p ↓	AC9 ↑	AMPK ↑	down ↓	Rat, CCI, SNI	[57]
lncRNA CCA11 ↑	miR-155 ↓	-	SGK3 ↑	down ↓	Rat, bilateral CCI	[58]
-	miR-155 ↓	-	SGK3 ↑	down ↓	SD rat, bilateral CCI	[59]
-	miR-15a/16 ↓	-	GRK2 ↑	down ↓	Mouse, CCI	[55]

Abbreviations of nerve injury models: CCI, chronic constriction injury; DRG, dorsal root ganglia; SD rat, Sprague Dawley rat; SNI, spared nerve injury. Upward arrows (↑) indicate upregulation; downward arrows (↓) indicate downregulation.

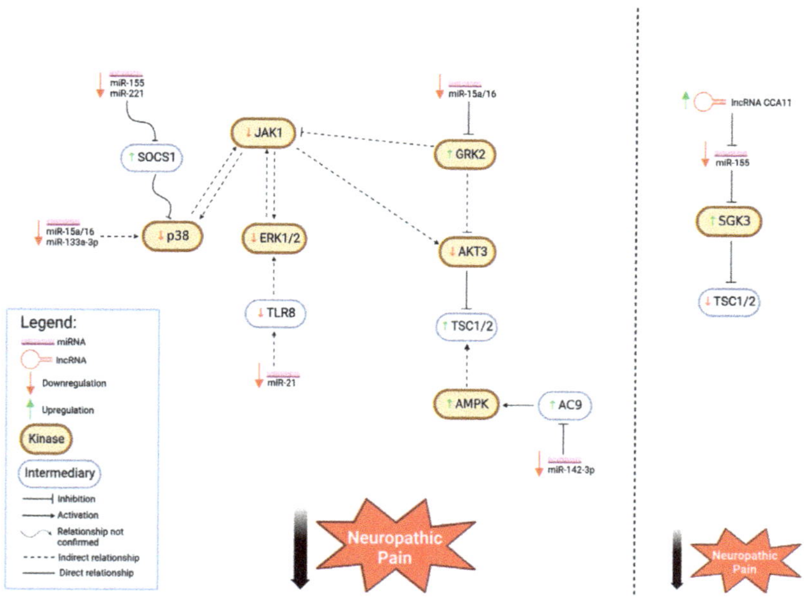

Figure 2. Downregulation of miRNA, via kinase mediation, alleviates neuropathic pain. Downregulation of miRNAs is indicated by a downward red arrow. Changes in the levels of kinases and/or other factors

can be either increased, as indicated by an upward green arrow, or decreased, as indicated by a downward red arrow. Note: the SGK3-mediated pathway is distinct from the indicated kinase network due to its opposing effect on TSC1/2 compared to AKT3, which is another upstream regulator of TSC1/2. Depending on the upstream regulator involved, TSC1/2 expression can be either upregulated or downregulated, correlating with pain attenuation. Ultimately, the combined effects of changes in miRNA levels and other regulatory factors contribute to the reduction in neuropathic pain.

Tables 1 and 2 list the miRNA–kinase–neuropathic pain cases, indicating the directionality of changes for miRNAs, kinases, and neuropathic pain behaviors. Table 3 lists the kinase names with their abbreviations discussed in this article, as well as alternative abbreviations often encountered in the literature. By categorizing these patterns of miRNA involvement in neuropathic pain, we can begin to discern relevant regulatory mechanisms and potential therapeutic targets. The following sections will explore specific miRNA-protein kinase interactions, highlighting the functional roles of protein kinases in neuropathic pain and implications for future research and clinical applications.

Table 3. Protein kinase names and abbreviations.

Protein Kinase Abbreviation	Alternative Abbreviations	Full Name
MAP3K4	MTK1; MEKK4; MAPKKK4; PRO0412; MKKK4	Mitogen-activated protein kinase kinase kinase 4
JNK	MAPK8; JNK1; PRKM8; SAPK1; JNK-46; JNK1A2; SAPK1c; JNK21B1/2	Jun N-terminal kinase
MKK4	MAP2K4; JNKK; MEK4; SEK1; SKK1; JNKK1; SERK1; MAPKK4; PRKMK4; SAPKK1	Mitogen-activated protein kinase kinase 4
MEK1/2	MAP2K1/2	Mitogen-activated protein kinase 1/2
ERK1/2	MAPK1/2	Extracellular signal-regulated kinase 1/2
JAK1	JTK3; AIIDE; JAK1A; JAK1B	Janus kinase 1
MAP2K	MKK	Mitogen-activated protein kinase kinase
AKT3	MPFH; PKBG; MPPH2; PRKBG; STK-2; PKB-GAMMA; RAC-gamma; RAC-PK-gamma	AKT Serine/Threonine kinase
mTOR	SKS; FRAP; FRAP1; FRAP2; RAFT1; RAPT1	Mechanistic target of rapamycin kinase
IRAK1	IRAK; pelle	Interleukin 1 receptor-associated kinase
MAPK6	ERK3; PRKM6; p97MAPK; HsT17250	Mitogen-activated protein kinase 6
GRK2	BARK1; ADRBK1; BETA-ARK1	G protein-coupled receptor kinase 2
AMPK	PRKAA1; AMPKa1; AMPK alpha 1	Adenosine monophosphate-activated protein kinase
SGK3	CISK; SGK2; SGKL	Serum/glucocorticoid-regulated kinase family member 3

3. Protein Kinase Involvement in miRNA Upregulation Leading to Alleviation of Neuropathic Pain

Multiple kinases, through direct or indirect interactions, can form a complex regulatory network (Figure 1). A general pattern within this network reveals that kinase activity is co-regulated with the attenuation of neuropathic pain; specifically, key kinases are downregulated in parallel with reductions in pain behavior. This coordinated downregulation suggests a functional relationship between these kinases and pain modulation. Importantly, miRNAs play a crucial role in this regulatory framework, as their upregulation in these cases contributes to the suppression of specific kinases involved in pain signaling. By targeting and downregulating these kinases, upregulated miRNAs help reinforce the pathway leading to neuropathic pain relief. This intricate interplay between miRNA expression and kinase regulation highlights a potential avenue for therapeutic intervention in neuropathic pain management.

3.1. Mitogen-Activated Protein Kinase Kinase Kinase 4 (MAP3K4)

Mitogen-activated protein kinase kinase kinase 4 (MAP3K4) plays a crucial role in cellular signaling, particularly in the mitogen-activated protein kinase (MAPK) cascade [60–62]. In this cascade, it is upstream from many kinases such as MKK4/6/7, JNK, and p38 [60–62].

MAP3K4 activates the MAPK pathway by phosphorylating and activating MAP2Ks [60–62], specifically MKK4 and MKK7 [62], which are responsible for phosphorylating and activating downstream MAPKs such as JNK and p38 [63,64]; therefore, MAP3K4 serves as a key regulator of the MAPK signaling cascade.

MAP3K4 indirectly [60–62] influences MEK1/2 through cross-talk between the JNK/p38 MAPK pathways and the ERK pathway [60,65–68]. Additionally, MAP3K4 is an upstream kinase in the MAPK pathway, triggering the cascade, which eventually leads to MEK1/2 activation [62,69].

Cross-talk between JAK1 and MAP3K4 occurs through the activation of MAPK pathways [60,70]. JAK1-STAT signaling can influence the activation of MAPK cascades indirectly by promoting the expression of genes that regulate proteins involved in MAPK activation [71]. Specifically, MAP3K4 can activate JNK [62,72], which can contribute to the inflammatory response initiated by JAK1 activation [73–75]. This cross-talk between cytokine signaling and MAPK pathways defines the complex relationship between the two kinases.

In the CCI model of rats, increased miR-183 levels suppressed MAP3K4 activities, resulting in attenuated neuropathic pain [39].

3.2. AKT Serine/Threonine Kinase 3 (AKT3)

AKT serine/threonine kinase 3 (AKT3) is a key regulator in the phosphoinositide 3-kinase (PI3K)-AKT-mTOR pathway, influencing cell growth, survival, and metabolism, particularly in brain development and neuronal protection [76].

In the neuropathic pain model of CCI, increased levels of miR-15a [40], miR-150 [41], and miR-20b-5p [42] suppressed AKT3 kinase activities, leading to attenuated neuropathic pain.

3.3. Mechanistic Target of Rapamycin Kinase (mTOR)

The mechanistic target of rapamycin kinase (mTOR) mediates cellular responses to stressors such as DNA damage and nutrient deprivation [77]. mTOR is a component of two distinct complexes: mechanistic target of rapamycin complex 1 (mTORC1), which controls protein synthesis, cell growth, and proliferation [78], and mechanistic target of rapamycin complex 1 (mTORC2), which is a regulator of the actin cytoskeleton and promotes cell survival and cell cycle progression [78].

AKT3 promotes mTOR activation by inhibiting tuberous sclerosis complex subunit 1/2 (TSC1/2) through phosphorylation [79–82]. This, in turn, allows Ras homolog enriched in brain (Rheb) to remain active and stimulate mTOR signaling [83,84]; thus, the downregulation of AKT3 positively correlates with mTOR downregulation and neuropathic pain attenuation.

In the neuropathic pain model of CCI, increased levels of miR-101 [43] and miR-183 [44] suppressed mTOR expression, attenuating neuropathic pain.

3.4. Mitogen-Activated Protein Kinase 1/2 (MEK1/2)

Mitogen-activated protein kinase 1/2 (MEK1/2) is a dual-specificity kinase that functions as a critical component of the MAPK/ERK signaling pathway [85]. They act as intermediaries between MAP3Ks, such as MAP3K4, by phosphorylating and activating ERK1/2 [86], which in turn regulates cell proliferation, differentiation, and survival [87].

MAP3K4 is upstream of MEK1/2 in certain signaling cascades [88], and activates MEK1/2 through intermediates such as MAP2Ks [86], contributing to ERK1/2 activation.

MEK1/2 are the direct activators of ERK1/2 [85]. Upon activation by upstream kinases, MEK1/2 phosphorylates ERK1/2 on specific threonine and tyrosine residues, leading to ERK1/2 activation and subsequent cellular responses [89].

Paired-box gene 2 (PAX2) is a transcription factor that promotes the expression of upstream activators such as receptor tyrosine kinases (RTKs) [90,91], leading to the activation of Ras and then Raf, ultimately leading to the phosphorylation and activation of MEK1/2 [92].

In the SCI rat model, increased levels of miR-362-3p suppress PAX2, which in turn leads to the suppression of MEK1/2 and, therefore, reduced neuropathic pain levels [45]; similarly, in the model of CCI of DRG in rats, increased miR-206 suppresses MEK and neuropathic pain [46], though it is not indicated that PAX2 is involved in this mechanism, nor is the specific isoform of MEK specified.

3.5. Extracellular Signal-Regulated Kinase 1/2 (ERK1/2)

Extracellular signal-regulated kinase (ERK1/2) is located downstream from MEK in the MAPK cascade and is a key enzyme in the ERK signaling pathway [88]. MEK1/2 phosphorylates ERK1/2 on threonine (T) and tyrosine (Y) residues within the threonine-glutamic acid-tyrosine (TEY) motif [93], activating ERK1/2.

Upon activation, toll-like receptor 8 (TLR8) recruits myeloid differentiation primary response 88 (MyD88) [94] which facilitates the activation of another protein kinase, IRAK1 [94–96], associates with the tumor necrosis factor (TNF) receptor-associated factor 6 (TRAF6) [97], aiding in the activation of MAPK signaling. This, in turn, leads to the activation of MEK1/2, which then phosphorylates and activates ERK1/2.

C-X-C motif chemokine ligand 13 (CXCL13) is a chemokine that binds to the C-X-C motif chemokine receptor 5 (CXCR5) [98], which is a G-protein-coupled receptor (GPCR). Upon CXCL13 activation, CXCR5 activates G-protein signaling, leading to MEK1/2 activation [99], which leads to ERK1/2 phosphorylation and activation [95].

In the neuropathic pain model of SNL-induced DRG of rats, increased levels of miR-143 [48] suppressed ERK1/2 activity, leading to attenuated neuropathic pain; similarly, in the neuropathic pain model of SNL in mice, increased miR-186-5p [47] suppressed CXCL13 and subsequently suppressed CXCR5 and ERK, leading to attenuated neuropathic pain. In this specific model, though, it is not specified which isoform of ERK is suppressed, but rather, the general type of kinase, ERK, is identified.

3.6. Mitogen-Activated Protein Kinase 6 (MAPK6)

Mitogen-activated protein kinase 6 (MAPK6) is a non-canonical MAPK, meaning it has unique activation mechanisms and does not follow the traditional rapidly accelerated fibrosarcoma (RAF)-MEK-ERK cascade [100–102]. Unlike ERK1/2's activation through the TEY motif, MAPK6's corresponding motif is SEG [100,103–105]. Not much is known about the role of MAPK6, but it does engage in regulatory relationships within the broad MAPK signaling network [106,107]. MAPK6 is activated through protein phosphorylation cascades and acts as an integration point for multiple biochemical signals [104,107].

In the neuropathic pain model of CCI, the upregulation of miR-26a-5p suppressed MAPK6 expression, resulting in attenuated neuropathic pain [49].

3.7. Interleukin 1 Receptor-Associated Kinase (IRAK1)

Interleukin 1 receptor-associated kinase (IRAK1) plays an important role in the regulation of the expression of inflammatory genes. Upon activation by their respective ligands, toll-like receptors (TLRs) and interleukin-1 (IL-1) recruit MyD88 [97], leading to the phosphorylation and activation of IRAK1 [96,108]. Once activated, IRAK1 phosphorylates and activates TRAF6, which then activates several kinases in the MAPK family, including ERK1/2, p38, and JNK [94].

In the neuropathic pain model of CCI, increased expression of miR-146a-5p suppressed IRAK1 signaling, leading to attenuated neuropathic pain [50].

3.8. Janus Kinase 1 (JAK1)

Janus kinase 1 (JAK1) is a protein that helps transmit signals for cytokines and growth factors [109], making it a key part of immune function. JAK1 is activated by cytokines, stimulating phosphoinositide 3-kinase (PI3K), which in turn activates AKT3 [109].

In the neuropathic pain model of SNL in rats, decreased levels of the lncRNA (long non-coding RNA) LINC00052 (long intergenic non-coding RNA 00052) have been found to increase levels of miR-448 [51], leading to decreased JAK1 levels and attenuated neuropathic pain.

3.9. G Protein-Coupled Receptor Kinase 2 (GRK2)

G protein-coupled receptor kinase 2 (GRK2) plays a key role in GPCR signaling [110,111], indirectly inhibiting JAK1 by phosphorylating upstream GPCRs [112–114], which disrupts the JAK signaling pathway and thereby reducing the downstream signaling cascade associated with JAK1 activity [112,115].

GRK2 negatively regulates AKT3 signaling through interactions with phosphatases such as protein phosphatase 2A (PP2A), leading to AKT3 dephosphorylation and, therefore, inactivation [110,111].

In the neuropathic pain model of de novo GRK2 knockout mice, increased levels of miR-124 [38] increased GRK2 expression, leading to attenuated neuropathic pain.

4. Protein Kinase Involvement in miRNA Downregulation Leading to Alleviation of Neuropathic Pain

Figure 2 illustrates the protein kinases involved in the second motif, along with their direct and indirect interaction partners. This kinase network presents a more intricate regulatory landscape in the context of neuropathic pain behaviors. Unlike the uniform co-regulation observed in other networks, kinases in these cases show a more nuanced interplay, where some kinases exhibit a positive correlation with neuropathic pain, meaning their upregulation is associated with increased pain behaviors, while others display a negative correlation in which their downregulation aligns with pain attenuation. These opposing regulatory pathways suggest that different subsets of kinases may contribute to distinct mechanisms underlying pain modulation, potentially reflecting the balance between pro-inflammatory and anti-inflammatory signaling pathways. Understanding these complex interactions may provide deeper insight into the molecular underpinnings of neuropathic pain and inform the development of more targeted therapeutic strategies.

4.1. Mitogen-Activated Protein Kinase (MAPK): p38

Mitogen-activated protein kinase (MAPK) is a protein kinase family that controls how cells respond to stimuli [104]. p38 is part of the MAPK family and regulates many important cellular processes [116] such as inflammation, cell growth, apoptosis, and tissue homeostasis [117]. Four p38 isoforms have been identified (p38α, p38β, p38γ, and p38δ) [117], though it is not indicated which isoform specifically is involved in the neuropathic pain network.

JAK1 and MAPK pathways interact in a complex but coordinated manner. JAK1, through cytokine receptor activation [118,119], can indirectly activate MAPK signaling [120,121]; additionally, MAPK signaling can modulate JAK1 activity [122,123], creating feedback loops that fine-tune cellular functions.

Suppressor of cytokine signaling 1 (SOCS1) inhibits MAPK signaling by indirectly blocking the activation of the upstream kinases, primarily by targeting and inhibiting the JAK family of kinases, which are crucial for the phosphorylation cascade leading to MAPK activation [124].

In the CCI model of rats, decreased levels of both miR-155 [52] and miR-221 [53] led to increased levels of SOCS1. This decreased p38 expression levels, attenuating neuropathic pain.

In the diabetes mellitus (DM) sciatic nerve of model rats, decreased levels of miR-133-3p [54] suppressed p38, attenuating neuropathic pain; additionally, in the CCI model of mice, decreased levels of miR-15a/16 [55] also suppressed p38, attenuating neuropathic pain.

4.2. Extracellular Signal-Regulated Kinase (ERK)

Extracellular signal-regulated kinase (ERK) is located downstream from MEK in the MAPK cascade and is a key enzyme within the ERK signaling pathway [88].

In the DRG Tlr8 knockout mouse model, decreased expression of miR-21 [56] downregulated Tlr8 and, consequently, led to decreased levels of ERK as well as attenuated neuropathic pain.

4.3. Adenosine Monophosphate-Activated Protein Kinase (AMPK)

Adenosine monophosphate-activated protein kinase (AMPK) is a protein kinase that plays a crucial role in regulating energy metabolism [125]. AMPK and AKT3 both work on TSC1/2 with opposing effects. AMPK activates TSC1/2, which leads to mTOR inhibition [126,127], while AKT3 inhibits TSC1/2, which leads to mTOR activation [128,129]. This balance between AMPK and AKT3 through TSC1/2 ensures regulated cellular responses.

In the CCI SNI model of rats, decreased levels of miR-142-3p led to an increase in AC9 levels, which in turn led to a decrease in cyclic AMP (cAMP) [57]. This decrease in cAMP increased AMPK expression, attenuating neuropathic pain.

4.4. Serum/Glucocorticoid Regulated Kinase Family Member 3 (SGK3)

Serum/glucocorticoid regulated kinase family member 3 (SGK3) phosphorylates several target proteins and has a role in neutral amino acid transport and activation of potassium and chloride channels [130,131]. SGK3 phosphorylates TSC, inhibiting TSC1/2 [132,133]. This inhibition leads to the activation of Rheb [134,135], which stimulates mTOR [133,134].

In the neuropathic pain models of bilateral CCI, upregulation of lncRNA CCA11 [58] suppressed miR-155 [59], upregulating SGK3 and attenuating neuropathic pain.

4.5. G Protein-Coupled Receptor Kinase 2 (GRK2)

G protein-coupled receptor kinase 2 (GRK2) plays a key role in GPCR signaling [110,111]. GRK2 negatively regulates AKT3 signaling through interactions with phosphatases such as protein phosphatase 2A (PP2A) [136–138], leading to AKT3 dephosphorylation and inactivation.

In the CCI model of mice, decreased levels of miR-15a/16 [55] led to increased GRK2 expression, attenuating neuropathic pain.

5. Concluding Remarks

In this review, we summarize key findings from the literature demonstrating causal relationships between miRNA regulation and neuropathic pain behaviors, with a specific focus on protein kinases as mediators of these effects. Two relationship patterns characterize miRNA modulation of neuropathic pain: (1) upregulation of miRNA attenuates neuropathic pain, largely through the suppression of pro-nociceptive protein kinases, and (2) downregulation of miRNA leads to pain attenuation, either by relieving the receptive regulation of anti-neuropathic signaling pathways, or by directly promoting their activation. These relationship patterns highlight the complex regulatory networks underlying the mechanisms for neuropathic pain, where protein kinases serve as critical molecular mediators in the cellular signaling cascades.

Given the high prevalence of neuropathic pain and a lack of effective therapies [6,7,9–17], there is an urgent need to develop novel treatment strategies that target the underlying molecular mechanisms of pain pathophysiology. Our discussion underscores the potential of miRNA-based approaches in this context, as miRNAs serve as upstream regulators and are capable of modulating multiple pain-related pathways simultaneously. The role of protein kinases as mediators of miRNA also points to the potential of targeting intracellular signaling molecules as a pain management strategy, offering the opportunity to identify kinase-specific interventions that may work independently or synergistically with miRNA-targeted therapies.

The therapeutic potential of miRNA modulation is increasingly being recognized [139–141]; specifically, both miRNA mimics and miRNA inhibitors can be used to manipulate miRNA levels, thus achieving the in vivo effect of either miRNA upregulation or miRNA downregulation [142–144]. In the context of neuropathic pain, the evidence summarized in this review suggests that both miRNA activation and inhibition could have therapeutic value, depending on the specific miRNA and its downstream targets. For miRNAs whose upregulation can lead to alleviation of neuropathic pain (Table 1, Figure 1), desirable therapeutic outcomes may be achieved with miRNA mimics, i.e., synthetic double-stranded RNA molecules that mimic the function of these endogenous miRNAs; similarly, for miRNAs whose downregulation can attenuate neuropathic pain (Table 2, Figure 2), favorable clinical results may be attained with miRNA inhibitors, i.e., single-stranded RNAs that are complementary to endogenous miRNAs, thus achieving the effect of gene silencing by specifically inhibiting these endogenous miRNAs. The ability to selectively regulate miRNA activity using miRNA modulators may open exciting avenues for the development of precision medicine approaches to neuropathic pain management.

Author Contributions: Conceptualization, L.Y.; data analysis, L.C. and Z.Č.; writing—original draft preparation, L.C., Z.Č. and L.Y.; writing—review and editing, L.C., Z.Č. and L.Y.; supervision, L.Y.; project administration, L.Y.; funding acquisition, L.Y. All authors have read and agreed to the published version of the manuscript.

Funding: This work was supported in part by grants from the National Institutes of Health of the United States (NS108887 and NS120617).

Institutional Review Board Statement: Not applicable.

Informed Consent Statement: Not applicable.

Data Availability Statement: No new data were created or analyzed in this study.

Acknowledgments: We thank Anuja Bahulekar, Esha Paghdal, Akash Patel, Kishan Patel, Shea Patel, and Yash Patel for helpful comments.

Conflicts of Interest: The authors declare no conflicts of interest.

References

1. Colloca, L.; Ludman, T.; Bouhassira, D.; Baron, R.; Dickenson, A.H.; Yarnitsky, D.; Freeman, R.; Truini, A.; Attal, N.; Finnerup, N.B.; et al. Neuropathic pain. *Nat. Rev. Dis. Primers* **2017**, *3*, 17002. [CrossRef]
2. Scholz, J.; Finnerup, N.B.; Attal, N.; Aziz, Q.; Baron, R.; Bennett, M.I.; Benoliel, R.; Cohen, M.; Cruccu, G.; Davis, K.D.; et al. The IASP classification of chronic pain for ICD-11: Chronic neuropathic pain. *Pain* **2019**, *160*, 53–59. [CrossRef] [PubMed]
3. Leone, C.M.; Truini, A. Understanding neuropathic pain: The role of neurophysiological tests in unveiling underlying mechanisms. *J. Anesth. Analg. Crit. Care* **2024**, *4*, 77. [CrossRef] [PubMed]
4. Ma, Y.C.; Kang, Z.B.; Shi, Y.Q.; Ji, W.Y.; Zhou, W.M.; Nan, W. The Complexity of Neuropathic Pain and Central Sensitization: Exploring Mechanisms and Therapeutic Prospects. *J. Integr. Neurosci.* **2024**, *23*, 89. [CrossRef] [PubMed]
5. Lima Pessôa, B.; Hauwanga, W.N.; Thomas, A.; Valentim, G.; McBenedict, B. A Comprehensive Narrative Review of Neuropathic Pain: From Pathophysiology to Surgical Treatment. *Cureus* **2024**, *16*, e58025. [CrossRef] [PubMed]
6. van Hecke, O.; Austin, S.K.; Khan, R.A.; Smith, B.H.; Torrance, N. Neuropathic pain in the general population: A systematic review of epidemiological studies. *Pain* **2014**, *155*, 654–662. [CrossRef] [PubMed]
7. Cavalli, E.; Mammana, S.; Nicoletti, F.; Bramanti, P.; Mazzon, E. The neuropathic pain: An overview of the current treatment and future therapeutic approaches. *Int. J. Immunopathol. Pharmacol.* **2019**, *33*, 2058738419838383. [CrossRef] [PubMed]
8. Bouhassira, D.; Lanteri-Minet, M.; Attal, N.; Laurent, B.; Touboul, C. Prevalence of chronic pain with neuropathic characteristics in the general population. *Pain* **2008**, *136*, 380–387. [CrossRef] [PubMed]
9. Baron, R.; Binder, A.; Wasner, G. Neuropathic pain: Diagnosis, pathophysiological mechanisms, and treatment. *Lancet Neurol.* **2010**, *9*, 807–819. [CrossRef] [PubMed]
10. Attal, N.; Bouhassira, D.; Baron, R. Diagnosis and assessment of neuropathic pain through questionnaires. *Lancet Neurol.* **2018**, *17*, 456–466. [CrossRef] [PubMed]
11. Feldman, A.; Weaver, J. Pharmacologic and Nonpharmacologic Management of Neuropathic Pain. *Semin. Neurol.* **2025**, *45*, 145–156. [CrossRef] [PubMed]
12. Dworkin, R.H.; O'Connor, A.B.; Backonja, M.; Farrar, J.T.; Finnerup, N.B.; Jensen, T.S.; Kalso, E.A.; Loeser, J.D.; Miaskowski, C.; Nurmikko, T.J.; et al. Pharmacologic management of neuropathic pain: Evidence-based recommendations. *Pain* **2007**, *132*, 237–251. [CrossRef]
13. O'Connor, A.B.; Dworkin, R.H. Treatment of neuropathic pain: An overview of recent guidelines. *Am. J. Med.* **2009**, *122*, S22–S32. [CrossRef] [PubMed]
14. Dworkin, R.H.; O'Connor, A.B.; Kent, J.; Mackey, S.C.; Raja, S.N.; Stacey, B.R.; Levy, R.M.; Backonja, M.; Baron, R.; Harke, H.; et al. Interventional management of neuropathic pain: NeuPSIG recommendations. *Pain* **2013**, *154*, 2249–2261. [CrossRef] [PubMed]
15. Andrejic, N.; Božovic, I.; Moradi, H.; Tataei, R.; Knezevic, N.N. Neuropathic pain management: A focused review of current treatments and novel data from main ongoing clinical trials. *Expert. Opin. Investig. Drugs* **2025**. Online ahead of print. [CrossRef] [PubMed]

16. Sadegh, A.A.; Gehr, N.L.; Finnerup, N.B. A systematic review and meta-analysis of randomized controlled head-to-head trials of recommended drugs for neuropathic pain. *Pain. Rep.* **2024**, *9*, e1138. [CrossRef] [PubMed]
17. Moisset, X. Neuropathic pain: Evidence based recommendations. *Presse Med.* **2024**, *53*, 104232. [CrossRef] [PubMed]
18. Pacifico, P.; Menichella, D.M. Molecular mechanisms of neuropathic pain. *Int. Rev. Neurobiol.* **2024**, *179*, 279–309. [CrossRef] [PubMed]
19. Rugnath, R.; Orzechowicz, C.; Newell, C.; Carullo, V.; Rugnath, A. A Literature Review: The Mechanisms and Treatment of Neuropathic Pain-A Brief Discussion. *Biomedicines* **2024**, *12*, 204. [CrossRef] [PubMed]
20. Bartel, D.P. MicroRNAs: Genomics, biogenesis, mechanism, and function. *Cell* **2004**, *116*, 281–297. [CrossRef] [PubMed]
21. Filipowicz, W.; Bhattacharyya, S.N.; Sonenberg, N. Mechanisms of post-transcriptional regulation by microRNAs: Are the answers in sight? *Nat. Rev. Genet.* **2008**, *9*, 102–114. [CrossRef] [PubMed]
22. Guo, H.; Ingolia, N.T.; Weissman, J.S.; Bartel, D.P. Mammalian microRNAs predominantly act to decrease target mRNA levels. *Nature* **2010**, *466*, 835–840. [CrossRef] [PubMed]
23. Suzuki, H.I. Roles of MicroRNAs in Disease Biology. *JMA J.* **2023**, *6*, 104–113. [CrossRef] [PubMed]
24. Bredy, T.W.; Lin, Q.; Wei, W.; Baker-Andresen, D.; Mattick, J.S. MicroRNA regulation of neural plasticity and memory. *Neurobiol. Learn. Mem.* **2011**, *96*, 89–94. [CrossRef] [PubMed]
25. Kalpachidou, T.; Kummer, K.K.; Kress, M. Non-coding RNAs in neuropathic pain. *Neuronal Signal* **2020**, *4*, NS20190099. [CrossRef] [PubMed]
26. Cohen, P. The role of protein phosphorylation in human health and disease. The Sir Hans Krebs Medal Lecture. *Eur. J. Biochem.* **2001**, *268*, 5001–5010. [CrossRef] [PubMed]
27. Ardito, F.; Giuliani, M.; Perrone, D.; Troiano, G.; Lo Muzio, L. The crucial role of protein phosphorylation in cell signaling and its use as targeted therapy (Review). *Int. J. Mol. Med.* **2017**, *40*, 271–280. [CrossRef] [PubMed]
28. Houles, T.; Yoon, S.O.; Roux, P.P. The expanding landscape of canonical and non-canonical protein phosphorylation. *Trends Biochem. Sci.* **2024**, *49*, 986–999. [CrossRef] [PubMed]
29. Ji, R.R.; Gereau, R.W.t.; Malcangio, M.; Strichartz, G.R. MAP kinase and pain. *Brain Res. Rev.* **2009**, *60*, 135–148. [CrossRef] [PubMed]
30. Mai, L.; Zhu, X.; Huang, F.; He, H.; Fan, W. p38 mitogen-activated protein kinase and pain. *Life Sci.* **2020**, *256*, 117885. [CrossRef] [PubMed]
31. Song, G.; Yang, Z.; Guo, J.; Zheng, Y.; Su, X.; Wang, X. Interactions Among lncRNAs/circRNAs, miRNAs, and mRNAs in Neuropathic Pain. *Neurotherapeutics* **2020**, *17*, 917–931. [CrossRef] [PubMed]
32. Gada, Y.; Pandey, A.; Jadhav, N.; Ajgaonkar, S.; Mehta, D.; Nair, S. New Vistas in microRNA Regulatory Interactome in Neuropathic Pain. *Front. Pharmacol.* **2021**, *12*, 778014. [CrossRef] [PubMed]
33. Jiang, M.; Wang, Y.; Wang, J.; Feng, S.; Wang, X. The etiological roles of miRNAs, lncRNAs, and circRNAs in neuropathic pain: A narrative review. *J. Clin. Lab. Anal.* **2022**, *36*, e24592. [CrossRef] [PubMed]
34. Morchio, M.; Sher, E.; Collier, D.A.; Lambert, D.W.; Boissonade, F.M. The Role of miRNAs in Neuropathic Pain. *Biomedicines* **2023**, *11*, 775. [CrossRef] [PubMed]
35. Zhao, Y.Y.; Wu, Z.J.; Zhu, L.J.; Niu, T.X.; Liu, B.; Li, J. Emerging roles of miRNAs in neuropathic pain: From new findings to novel mechanisms. *Front. Mol. Neurosci.* **2023**, *16*, 1110975. [CrossRef] [PubMed]
36. Sampath, K.K.; Belcher, S.; Hales, J.; Thomson, O.P.; Farrell, G.; Gisselman, A.S.; Katare, R.; Tumilty, S. The role of micro-RNAs in neuropathic pain-a scoping review. *Pain. Rep.* **2023**, *8*, e1108. [CrossRef] [PubMed]
37. Golmakani, H.; Azimian, A.; Golmakani, E. Newly discovered functions of miRNAs in neuropathic pain: Transitioning from recent discoveries to innovative underlying mechanisms. *Mol. Pain.* **2024**, *20*, 17448069231225845. [CrossRef] [PubMed]
38. Willemen, H.L.; Huo, X.J.; Mao-Ying, Q.L.; Zijlstra, J.; Heijnen, C.J.; Kavelaars, A. MicroRNA-124 as a novel treatment for persistent hyperalgesia. *J. Neuroinflammation* **2012**, *9*, 143. [CrossRef] [PubMed]
39. Huang, L.; Wang, L. Upregulation of miR-183 represses neuropathic pain through inhibiton of MAP3K4 in CCI rat models. *J. Cell Physiol.* **2020**, *235*, 3815–3822. [CrossRef] [PubMed]
40. Cai, L.; Liu, X.; Guo, Q.; Huang, Q.; Zhang, Q.; Cao, Z. MiR-15a attenuates peripheral nerve injury-induced neuropathic pain by targeting AKT3 to regulate autophagy. *Genes. Genomics* **2020**, *42*, 77–85. [CrossRef] [PubMed]
41. Cai, W.; Zhang, Y.; Liu, Y.; Liu, H.; Zhang, Z.; Su, Z. Effects of miR-150 on neuropathic pain process via targeting AKT3. *Biochem. Biophys. Res. Commun.* **2019**, *517*, 532–537. [CrossRef] [PubMed]
42. You, H.; Zhang, L.; Chen, Z.; Liu, W.; Wang, H.; He, H. MiR-20b-5p relieves neuropathic pain by targeting Akt3 in a chronic constriction injury rat model. *Synapse* **2019**, *73*, e22125. [CrossRef] [PubMed]
43. Xie, T.; Zhang, J.; Kang, Z.; Liu, F.; Lin, Z. miR-101 down-regulates mTOR expression and attenuates neuropathic pain in chronic constriction injury rat models. *Neurosci. Res.* **2020**, *158*, 30–36. [CrossRef] [PubMed]

44. Xie, X.; Ma, L.; Xi, K.; Zhang, W.; Fan, D. MicroRNA-183 Suppresses Neuropathic Pain and Expression of AMPA Receptors by Targeting mTOR/VEGF Signaling Pathway. *Cell Physiol. Biochem.* **2017**, *41*, 181–192. [CrossRef] [PubMed]
45. Hu, Y.; Liu, Q.; Zhang, M.; Yan, Y.; Yu, H.; Ge, L. MicroRNA-362-3p attenuates motor deficit following spinal cord injury via targeting paired box gene 2. *J. Integr. Neurosci.* **2019**, *18*, 57–64. [CrossRef] [PubMed]
46. Sun, W.; Zhang, L.; Li, R. Overexpression of miR-206 ameliorates chronic constriction injury-induced neuropathic pain in rats via the MEK/ERK pathway by targeting brain-derived neurotrophic factor. *Neurosci. Lett.* **2017**, *646*, 68–74. [CrossRef] [PubMed]
47. Jiang, B.C.; Cao, D.L.; Zhang, X.; Zhang, Z.J.; He, L.N.; Li, C.H.; Zhang, W.W.; Wu, X.B.; Berta, T.; Ji, R.R.; et al. CXCL13 drives spinal astrocyte activation and neuropathic pain via CXCR5. *J. Clin. Investig.* **2016**, *126*, 745–761. [CrossRef] [PubMed]
48. Xu, B.; Cao, J.; Zhang, J.; Jia, S.; Wu, S.; Mo, K.; Wei, G.; Liang, L.; Miao, X.; Bekker, A.; et al. Role of MicroRNA-143 in Nerve Injury-Induced Upregulation of Dnmt3a Expression in Primary Sensory Neurons. *Front. Mol. Neurosci.* **2017**, *10*, 350. [CrossRef] [PubMed]
49. Zhang, Y.; Su, Z.; Liu, H.L.; Li, L.; Wei, M.; Ge, D.J.; Zhang, Z.J. Effects of miR-26a-5p on neuropathic pain development by targeting MAPK6 in in CCI rat models. *Biomed. Pharmacother.* **2018**, *107*, 644–649. [CrossRef] [PubMed]
50. Wang, Z.; Liu, F.; Wei, M.; Qiu, Y.; Ma, C.; Shen, L.; Huang, Y. Chronic constriction injury-induced microRNA-146a-5p alleviates neuropathic pain through suppression of IRAK1/TRAF6 signaling pathway. *J. Neuroinflammation* **2018**, *15*, 179. [CrossRef] [PubMed]
51. Wang, L.; Zhu, K.; Yang, B.; Cai, Y. Knockdown of Linc00052 alleviated spinal nerve ligation-triggered neuropathic pain through regulating miR-448 and JAK1. *J. Cell Physiol.* **2020**, *235*, 6528–6535. [CrossRef] [PubMed]
52. Tan, Y.; Yang, J.; Xiang, K.; Tan, Q.; Guo, Q. Suppression of microRNA-155 attenuates neuropathic pain by regulating SOCS1 signalling pathway. *Neurochem. Res.* **2015**, *40*, 550–560. [CrossRef] [PubMed]
53. Xia, L.; Zhang, Y.; Dong, T. Inhibition of MicroRNA-221 Alleviates Neuropathic Pain Through Targeting Suppressor of Cytokine Signaling 1. *J. Mol. Neurosci.* **2016**, *59*, 411–420. [CrossRef] [PubMed]
54. Chang, L.L.; Wang, H.C.; Tseng, K.Y.; Su, M.P.; Wang, J.Y.; Chuang, Y.T.; Wang, Y.H.; Cheng, K.I. Upregulation of miR-133a-3p in the Sciatic Nerve Contributes to Neuropathic Pain Development. *Mol. Neurobiol.* **2020**, *57*, 3931–3942. [CrossRef] [PubMed]
55. Li, T.; Wan, Y.; Sun, L.; Tao, S.; Chen, P.; Liu, C.; Wang, K.; Zhou, C.; Zhao, G. Inhibition of MicroRNA-15a/16 Expression Alleviates Neuropathic Pain Development through Upregulation of G Protein-Coupled Receptor Kinase 2. *Biomol. Ther.* **2019**, *27*, 414–422. [CrossRef] [PubMed]
56. Zhang, Z.J.; Guo, J.S.; Li, S.S.; Wu, X.B.; Cao, D.L.; Jiang, B.C.; Jing, P.B.; Bai, X.Q.; Li, C.H.; Wu, Z.H.; et al. TLR8 and its endogenous ligand miR-21 contribute to neuropathic pain in murine DRG. *J. Exp. Med.* **2018**, *215*, 3019–3037. [CrossRef] [PubMed]
57. Li, X.; Wang, S.; Yang, X.; Chu, H. miR-142-3p targets AC9 to regulate sciatic nerve injury-induced neuropathic pain by regulating the cAMP/AMPK signalling pathway. *Int. J. Mol. Med.* **2021**, *47*, 561–572. [CrossRef] [PubMed]
58. Dou, L.; Lin, H.; Wang, K.; Zhu, G.; Zou, X.; Chang, E.; Zhu, Y. Long non-coding RNA CCAT1 modulates neuropathic pain progression through sponging miR-155. *Oncotarget* **2017**, *8*, 89949–89957. [CrossRef] [PubMed]
59. Liu, S.; Zhu, B.; Sun, Y.; Xie, X. MiR-155 modulates the progression of neuropathic pain through targeting SGK3. *Int. J. Clin. Exp. Pathol.* **2015**, *8*, 14374–14382. [PubMed]
60. Takekawa, M.; Posas, F.; Saito, H. A human homolog of the yeast Ssk2/Ssk22 MAP kinase kinase kinases, MTK1, mediates stress-induced activation of the p38 and JNK pathways. *EMBO J.* **1997**, *16*, 4973–4982. [CrossRef] [PubMed]
61. Nguyen, K.; Tran, M.N.; Rivera, A.; Cheng, T.; Windsor, G.O.; Chabot, A.B.; Cavanaugh, J.E.; Collins-Burow, B.M.; Lee, S.B.; Drewry, D.H.; et al. MAP3K Family Review and Correlations with Patient Survival Outcomes in Various Cancer Types. *Front. Biosci.* **2022**, *27*, 167. [CrossRef] [PubMed]
62. Huang, Y.; Wang, G.; Zhang, N.; Zeng, X. MAP3K4 kinase action and dual role in cancer. *Discov. Oncol.* **2024**, *15*, 99. [CrossRef] [PubMed]
63. Tournier, C.; Dong, C.; Turner, T.K.; Jones, S.N.; Flavell, R.A.; Davis, R.J. MKK7 is an essential component of the JNK signal transduction pathway activated by proinflammatory cytokines. *Genes. Dev.* **2001**, *15*, 1419–1426. [CrossRef] [PubMed]
64. Ho, D.T.; Bardwell, A.J.; Abdollahi, M.; Bardwell, L. A docking site in MKK4 mediates high affinity binding to JNK MAPKs and competes with similar docking sites in JNK substrates. *J. Biol. Chem.* **2003**, *278*, 32662–32672. [CrossRef] [PubMed]
65. Zheng, C.F.; Guan, K.L. Cloning and characterization of two distinct human extracellular signal-regulated kinase activator kinases, MEK1 and MEK2. *J. Biol. Chem.* **1993**, *268*, 11435–11439. [CrossRef] [PubMed]
66. Blank, J.L.; Gerwins, P.; Elliott, E.M.; Sather, S.; Johnson, G.L. Molecular cloning of mitogen-activated protein/ERK kinase kinases (MEKK) 2 and 3. Regulation of sequential phosphorylation pathways involving mitogen-activated protein kinase and c-Jun kinase. *J. Biol. Chem.* **1996**, *271*, 5361–5368. [CrossRef] [PubMed]

67. Gerwins, P.; Blank, J.L.; Johnson, G.L. Cloning of a novel mitogen-activated protein kinase kinase kinase, MEKK4, that selectively regulates the c-Jun amino terminal kinase pathway. *J. Biol. Chem.* **1997**, *272*, 8288–8295. [CrossRef] [PubMed]
68. Xu, B.; Wilsbacher, J.L.; Collisson, T.; Cobb, M.H. The N-terminal ERK-binding site of MEK1 is required for efficient feedback phosphorylation by ERK2 in vitro and ERK activation in vivo. *J. Biol. Chem.* **1999**, *274*, 34029–34035. [CrossRef] [PubMed]
69. Shaul, Y.D.; Seger, R. The MEK/ERK cascade: From signaling specificity to diverse functions. *Biochim. Biophys. Acta* **2007**, *1773*, 1213–1226. [CrossRef] [PubMed]
70. Zhang, W.; Liu, H.T. MAPK signal pathways in the regulation of cell proliferation in mammalian cells. *Cell Res.* **2002**, *12*, 9–18. [CrossRef] [PubMed]
71. Hu, Q.; Bian, Q.; Rong, D.; Wang, L.; Song, J.; Huang, H.S.; Zeng, J.; Mei, J.; Wang, P.Y. JAK/STAT pathway: Extracellular signals, diseases, immunity, and therapeutic regimens. *Front. Bioeng. Biotechnol.* **2023**, *11*, 1110765. [CrossRef] [PubMed]
72. Chen, W.; White, M.A.; Cobb, M.H. Stimulus-specific requirements for MAP3 kinases in activating the JNK pathway. *J. Biol. Chem.* **2002**, *277*, 49105–49110. [CrossRef] [PubMed]
73. Guma, M.; Firestein, G.S. c-Jun N-Terminal Kinase in Inflammation and Rheumatic Diseases. *Open Rheumatol. J.* **2012**, *6*, 220–231. [CrossRef] [PubMed]
74. Jenkins, B.J. Transcriptional regulation of pattern recognition receptors by Jak/STAT signaling, and the implications for disease pathogenesis. *J. Interferon Cytokine Res.* **2014**, *34*, 750–758. [CrossRef] [PubMed]
75. Thompson, H.J.; Lutsiv, T. Natural Products in Precision Oncology: Plant-Based Small Molecule Inhibitors of Protein Kinases for Cancer Chemoprevention. *Nutrients* **2023**, *15*, 1192. [CrossRef] [PubMed]
76. Guo, N.; Wang, X.; Xu, M.; Bai, J.; Yu, H.; Le, Z. PI3K/AKT signaling pathway: Molecular mechanisms and therapeutic potential in depression. *Pharmacol. Res.* **2024**, *206*, 107300. [CrossRef] [PubMed]
77. Soliman, G.A. The role of mechanistic target of rapamycin (mTOR) complexes signaling in the immune responses. *Nutrients* **2013**, *5*, 2231–2257. [CrossRef] [PubMed]
78. Simcox, J.; Lamming, D.W. The central moTOR of metabolism. *Dev. Cell* **2022**, *57*, 691–706. [CrossRef] [PubMed]
79. Dan, H.C.; Ebbs, A.; Pasparakis, M.; Van Dyke, T.; Basseres, D.S.; Baldwin, A.S. Akt-dependent activation of mTORC1 complex involves phosphorylation of mTOR (mammalian target of rapamycin) by IkappaB kinase alpha (IKKalpha). *J. Biol. Chem.* **2014**, *289*, 25227–25240. [CrossRef] [PubMed]
80. Tee, A.R.; Sampson, J.R.; Pal, D.K.; Bateman, J.M. The role of mTOR signalling in neurogenesis, insights from tuberous sclerosis complex. *Semin. Cell Dev. Biol.* **2016**, *52*, 12–20. [CrossRef] [PubMed]
81. Dunkerly-Eyring, B.L.; Pan, S.; Pinilla-Vera, M.; McKoy, D.; Mishra, S.; Grajeda Martinez, M.I.; Oeing, C.U.; Ranek, M.J.; Kass, D.A. Single serine on TSC2 exerts biased control over mTORC1 activation mediated by ERK1/2 but not Akt. *Life Sci. Alliance* **2022**, *5*, 1169. [CrossRef] [PubMed]
82. Cormerais, Y.; Lapp, S.C.; Kalafut, K.C.; Cisse, M.Y.; Shin, J.; Stefadu, B.; Personnaz, J.; Schrotter, S.; D'Amore, A.; Martin, E.R.; et al. AKT-mediated phosphorylation of TSC2 controls stimulus- and tissue-specific mTORC1 signaling and organ growth. *bioRxiv* **2024**. [CrossRef] [PubMed]
83. Polchi, A.; Magini, A.; Meo, D.D.; Tancini, B.; Emiliani, C. mTOR Signaling and Neural Stem Cells: The Tuberous Sclerosis Complex Model. *Int. J. Mol. Sci.* **2018**, *19*, 1474. [CrossRef] [PubMed]
84. Liao, W.T.; Chiang, Y.J.; Yang-Yen, H.F.; Hsu, L.C.; Chang, Z.F.; Yen, J.J.Y. CBAP regulates the function of Akt-associated TSC protein complexes to modulate mTORC1 signaling. *J. Biol. Chem.* **2023**, *299*, 105455. [CrossRef] [PubMed]
85. Wang, H.; Chi, L.; Yu, F.; Dai, H.; Si, X.; Gao, C.; Wang, Z.; Liu, L.; Zheng, J.; Ke, Y.; et al. The overview of Mitogen-activated extracellular signal-regulated kinase (MEK)-based dual inhibitor in the treatment of cancers. *Bioorg Med. Chem.* **2022**, *70*, 116922. [CrossRef] [PubMed]
86. Wortzel, I.; Seger, R. The ERK Cascade: Distinct Functions within Various Subcellular Organelles. *Genes. Cancer* **2011**, *2*, 195–209. [CrossRef] [PubMed]
87. Sun, Y.; Liu, W.Z.; Liu, T.; Feng, X.; Yang, N.; Zhou, H.F. Signaling pathway of MAPK/ERK in cell proliferation, differentiation, migration, senescence and apoptosis. *J. Recept. Signal Transduct. Res.* **2015**, *35*, 600–604. [CrossRef] [PubMed]
88. Seger, R.; Krebs, E.G. The MAPK signaling cascade. *FASEB J.* **1995**, *9*, 726–735. [CrossRef]
89. Roskoski, R., Jr. ERK1/2 MAP kinases: Structure, function, and regulation. *Pharmacol. Res.* **2012**, *66*, 105–143. [CrossRef] [PubMed]
90. Sanyanusin, P.; Norrish, J.H.; Ward, T.A.; Nebel, A.; McNoe, L.A.; Eccles, M.R. Genomic structure of the human PAX2 gene. *Genomics* **1996**, *35*, 258–261. [CrossRef] [PubMed]
91. Baldewijns, M.M.; van Vlodrop, I.J.; Vermeulen, P.B.; Soetekouw, P.M.; van Engeland, M.; de Bruine, A.P. VHL and HIF signalling in renal cell carcinogenesis. *J. Pathol.* **2010**, *221*, 125–138. [CrossRef] [PubMed]
92. Simon, M.A. Receptor tyrosine kinases: Specific outcomes from general signals. *Cell* **2000**, *103*, 13–15. [CrossRef] [PubMed]

93. Lai, S.; Pelech, S. Regulatory roles of conserved phosphorylation sites in the activation T-loop of the MAP kinase ERK1. *Mol. Biol. Cell* **2016**, *27*, 1040–1050. [CrossRef] [PubMed]
94. Flannery, S.; Bowie, A.G. The interleukin-1 receptor-associated kinases: Critical regulators of innate immune signalling. *Biochem. Pharmacol.* **2010**, *80*, 1981–1991. [CrossRef] [PubMed]
95. Guegan, J.P.; Fremin, C.; Baffet, G. The MAPK MEK1/2-ERK1/2 Pathway and Its Implication in Hepatocyte Cell Cycle Control. *Int. J. Hepatol.* **2012**, *2012*, 328372. [CrossRef] [PubMed]
96. Vollmer, S.; Strickson, S.; Zhang, T.; Gray, N.; Lee, K.L.; Rao, V.R.; Cohen, P. The mechanism of activation of IRAK1 and IRAK4 by interleukin-1 and Toll-like receptor agonists. *Biochem. J.* **2017**, *474*, 2027–2038. [CrossRef] [PubMed]
97. Chen, L.; Zheng, L.; Chen, P.; Liang, G. Myeloid Differentiation Primary Response Protein 88 (MyD88): The Central Hub of TLR/IL-1R Signaling. *J. Med. Chem.* **2020**, *63*, 13316–13329. [CrossRef] [PubMed]
98. Hsieh, C.H.; Jian, C.Z.; Lin, L.I.; Low, G.S.; Ou, P.Y.; Hsu, C.; Ou, D.L. Potential Role of CXCL13/CXCR5 Signaling in Immune Checkpoint Inhibitor Treatment in Cancer. *Cancers* **2022**, *14*, 294. [CrossRef] [PubMed]
99. Wang, B.; Wang, M.; Ao, D.; Wei, X. CXCL13-CXCR5 axis: Regulation in inflammatory diseases and cancer. *Biochim. Biophys. Acta Rev. Cancer* **2022**, *1877*, 188799. [CrossRef] [PubMed]
100. De la Mota-Peynado, A.; Chernoff, J.; Beeser, A. Identification of the atypical MAPK Erk3 as a novel substrate for p21-activated kinase (Pak) activity. *J. Biol. Chem.* **2011**, *286*, 13603–13611. [CrossRef] [PubMed]
101. Chen, C.; Nelson, L.J.; Avila, M.A.; Cubero, F.J. Mitogen-Activated Protein Kinases (MAPKs) and Cholangiocarcinoma: The Missing Link. *Cells* **2019**, *8*, 172. [CrossRef] [PubMed]
102. Hegazy, M.; Elkady, M.A.; Yehia, A.M.; Elsakka, E.G.E.; Abulsoud, A.I.; Abdelmaksoud, N.M.; Elshafei, A.; Abdelghany, T.M.; Elkhawaga, S.Y.; Ismail, A.; et al. The role of miRNAs in laryngeal cancer pathogenesis and therapeutic resistance—A focus on signaling pathways interplay. *Pathol. Res. Pract.* **2023**, *246*, 154510. [CrossRef] [PubMed]
103. McKay, M.M.; Morrison, D.K. Integrating signals from RTKs to ERK/MAPK. *Oncogene* **2007**, *26*, 3113–3121. [CrossRef] [PubMed]
104. Cargnello, M.; Roux, P.P. Activation and function of the MAPKs and their substrates, the MAPK-activated protein kinases. *Microbiol. Mol. Biol. Rev.* **2011**, *75*, 50–83. [CrossRef] [PubMed]
105. Lau, A.T.Y.; Xu, Y.M. Regulation of human mitogen-activated protein kinase 15 (extracellular signal-regulated kinase 7/8) and its functions: A recent update. *J. Cell Physiol.* **2018**, *234*, 75–88. [CrossRef] [PubMed]
106. Dietz, K.J.; Jacquot, J.P.; Harris, G. Hubs and bottlenecks in plant molecular signalling networks. *New Phytol.* **2010**, *188*, 919–938. [CrossRef] [PubMed]
107. Bogucka, K.; Pompaiah, M.; Marini, F.; Binder, H.; Harms, G.; Kaulich, M.; Klein, M.; Michel, C.; Radsak, M.P.; Rosigkeit, S.; et al. ERK3/MAPK6 controls IL-8 production and chemotaxis. *Elife* **2020**, *9*, e52511. [CrossRef] [PubMed]
108. Xiang, W.; Chao, Z.Y.; Feng, D.Y. Role of Toll-like receptor/MYD88 signaling in neurodegenerative diseases. *Rev. Neurosci.* **2015**, *26*, 407–414. [CrossRef] [PubMed]
109. Morris, R.; Kershaw, N.J.; Babon, J.J. The molecular details of cytokine signaling via the JAK/STAT pathway. *Protein Sci.* **2018**, *27*, 1984–2009. [CrossRef] [PubMed]
110. Ribas, C.; Penela, P.; Murga, C.; Salcedo, A.; Garcia-Hoz, C.; Jurado-Pueyo, M.; Aymerich, I.; Mayor, F., Jr. The G protein-coupled receptor kinase (GRK) interactome: Role of GRKs in GPCR regulation and signaling. *Biochim. Biophys. Acta* **2007**, *1768*, 913–922. [CrossRef] [PubMed]
111. Gurevich, V.V.; Gurevich, E.V. GPCR Signaling Regulation: The Role of GRKs and Arrestins. *Front. Pharmacol.* **2019**, *10*, 125. [CrossRef] [PubMed]
112. Tripathi, D.K.; Poluri, K.M. Molecular insights into kinase mediated signaling pathways of chemokines and their cognate G protein coupled receptors. *Front. Biosci.* **2020**, *25*, 1361–1385. [CrossRef] [PubMed]
113. Tao, J.; Jiang, C.; Guo, P.; Chen, H.; Zhu, Z.; Su, T.; Zhou, W.; Tai, Y.; Han, C.; Ma, Y.; et al. A novel GRK2 inhibitor alleviates experimental arthritis through restraining Th17 cell differentiation. *Biomed. Pharmacother.* **2023**, *157*, 113997. [CrossRef] [PubMed]
114. Tian, T.; Wei, M.; Guan, Y.; Rao, L.; Luo, T.; Han, C.; Wei, W.; Ma, Y. Paeoniflorin-6′-O-benzene sulfonate inhibits keratinocyte proliferation by restoring GRK2-JAK1 colocalization in mouse model of psoriasis. *Cell Signal* **2025**, *131*, 111706. [CrossRef] [PubMed]
115. Palikhe, S.; Ohashi, W.; Sakamoto, T.; Hattori, K.; Kawakami, M.; Andoh, T.; Yamazaki, H.; Hattori, Y. Regulatory Role of GRK2 in the TLR Signaling-Mediated iNOS Induction Pathway in Microglial Cells. *Front. Pharmacol.* **2019**, *10*, 59. [CrossRef] [PubMed]
116. Zarubin, T.; Han, J. Activation and signaling of the p38 MAP kinase pathway. *Cell Res.* **2005**, *15*, 11–18. [CrossRef] [PubMed]
117. Roux, P.P.; Blenis, J. ERK and p38 MAPK-activated protein kinases: A family of protein kinases with diverse biological functions. *Microbiol. Mol. Biol. Rev.* **2004**, *68*, 320–344. [CrossRef] [PubMed]
118. Yoshimura, A.; Naka, T.; Kubo, M. SOCS proteins, cytokine signalling and immune regulation. *Nat. Rev. Immunol.* **2007**, *7*, 454–465. [CrossRef] [PubMed]

119. Liang, Y.B.; Tang, H.; Chen, Z.B.; Zeng, L.J.; Wu, J.G.; Yang, W.; Li, Z.Y.; Ma, Z.F. Downregulated SOCS1 expression activates the JAK1/STAT1 pathway and promotes polarization of macrophages into M1 type. *Mol. Med. Rep.* **2017**, *16*, 6405–6411. [CrossRef] [PubMed]
120. Rane, S.G.; Reddy, E.P. Janus kinases: Components of multiple signaling pathways. *Oncogene* **2000**, *19*, 5662–5679. [CrossRef] [PubMed]
121. Bousoik, E.; Montazeri Aliabadi, H. "Do We Know Jack" About JAK? A Closer Look at JAK/STAT Signaling Pathway. *Front. Oncol.* **2018**, *8*, 287. [CrossRef] [PubMed]
122. Rawlings, J.S.; Rosler, K.M.; Harrison, D.A. The JAK/STAT signaling pathway. *J. Cell Sci.* **2004**, *117*, 1281–1283. [CrossRef] [PubMed]
123. Gorina, R.; Font-Nieves, M.; Marquez-Kisinousky, L.; Santalucia, T.; Planas, A.M. Astrocyte TLR4 activation induces a proinflammatory environment through the interplay between MyD88-dependent NFkappaB signaling, MAPK, and Jak1/Stat1 pathways. *Glia* **2011**, *59*, 242–255. [CrossRef] [PubMed]
124. Liau, N.P.D.; Laktyushin, A.; Lucet, I.S.; Murphy, J.M.; Yao, S.; Whitlock, E.; Callaghan, K.; Nicola, N.A.; Kershaw, N.J.; Babon, J.J. The molecular basis of JAK/STAT inhibition by SOCS1. *Nat. Commun.* **2018**, *9*, 1558. [CrossRef] [PubMed]
125. Ke, R.; Xu, Q.; Li, C.; Luo, L.; Huang, D. Mechanisms of AMPK in the maintenance of ATP balance during energy metabolism. *Cell Biol. Int.* **2018**, *42*, 384–392. [CrossRef] [PubMed]
126. Findlay, G.M.; Harrington, L.S.; Lamb, R.F. TSC1-2 tumour suppressor and regulation of mTOR signalling: Linking cell growth and proliferation? *Curr. Opin. Genet. Dev.* **2005**, *15*, 69–76. [CrossRef] [PubMed]
127. Chen, W.; Pan, Y.; Wang, S.; Liu, Y.; Chen, G.; Zhou, L.; Ni, W.; Wang, A.; Lu, Y. Cryptotanshinone activates AMPK-TSC2 axis leading to inhibition of mTORC1 signaling in cancer cells. *BMC Cancer* **2017**, *17*, 34. [CrossRef] [PubMed]
128. Lin, H.P.; Lin, C.Y.; Huo, C.; Jan, Y.J.; Tseng, J.C.; Jiang, S.S.; Kuo, Y.Y.; Chen, S.C.; Wang, C.T.; Chan, T.M.; et al. AKT3 promotes prostate cancer proliferation cells through regulation of Akt, B-Raf, and TSC1/TSC2. *Oncotarget* **2015**, *6*, 27097–27112. [CrossRef] [PubMed]
129. Sharma, A.; Mehan, S. Targeting PI3K-AKT/mTOR signaling in the prevention of autism. *Neurochem. Int.* **2021**, *147*, 105067. [CrossRef] [PubMed]
130. Lang, F.; Cohen, P. Regulation and physiological roles of serum- and glucocorticoid-induced protein kinase isoforms. *Sci. STKE* **2001**, *2001*, re17. [CrossRef] [PubMed]
131. Palmada, M.; Speil, A.; Jeyaraj, S.; Bohmer, C.; Lang, F. The serine/threonine kinases SGK1, 3 and PKB stimulate the amino acid transporter ASCT2. *Biochem. Biophys. Res. Commun.* **2005**, *331*, 272–277. [CrossRef] [PubMed]
132. Efeyan, A.; Sabatini, D.M. mTOR and cancer: Many loops in one pathway. *Curr. Opin. Cell Biol.* **2010**, *22*, 169–176. [CrossRef] [PubMed]
133. Bago, R.; Sommer, E.; Castel, P.; Crafter, C.; Bailey, F.P.; Shpiro, N.; Baselga, J.; Cross, D.; Eyers, P.A.; Alessi, D.R. The hVps34-SGK3 pathway alleviates sustained PI3K/Akt inhibition by stimulating mTORC1 and tumour growth. *EMBO J.* **2016**, *35*, 1902–1922. [CrossRef] [PubMed]
134. Nobukini, T.; Thomas, G. The mTOR/S6K signalling pathway: The role of the TSC1/2 tumour suppressor complex and the proto-oncogene Rheb. *Novartis Found. Symp.* **2004**, *262*, 148–154, discussion 54-9, 265–268. [PubMed]
135. Yang, Q.; Inoki, K.; Kim, E.; Guan, K.L. TSC1/TSC2 and Rheb have different effects on TORC1 and TORC2 activity. *Proc. Natl. Acad. Sci. USA* **2006**, *103*, 6811–6816. [CrossRef] [PubMed]
136. Ugi, S.; Imamura, T.; Maegawa, H.; Egawa, K.; Yoshizaki, T.; Shi, K.; Obata, T.; Ebina, Y.; Kashiwagi, A.; Olefsky, J.M. Protein phosphatase 2A negatively regulates insulin's metabolic signaling pathway by inhibiting Akt (protein kinase B) activity in 3T3-L1 adipocytes. *Mol. Cell Biol.* **2004**, *24*, 8778–8789. [CrossRef] [PubMed]
137. Andrabi, S.; Gjoerup, O.V.; Kean, J.A.; Roberts, T.M.; Schaffhausen, B. Protein phosphatase 2A regulates life and death decisions via Akt in a context-dependent manner. *Proc. Natl. Acad. Sci. USA* **2007**, *104*, 19011–19016. [CrossRef] [PubMed]
138. Bao, F.; Hao, P.; An, S.; Yang, Y.; Liu, Y.; Hao, Q.; Ejaz, M.; Guo, X.X.; Xu, T.R. Akt scaffold proteins: The key to controlling specificity of Akt signaling. *Am. J. Physiol. Cell Physiol.* **2021**, *321*, C429–C442. [CrossRef] [PubMed]
139. Schmidt, M.F. miRNA Targeting Drugs: The Next Blockbusters? *Methods Mol. Biol.* **2017**, *1517*, 3–22. [CrossRef] [PubMed]
140. Bonneau, E.; Neveu, B.; Kostantin, E.; Tsongalis, G.J.; De Guire, V. How close are miRNAs from clinical practice? A perspective on the diagnostic and therapeutic market. *Ejifcc* **2019**, *30*, 114–127. [PubMed]
141. Sun, X.; Setrerrahmane, S.; Li, C.; Hu, J.; Xu, H. Nucleic acid drugs: Recent progress and future perspectives. *Signal Transduct. Target. Ther.* **2024**, *9*, 316. [CrossRef] [PubMed]
142. Chakraborty, C.; Sharma, A.R.; Sharma, G.; Lee, S.S. Therapeutic advances of miRNAs: A preclinical and clinical update. *J. Adv. Res.* **2021**, *28*, 127–138. [CrossRef] [PubMed]

143. Holjencin, C.; Jakymiw, A. MicroRNAs and Their Big Therapeutic Impacts: Delivery Strategies for Cancer Intervention. *Cells* **2022**, *11*, 2332. [CrossRef] [PubMed]
144. Kim, T.; Croce, C.M. MicroRNA: Trends in clinical trials of cancer diagnosis and therapy strategies. *Exp. Mol. Med.* **2023**, *55*, 1314–1321. [CrossRef] [PubMed]

cells

MDPI

Review

Molecular Mechanisms and Pathways in Visceral Pain

Qiqi Zhou [1,2,*] and George Nicholas Verne [1,2]

[1] College of Medicine, University of Tennessee, Memphis, TN 38163, USA; gverne@uthsc.edu
[2] Lt. Col. Luke Weathers, Jr. VA Medical Center, Memphis, TN 38105, USA
* Correspondence: qzhou12@uthsc.edu

Abstract: Chronic visceral pain, a significant contributor to morbidity in the United States, affects millions and results in substantial economic costs. Despite its impact, the mechanisms underlying disorders of gut–brain interaction (DGBIs), such as irritable bowel syndrome (IBS), remain poorly understood. Visceral hypersensitivity, a hallmark of chronic visceral pain, involves an enhanced pain response in internal organs to normal stimuli. Various factors like inflammation, intestinal hyperpermeability, and epigenetic modifications influence its presentation. Emerging evidence suggests that persistent colonic stimuli, disrupted gut barriers, and altered non-coding RNA (ncRNA) expression contribute to the pathophysiology of visceral pain. Additionally, cross-sensitization of afferent pathways shared by pelvic organs underpins the overlap of chronic pelvic pain disorders, such as interstitial cystitis and IBS. Central sensitization and viscerosomatic convergence further exacerbate pain, with evidence showing IBS patients exhibit hypersensitivity to both visceral and somatic stimuli. The molecular mechanisms of visceral pain involve critical mediators such as cytokines, prostaglandins, and neuropeptides, alongside ion channels like transient receptor potential vanilloid 1 (TRPV1) and acid-sensing ion channels (ASICs). These molecular insights indicate potential therapeutic targets and highlight the possible use of TRPV1 antagonists and ASIC inhibitors to mitigate visceral pain. This review explores the neurophysiological pathways of visceral pain, focusing on peripheral and central sensitization mechanisms, to advance the development of targeted treatments for chronic pain syndromes, particularly IBS and related disorders.

Keywords: disorders of gut–brain interaction; irritable bowel syndrome; visceral hypersensitivity; ncRNA

1. Chronic Visceral Pain: Mechanisms and Overlapping Pathologies

Chronic visceral pain is among the most debilitating functional disorders, impacting an estimated 100 million individuals in the United States. This condition, which underlies disorders of gut–brain interaction (DGBIs) such as irritable bowel syndrome (IBS), imposes significant healthcare and economic burdens, with costs exceeding USD 700 billion annually. Beyond the financial implications, the condition dramatically impairs the quality of life, manifesting in persistent discomfort, emotional distress, and reduced productivity. Despite its widespread prevalence and severe repercussions, the underlying mechanisms driving DGBIs and chronic visceral pain remain poorly understood, posing challenges to the development of effective therapeutic interventions [1–4].

While significant strides have been made in understanding DGBIs, the pathophysiology of visceral pain remains elusive [5]. Several mechanisms may contribute to this condition, including neuroinflammation, peripheral sensitization, and gut barrier dysfunction. Neuroinflammation is thought to exacerbate pain signals by altering neural pathways, while peripheral sensitization increases the responsiveness of nociceptors to stimuli. Meanwhile, gut barrier dysfunction may facilitate the translocation of antigens and pathogens that trigger immune responses and sensitize visceral nerves [6]. However, these mechanisms are often studied in isolation, and the field still lacks an integrated framework that cohesively explains their interactions. Addressing this gap is crucial for advancing targeted therapies and improving outcomes for individuals living with these chronic conditions.

This review aims to explore the molecular and physiological underpinnings of visceral pain, emphasizing the neurophysiological pathways and the contributions of central and peripheral sensitization. By bridging basic science with clinical implications, it seeks to provide a comprehensive perspective on the mechanisms of visceral pain, laying the groundwork for the development of targeted and effective therapeutic strategies. Advancing our understanding of these mechanisms is essential for addressing the unmet needs of patients and improving their quality of life.

2. Pathophysiology of Visceral Pain and Hypersensitivity

Chronic abdominal pain, which is the leading cause of visits to gastroenterology clinics, is most commonly attributed to DGBIs, particularly IBS [7]. While IBS is characterized by abdominal pain associated with altered bowel habits, it is also frequently accompanied by an exaggerated pain response to otherwise normal stimuli in the gut and other systems [4], known as visceral hypersensitivity or hyperalgesia. In these cases, even mild stimuli (e.g., distention or mild contractions) are perceived as intensely painful, a phenomenon often accompanied by allodynia (pain in response to non-painful stimuli). This extended response goes beyond the gastrointestinal (GI) tract, implicating multiple interconnected pathways. Heightened sensitivity represents a maladaptive neural response to environmental factors, previous injury, or stress, and highlights the complex interaction between peripheral mechanisms, such as those driven by gut inflammation, and central mechanisms within the nervous system.

Emerging evidence has identified several critical mechanisms that contribute to visceral hypersensitivity and pain, each contributing to a multi-level interaction between the peripheral and central nervous systems:

1. Neuronal sensitization: Persistent, abnormal stimuli from the colon can lead to prolonged hypersensitivity by sensitizing spinal neurons [8,9]. This process involves viscerosomatic convergence, where nociceptive input from visceral structures (e.g., the gut and other internal organs) overlaps with somatic input (e.g., from the skin, muscles, and soft tissues), leading to a compound response. The sustained input from the colon can "prime" spinal neurons, making them more responsive to future stimuli [9].
2. Increased intestinal permeability: The gut barrier is crucial for maintaining immune homeostasis and protecting against harmful stimuli [10]. However, barrier disruptions, such as those seen in conditions like IBS, can contribute to pain by allowing pathogenic or inflammatory mediators (e.g., cytokines, bacteria) to interact with afferent nerve fibers [6]. This "leaky gut" phenomenon may not only heighten nociception but also activate immune responses that further sensitize visceral pathways [10].
3. Epigenetic influences: Recent studies have illuminated the role of epigenetic regulation in the pathophysiology of visceral pain [11]. Altered expression of microRNAs (miRNAs) in GI tissues, potentially delivered via extracellular vesicles (EVs), may affect the expression of pain-related genes. These small RNA molecules can modulate

pain signaling pathways at both the peripheral and central levels, adding another layer of complexity to visceral hypersensitivity mechanisms.

These factors demonstrate the dynamic interplay between peripheral and central mechanisms and underscore the importance of studying both to fully understand chronic visceral pain [12]. This dual-layered process, involving both the gut and the brain, challenges current therapeutic approaches that often focus on only one aspect while excluding the other.

2.1. Mechanisms of Neuronal Sensitization

Ongoing research continues to unravel the complexities of visceral pain mechanisms, with a growing focus on peripheral and central sensitization. The interplay between these processes underscores the need for a comprehensive approach that addresses the full spectrum of factors contributing to chronic pain. The roles of silent nociceptors, immune–neural interactions, and epigenetic modifications are among the emerging areas of interest, offering a more nuanced understanding of visceral hypersensitivity and chronic pain syndromes.

2.1.1. Afferent Mechanisms of Visceral Pain

Primary visceral afferents in the gut are pivotal in the development and maintenance of chronic visceral hypersensitivity, as shown by studies in both human and animal models [13]. These specialized sensory neurons convey stimuli from the gut to the central nervous system (CNS), forming the basis for gut sensory processing [2]. Their receptors are distributed across the serosal, muscular, and mucosal layers of the GI tract, allowing them to detect mechanical (e.g., distension), chemical, and luminal stimuli. While most GI input remains below conscious perception, pathological triggers such as inflammation, trauma, or environmental stressors can sensitize the gut, heightening its response to luminal distension. This sensitization underlies the visceral hypersensitivity observed in IBS patients, contributing to symptoms such as bloating, abdominal pain, and altered bowel habits.

Silent nociceptors are primary visceral afferents that have a particularly significant role in this process. They are normally inactive but become mechanosensitive and spontaneously active following tissue injury, creating a feedback loop that perpetuates visceral hypersensitivity [12]. This transformation is critical in the transition from acute nociceptive pain to chronic visceral hypersensitivity, implicating both peripheral and central mechanisms in the maintenance of pain. For instance, acute introduction of bile salts into the colon amplifies mechanosensitive colonic afferent firing during distension, leading to exaggerated pain responses and illustrating the interplay between chemical and mechanical sensitization [8].

Recent insights suggest that while acute injury often results in temporary mechanosensitization, persistent hyperalgesia may develop from prolonged or recurrent tissue damage. Animal models of colitis and visceral hypersensitivity demonstrate that colonic irritation can trigger long-lasting sensitization of the gut, even in the absence of ongoing inflammation or structural abnormalities [13]. Ongoing afferent input from peripheral sources leads to various pathological outcomes, including spontaneous motor abnormalities, hyperalgesia, pain, and allodynia [14]. Additionally, transient inflammation of the colon that leads to colonic distension can initiate sustained visceral hypersensitivity, with heightened abdominal muscle contractility and hyperexcitability of viscerosensitive neurons in the lumbosacral spinal cord (L6-S1) [9]. Agents such as lidocaine, which block sensitized visceral afferents, reverse these changes [14]. Significant alterations also occurred in the signaling pathways within the spinal cord, including the upregulation of pro-inflammatory mediators, which sustain pain processing even in the absence of ongoing inflammation in the gut [12,14]. Clinical observations in conditions such as complex regional pain syndrome further corroborate findings from these models [15]. In these cases, administration of peripherally applied anesthetics alleviates widespread pain and hypersensitivity, highlighting the role of nociceptive input in maintaining central sensitization.

Notably, continuous impulses from nociceptive colonic afferent neurons may partially sustain widespread zones of hypersensitivity observed in conditions such as fibromyalgia, neuropathic pain, and IBS. This hypothesis was evaluated in a double-blind crossover trial using intracolonic lidocaine jelly in IBS patients [16]. The trial showed significant reductions in hypersensitivity to nociceptive colonic distension and thermal stimuli applied to the foot, underscoring the role of tonic afferent input from the gut in secondary somatic hypersensitivity. Similar findings in animal models of IBS revealed that intracolonic lidocaine normalized both colonic and thermal hypersensitivity without detectable systemic lidocaine levels, indicating localized effects [13]. These findings underscore the contribution of persistent peripheral input from the gut to both primary visceral and secondary somatic hypersensitivity, as exemplified in IBS patients.

Beyond local tissue injury, systemic and developmental factors contribute to chronic hyperalgesia. For instance, neonatal colonic irritation in animal models leads to persistent visceral hypersensitivity and central sensitization, effects not observed in adults [13]. This observation suggests that there is a critical window during development when the nervous system is particularly susceptible to long-term alterations. Additionally, these findings align with clinical observations that approximately 25% of adults develop IBS following an enteric infection [7]. Transient inflammation of the small bowel or colon during such infections can result in prolonged gut sensitization, even after the resolution of active infection.

Afferent sensitization is often accompanied by central amplification, where spinal cord neurons receiving continuous input from sensitized visceral afferents become more responsive to incoming pain signals, exacerbating the perception of pain. The interplay between these peripheral and central processes creates a vicious cycle of hypersensitivity, where even mild stimuli can trigger intense pain [15]. This intricate relationship between the gut and the nervous system is central to the pathogenesis of a variety of chronic pain conditions, including IBS, interstitial cystitis, and ureteric colic [17]. In each of these conditions, the normal processing of sensory information is disrupted, leading to the perception of pain in response to stimuli that would typically be non-painful. The mechanisms underlying these disorders are multifactorial, involving both altered visceral afferent signaling and changes in central pain processing pathways.

2.1.2. Central Sensitization and Viscerosomatic Convergence

Central sensitization, where spinal cord and brain neurons exhibit heightened excitability and hyperalgesic responses (increased pain sensitivity) to noxious stimuli, plays a central role in the development and persistence of chronic visceral pain [18,19]. This phenomenon amplifies and prolongs the perception of pain through increased glutamate release, N-methyl-D-aspartate (NMDA) receptor activation, and altered inhibitory signaling. Persistent activation of NMDA receptors and long-term potentiation (LTP) of synaptic transmission are critical components of spinal cord plasticity, promoting neuronal hyperexcitability and reinforcing synaptic strength [12,18,19]. Glial cell activation further exacerbates central sensitization. Microglia and astrocytes, activated by ongoing nociceptive input, release pro-inflammatory cytokines and chemokines (e.g., interleukin-1 beta (IL-1β), tumor necrosis factor-alpha (TNF-α)), perpetuating neuroinflammation and pain hypersensitivity [12]. This glial–neuronal crosstalk creates a self-reinforcing loop, contributing to the persistence of pain. Understanding these processes has informed therapeutic approaches targeting central sensitization, including receptor antagonists, ion channel blockers, and anti-inflammatory agents. In patients with IBS, this mechanism contributes to the heightened sensitivity to both visceral and somatic stimuli, suggesting that central sensitization is not merely a peripheral phenomenon but a systemic disorder [4].

Central sensitization is closely linked to viscerosomatic convergence, wherein nociceptive pathways from both the gut and somatic tissues share common spinal segments, particularly in the lumbosacral region. Experimental evidence has shown that sensory neurons in these regions can transmit pain signals originating from both visceral and somatic structures [4,17–19]. Studies further confirm that the overlap of these nociceptive pathways contributes to somatic hyperalgesia, where patients exhibit increased pain perception in areas innervated by the same spinal segments [2,15]. This overlap is thought to be responsible for the referred pain often reported by IBS patients whose discomfort extends beyond the gut to other parts of the body, such as the lower back or pelvis, and reflects the broader involvement of central mechanisms in IBS pathophysiology [20–22].

2.1.3. Overlap with Chronic Pelvic Pain Disorders

The pathophysiology of chronic visceral pain is not restricted to the GI system; it frequently overlaps with other chronic pelvic pain disorders, including conditions like interstitial cystitis, pelvic floor dysfunction, and chronic pelvic pain syndrome [21]. Animal models have demonstrated that localized inflammatory injury in one visceral organ (e.g., the colon) can lead to sensitization of afferent pathways shared by other pelvic organs. This cross-sensitization between different pelvic organs, such as the bladder and intestines, provides a mechanistic basis for the clinical overlap of symptoms in patients with disorders like IBS and interstitial cystitis.

The concept of neural crosstalk, where afferent fibers from different organs converge at the spinal cord, reinforces the idea that chronic pelvic pain is a manifestation of a broader systemic disorder involving aberrant pain processing at both peripheral and central levels. Studies have further demonstrated that the primary visceral afferent pathways, which carry pain signals from the intestines, also contribute to altered sensations in other pelvic viscera (e.g., bladder) [2,8]. This suggests that chronic pelvic pain syndromes are likely to share common neural mechanisms that extend beyond the GI system.

2.2. *Intestinal Barrier Dysfunction and Visceral Pain*

In health, the GI tract serves as a barrier against harmful substances and supports nutrient absorption. However, disruptions in this barrier can lead to hyperpermeability, bacterial translocation, sepsis, and systemic inflammation, which have been implicated in conditions like celiac disease, inflammatory bowel disease, and IBS [6,7]. Post-infectious IBS (PI-IBS), particularly its diarrhea-predominant subtype (PI-IBS-D), is associated with chronic mucosal inflammation, elevated cytokine levels, and mast cell activation [10,16]. These factors increase intestinal permeability, allowing bacteria and antigens to penetrate the mucosal layer, triggering immune responses that exacerbate pain and diarrhea. This heightened permeability establishes a feedback loop wherein inflammatory mediators, bacteria, and antigens sustain afferent signaling to the spinal cord, potentially sensitizing spinal segments and perpetuating central sensitization (Figure 1). Targeting this interplay between intestinal permeability and spinal sensitization offers a promising avenue for treating PI-IBS-D and other disorders characterized by visceral pain [22].

Loss of the downregulated in adenoma (DRA; SLC26A3) protein compromises the intestinal epithelial barrier by reducing tight junction (TJ) and adherens junction proteins, such as ZO-1, occludin, and E-cadherin, thereby increasing colonic permeability [23]. DRA deficiency is associated with gut dysbiosis and microbial changes that partially affect TJ protein expression. Enhanced binding of the CUG triplet repeat RNA-binding protein 1 (CUGBP1) to occludin and E-cadherin genes in DRA knockout (KO) mice suggests posttranscriptional mechanisms contribute to barrier dysfunction and IBD pathogenesis [24,25].

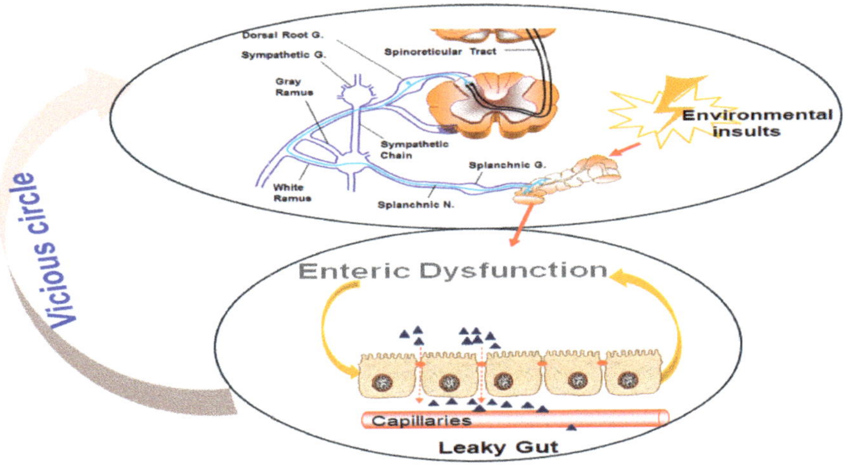

Figure 1. The vicious cycle of environmental insults, enteric dysfunction, and visceral hypersensitivity. (**Upper panel**) A neural pathway involving the splanchnic nerves and sympathetic chain, which respond to environmental insults, triggers nociceptive signaling. These signals are relayed through the DRG and spinoreticular tract, further exacerbating the stress response. (**Lower panel**) Enteric dysfunction is characterized by leaky gut syndrome, where compromised intestinal epithelial barrier integrity allows increased permeability to harmful molecules. This dysfunction amplifies systemic inflammation and sensitizes neural pathways, forming a feedback loop (vicious cycle) that perpetuates visceral hypersensitivity and chronic gut dysfunction.

Elevated vault RNA (vtRNA)1-1 levels, observed in EVs from shock patients and septic mice, impair intestinal epithelial renewal and barrier function by reducing intercellular junction proteins and Paneth cells. Mechanistically, vtRNA1-1 interacts with CUGBP1, increasing its association with mRNAs of Claudin-1, a TJ protein essential for maintaining intestinal barrier integrity, and occludin, thereby inhibiting their expression and exacerbating gut barrier dysfunction. These findings reveal a novel vtRNA1-1/CUGBP1 axis contributing to gut mucosal disruption during critical illness, providing insights into potential therapeutic targets for intestinal barrier restoration [26].

2.3. Epigenetic Mechanisms in Visceral Pain

Emerging evidence underscores the crucial role of epigenetic and non-coding RNA (ncRNA) mechanisms, particularly microRNAs (miRNAs), in the regulation of visceral pain. miRNAs are small, ncRNA molecules that fine-tune gene expression and affect a wide range of cellular processes by either degrading target messenger RNAs (mRNAs) or repressing their translation [27]. Through these mechanisms, miRNAs modulate critical cellular functions such as differentiation, proliferation, apoptosis, and neuroplasticity, which are essential for maintaining homeostasis in both the peripheral and central nervous systems.

miRNAs help regulate visceral pain through the molecular pathways that govern pain signaling and neuroplasticity. For instance, the miR-29 family, which is involved in physiological regulation processes, modulates enzymes that control DNA methylation, a key epigenetic modification that can alter gene expression. Through this regulation, miR-29 affects critical pathways involved in intestinal permeability, an essential factor in the development of chronic GI disorders. Increased intestinal permeability, often referred to as "leaky gut," is a hallmark of conditions like IBS, where the gut barrier function is compromised, allowing noxious stimuli and inflammatory mediators to interact with sensory nerves and contribute to visceral hypersensitivity [28]. Additionally, in IBS-D, upregulated miR-29a/b levels have been linked to reduced expression of Claudin-1 [29]. This downregulation contributes to increased permeability and chronic hypersensitivity. Silencing the miR-29 cluster has been shown to reverse intestinal hyperpermeability, indicating that miR-29 represents a potential therapeutic target.

The circular RNA (circRNA) Cdr1as acts as a repressor of intestinal epithelial regeneration and defense, with levels increasing in conditions such as colitis and sepsis in both mice and humans. Ablation of Cdr1as enhances intestinal mucosal renewal, promotes injury-induced epithelial regeneration, and provides protection against colitis. The inhibitory effects of Cdr1as on epithelial repair are mediated, at least partially, through interactions with miR-195, highlighting its role in impaired mucosal renewal [30].

IL-1β increases intestinal permeability by upregulating the miRNA MIR200C-3p, which suppresses occludin expression in enterocytes, disrupting the TJ barrier. In mice and patients with colitis, elevated IL-1β and MIR200C-3p levels correlate with reduced occludin and increased TJ permeability, a hallmark of intestinal inflammation. Targeting MIR200C-3p with an antagonist preserves occludin expression, reduces permeability, and mitigates barrier dysfunction, offering potential therapeutic benefits for colitis [31].

miRNAs also play a significant role in the regulation of pain pathways within the CNS. miRNAs can modulate the expression of genes involved in neuroplasticity, the ability of the nervous system to reorganize itself in response to injury or environmental changes [27]. In the context of chronic visceral pain, altered miRNA expression may contribute to the sensitization of spinal cord neurons, amplifying pain signals and promoting the development of central sensitization in chronic pain disorders, and can lead to the persistence of pain long after the initial injury or inflammation has subsided. Moreover, miRNAs can also influence the inflammatory milieu within the gut and CNS. By regulating the expression of cytokines, chemokines, and other inflammatory mediators, miRNAs contribute to the modulation of the immune response, which is often dysregulated in individuals with chronic visceral pain [10]. This dysregulation may exacerbate pain and prolong the sensitization of both peripheral and central pain pathways. miRNAs also play a role in modulating nociception and pain thresholds. For instance, miR-328 and miR-320 downregulate the neurokinin-1 receptor (NK1R) in bladder pain syndrome, while let-7 miRNAs influence opioid tolerance by targeting mu-opioid receptors [32,33].

The complexity of miRNA regulation suggests that these molecules could serve as both biomarkers for diagnosing chronic visceral pain disorders and therapeutic targets for novel interventions [33]. By selectively modulating specific miRNAs, it may be possible to restore normal gene expression patterns and alleviate the persistent pain associated with conditions such as IBS, interstitial cystitis, and other FGIDs.

miRNAs thus represent a crucial layer of regulation in the pathophysiology of chronic visceral pain. Their ability to influence gene expression and cellular processes, particularly in the context of intestinal permeability, pain signaling, and neuroplasticity, positions them as key players in the development and persistence of visceral hypersensitivity. Further research into the specific roles of individual miRNAs in these pathways may provide valuable insights into the molecular mechanisms underlying chronic visceral pain and open up new avenues for targeted therapeutic strategies.

3. Key Molecular Mediators and Possible Targets for Treatment of Visceral Pain

Current treatments for chronic visceral pain have limited efficacy [7], which highlights the pressing need for a deeper understanding of its underlying mechanisms. Insights into primary afferent physiology, neural crosstalk, and the integration of peripheral and central pain pathways hold significant promise for therapeutic innovation. Addressing the multifaceted nature of visceral pain through comprehensive research can pave the way for novel treatments, ultimately improving outcomes for patients with IBS and other DGBIs.

Recent advancements have identified potential therapeutic targets, such as specific ion channels, receptors, and signaling molecules that mediate visceral hypersensitivity [10]. By modulating these pathways, it may be possible to develop more effective treatments for chronic visceral pain, aimed not only at alleviating symptoms but also at addressing the underlying mechanisms of pain generation and maintenance. Furthermore, the growing understanding of the role of the CNS in amplifying visceral pain suggests that treatments targeting central sensitization, such as neuromodulatory therapies, may offer promise for patients suffering from conditions like IBS. In addition, approaches aimed at restoring gut barrier function or modulating the gut microbiota may hold potential for preventing or reversing the peripheral sensitization of visceral afferents.

3.1. Neurotransmitters and Neuromodulators in Pain Signaling

3.1.1. Glutamate and Gamma-Aminobutyric Acid (GABA)

Glutamate and GABA are fundamental to excitatory and inhibitory signaling in the CNS, particularly in pain modulation [15]. Glutamate, the primary excitatory neurotransmitter, is crucial for transmitting nociceptive signals. It works through ionotropic receptors, such as the NMDA and α-amino-3-hydroxy-5-methyl-4-isoxazolepropionic acid (AMPA) receptors, which mediate rapid synaptic transmission, and metabotropic glutamate receptors (mGluRs), which regulate synaptic plasticity. Enhanced glutamatergic signaling in the spinal cord is vital for central sensitization, a process characterized by heightened nociceptive neuron responsiveness. This phenomenon, involving sustained activity-dependent synaptic changes, including phosphorylation of NMDA receptors and increased AMPA receptor trafficking, has been shown to amplify excitatory signaling [14]. These mechanisms reflect pathological neural plasticity in response to persistent nociceptive inputs, thereby contributing to visceral hypersensitivity.

In contrast, GABA serves as the principal inhibitory neurotransmitter that maintains neural homeostasis by counterbalancing excitatory signals. GABAergic inhibition occurs through GABA-A (ionotropic) and GABA-B (metabotropic) receptors that suppress nociceptive transmission by hyperpolarizing postsynaptic neurons and reducing neurotransmitter release. Dysfunction in GABAergic pathways, such as downregulated receptor expression or altered chloride homeostasis, leads to increased pain sensitivity. GABAergic neurons in the ventral spinal cord are also targets for μ-opioid receptor-mediated presynaptic inhibition, indicating that GABA signaling and analgesic mechanisms interact [15].

These disruptions likely contribute to conditions like IBS, where impaired inhibitory control heightens pain perception. Together, glutamate-driven excitation and GABA-mediated inhibition are key regulators of nociceptive signaling and represent therapeutic targets for chronic pain management.

3.1.2. Substance P and Calcitonin Gene-Related Peptide (CGRP)

In nociceptive signaling, substance P is a critical neuropeptide that is produced in response to noxious stimuli [34–36]. Released by primary sensory neurons, it binds to neurokinin-1 (NK1) receptors to amplify pain transmission. Substance P also mediates neurogenic inflammation, contributing to the sensitization associated with chronic pain. By facilitating communication between the peripheral and central nervous systems, substance P plays a dual role in pain propagation and inflammation, making it a key target in managing visceral pain syndromes.

CGRP, often co-released with substance P, complements its effects by targeting CGRP receptors. It induces vasodilation and promotes inflammatory responses, integral to both peripheral and central sensitization. CGRP's role in migraine-related visceral symptoms and GI pain disorders underscores its importance in the vascular-neural interplay underlying visceral pain. CGRP and substance P can also combine synergistically to contribute to pain and inflammation [34–36]. These insights support their potential as therapeutic targets for visceral pain management.

3.1.3. Serotonin (5-HT): Receptor Subtypes

Serotonin (5-HT) is a key regulator of GI function, with receptor subtypes playing distinct roles in physiological and pathological processes [36,37]. 5-HT3 receptors are ligand-gated ion channels that mediate rapid excitatory neurotransmission. Their dysregulation contributes to visceral hypersensitivity in GI pain conditions like IBS. Therapeutic use of 5-HT3 antagonists has proven effective in alleviating pain and discomfort in DGBIs. Conversely, 5-HT4 receptors, part of the G-protein-coupled receptor family, regulate GI motility and sensitivity. Agonists targeting 5-HT4 receptors improve gut motility and reduce pain, highlighting their therapeutic potential.

Dysregulated serotonin signaling is a hallmark of visceral pain syndromes, linked to hypersensitivity and altered gut motility. Serotonin has a multifaceted role in GI physiology and pathology, providing a foundation for targeted drug development. Modulating serotonin signaling may be a promising approach to improve symptoms and quality of life in DGBIs [36,37].

3.1.4. Transient Receptor Potential Channels (TRP)

TRPV1 is a nonselective cation channel crucial for detecting noxious stimuli such as extreme heat, acidity, and capsaicin. Expressed primarily in peripheral nociceptive neurons, TRPV1 contributes to inflammatory pain responses and hyperalgesia by lowering its activation threshold during inflammation. TRPV1's role in pain pathways has been well defined, suggesting it as a possible target for therapies aimed at alleviating visceral pain [38,39].

Similarly, environmental irritants (e.g., mustard oil) and endogenous inflammatory mediators activate the Transient Receptor Potential Ankyrin 1 (TRPA1) channel, which contributes to inflammatory hyperalgesia. TRPA1 antagonists reduce pain behaviors in animal models, offering novel strategies for managing visceral pain. TRPV1 and TRPA1 have a dynamic interplay, playing complementary roles in pain processing [38].

3.1.5. Voltage-Gated Sodium Channels

Voltage-gated sodium channels (NaV), particularly NaV1.7 and NaV1.8, are pivotal in neuronal excitability and pain signaling [40]. These channels, predominantly expressed in sensory neurons, facilitate action potential propagation. Pathological conditions like tissue injury or inflammation enhance NaV channel activity, resulting in neuronal hyperexcitability and chronic pain syndromes. NaV1.1 is involved in mechanical pain, as shown by the use of selective spider toxin inhibitors that significantly reduce pain responses [41]. These findings underscore NaV channels as key targets for pain management.

3.1.6. Catechol-O-Methyltransferase (COMT)

COMT modulates nociceptive signaling by degrading catecholamines such as dopamine, epinephrine, and norepinephrine. Genetic variations in COMT expression influence pain perception and susceptibility to chronic pain conditions such as fibromyalgia and migraines [42]. The V158M single-nucleotide polymorphism (SNP) in the COMT gene has been linked to heightened pain sensitivity, underscoring its role in individual pain variability [43].

Recent studies have shown that loss or reduced expression of COMT not only affects catecholamine metabolism but also contributes to increased production of pro-inflammatory cytokines, particularly TNF-α, via a downstream miRNA-dependent mechanism. Specifically, decreased COMT activity leads to upregulation of miR-155, which in turn promotes TNF-α expression in enteric neurons and macrophages [44] (Figure 2). TNF-α then acts on nociceptive pathways to sensitize colonic and dorsal root ganglion (DRG) neurons, thereby enhancing visceral pain signaling [45].

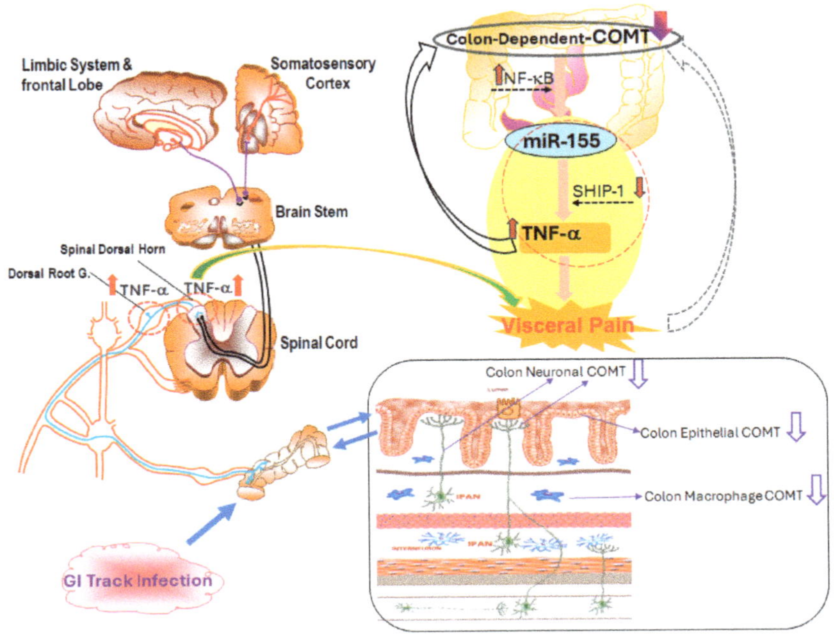

Figure 2. Mechanistic overview of colon-dependent catechol-O-methyltransferase (COMT) and its role in visceral pain. Both central and peripheral pathways are involved in visceral pain following GI tract infection. Gut barrier dysfunction, immune activation, and neural hypersensitivity all contribute to the pathogenesis of visceral pain. In epithelial, neuronal, and macrophage cells, colon-dependent COMT expression is reduced, resulting in dysregulated inflammatory responses. Upregulation of nuclear factor kappa-light-chain-enhancer of activated B cells (NF-κB) and miR-155, coupled with the suppression of Src homology 2 domain-containing inositol phosphatase 1 (SHIP-1), leads to increased TNF-α production, amplifying nociceptive signaling. Pro-inflammatory cytokines such as TNF-α activate NF-κB pathways within DRG and spinal dorsal horn neurons, further sensitizing pain pathways and transmitting signals to the brain stem, limbic system, and somatosensory cortex.

This COMT-miR-155-TNF-α axis provides a mechanistic bridge between genetic susceptibility and immune-driven inflammation in post-infectious IBS with diarrhea (PI-IBS-D), offering a potential therapeutic target for interrupting the cycle of chronic abdominal pain.

3.1.7. Ion Channels

In addition to key mediators of visceral pain pathophysiology, such as cytokines, prostaglandins, and neuropeptides, which are critical for activating and sensitizing nociceptors [39], ion channels like transient receptor potential vanilloid 1 (TRPV1) and acid-sensing ion channels (ASICs) are important for detecting noxious stimuli [35]. Because altered expression and function of these channels have been strongly implicated in the development of visceral hypersensitivity [38,39], these molecular mediators may be potential targets for therapeutic intervention. Furthermore, targeted therapies aimed at modulating the activity of specific ion channels and receptors involved in nociception could provide substantial benefits over traditional analgesics, including enhanced efficacy with fewer side effects. For example, TRPV1 antagonists and ASIC inhibitors are being explored as potential options for mitigating visceral pain by reducing afferent sensitization [38].

In particular, visceral hypersensitivity involving TRPV1-mediated nociceptive signaling and inflammation is a key mechanism underlying abdominal pain in patients with IBS with constipation (IBS-C). Tenapanor, a small-molecule inhibitor of the sodium/hydrogen exchanger isoform 3 (NHE3), inhibits absorption of sodium and phosphate in the GI tract. Treatment with tenapanor improves GI motility, decreases intestinal permeability and inflammation, and normalizes TRPV1 signaling, which may collectively reduce visceral hypersensitivity and associated abdominal pain. These findings suggest that targeting visceral hypersensitivity could be a promising approach for alleviating abdominal pain in IBS-C patients [46,47].

Table 1 provides a comprehensive overview of key neurotransmitters and neuromodulators involved in pain signaling.

Table 1. Neurotransmitters and Neuromodulators in Pain Signaling: Roles, Mechanisms, and Therapeutic Implications.

Neurotransmitter/ Neuromodulator	Mechanism	Role in Pain Signaling	Therapeutic Implications
Glutamate [22]	Primary excitatory neurotransmitter in the CNS; acts through NMDA, AMPA, and mGluRs	Mediates excitatory nociceptive signaling, contributing to central sensitization	Targeting glutamate receptors (NMDA, AMPA) and GABA receptors can help manage chronic pain and hypersensitivity, particularly in IBS
GABA [15]	Principal inhibitory neurotransmitter; acts through GABA-A and GABA-B receptors	Maintains neural homeostasis by inhibiting excessive excitatory signaling	
Substance P [36]	Released by primary sensory neurons; binding to NK1 receptors amplifies pain signals and mediates inflammation	Enhance pain transmission and contribute to neurogenic inflammation, facilitating peripheral and central sensitization	Potential targets for managing visceral pain syndromes, especially in IBS and related disorders
CGRP [35]	Promotes vasodilation and inflammation, often co-released with substance P		
Serotonin (5-HT) [36,37]	5-HT3 receptors: mediate excitatory transmission and contribute to visceral pain	5-HT signaling regulates gut function and is implicated in conditions like IBS Dysregulated 5-HT3 signaling contributes to visceral hypersensitivity	Modulation of serotonin receptors, particularly 5-HT3 antagonists and 5- HT4 agonists, offers potential therapeutic strategies for gut disorders
	5-HT4 receptors: regulate GI motility and sensitivity	5-HT signaling regulates gut function and is implicated in conditions like IBS 5-HT4 agonists improve motility and reduce pain	

Table 1. *Cont.*

Neurotransmitter/ Neuromodulator	Mechanism	Role in Pain Signaling	Therapeutic Implications
TRPV1 channels [38,39]	Activated by noxious stimuli, such as heat and acid	Contribute to inflammatory pain responses and hyperalgesia Lowers its activation threshold during inflammation	Antagonists show promise as novel therapies for visceral pain and hyperalgesia, particularly in inflammatory conditions
TRPA1 channels [38]	Activated by environmental irritants and inflammatory mediators	Contribute to inflammatory pain responses and hyperalgesia participates in inflammatory pain	
Voltage-gated sodium channels [40,41]	NaV1.7 and NaV1.8 channels facilitate action potential propagation in sensory neurons, with activity enhanced in pathological conditions like inflammation	Essential for pain signaling and contribute to neuronal hyperexcitability in chronic pain conditions	NaV channel blockers are potential pain management therapies, particularly for neuropathic and inflammatory pain
COMT [43,44]	COMT degrades catecholamines (dopamine, norepinephrine), modulating pain sensitivity	Genetic variants in COMT influence pain perception; higher COMT activity is associated with reduced pain sensitivity Inflammation can modulate COMT's effects on pain signaling	Targeting COMT in combination with other therapies may reduce chronic pain, particularly in conditions like fibromyalgia and IBS

The dual involvement of peripheral and central mechanisms in chronic visceral hypersensitivity underscores the complexity of these conditions and highlights the need for a multifaceted therapeutic approach. As research into the underlying mechanisms of visceral pain continues to evolve, there is increasing hope that novel interventions targeting both peripheral and central pathways will offer relief to patients suffering from these debilitating disorders, ultimately improving their quality of life.

4. Conclusions

Chronic visceral pain, particularly in the context of disorders such as IBS, is a complex, multifactorial condition that involves both peripheral and central mechanisms. Several well-supported mechanisms have been implicated in the development and persistence of visceral pain and hypersensitivity, including (i) nociceptive input from the colon, which contributes to the induction and maintenance of hypersensitivity; (ii) increased intestinal permeability, driving or sustaining visceral nociceptive responses; and (iii) alterations in miRNAs and EVs within target tissues, which may influence local and systemic signaling pathways. The interplay of these factors emphasizes the need for an integrated approach to understanding and treating this pervasive condition.

Recent advancements suggest that epigenetic regulation plays a pivotal role in modulating stress-induced visceral pain. The pathophysiology of GI disorders, including IBS, is increasingly linked to aberrant expression of microRNAs (miRNAs) and other molecular mechanisms. Emerging models also highlight the involvement of DNA methylation and disrupted miRNA signaling pathways as key contributors to these processes. Despite these insights, the molecular basis and neurobiology of specific patient endophenotypes experiencing visceral pain remain incompletely understood, underscoring the need for further research to identify actionable therapeutic targets.

While these mechanisms offer valuable insights, unidentified pathways likely contribute to the complex interplay between peripheral and central processes in chronic visceral pain. Synergistic interactions among these mechanisms, along with transient physiological triggers such as gut inflammation or increased intestinal permeability, may initiate and perpetuate visceral hypersensitivity.

Future studies should aim to unravel these intricate interactions to provide a more comprehensive understanding of visceral nociception. Such efforts could pave the way for novel, precisely targeted therapies that improve outcomes for patients with chronic visceral pain, surpassing the efficacy of current treatment options. By integrating insights from molecular biology, neurobiology, and patient-specific endophenotypes, we can move toward more effective therapies that address not only the peripheral origins of pain but also the central mechanisms that perpetuate it to make significant strides in the management of chronic GI disorders.

Author Contributions: Q.Z. and G.N.V. conceptualized the paper and wrote the initial and final draft. Editing and writing assistance was provided by Deborah Solymar of Science Editors Network. All authors have read and agreed to the published version of the manuscript.

Funding: This research was supported by grants from the National Institute of Diabetes and Digestive and Kidney Diseases (NIDDK) (DK099052, DK118959) to Q.Z. and G.N.V. and a grant from the Department of Veterans Affairs (CX001477, CX002910) to Q.Z. and G.N.V.

Data Availability Statement: No new data were created or analyzed in this study.

Conflicts of Interest: The authors declare no conflicts of interest.

Abbreviations

The following abbreviations are used in this manuscript:

ASIC	Acid-sensing ion channel
CGRP	Calcitonin gene-related peptide
circRNA	Circular ribonucleic acid
CNS	Central nervous system
COMT	Catechol-O-methyltransferase
CUGBP1	CUG triplet repeat RNA-binding protein 1
DGBI	Disorder of gut–brain interaction
DRA	Downregulated in adenoma protein
DRG	Dorsal root ganglia
EV	Extracellular vesicle
FGF	Fibroblast growth factor
FGID	Functional gastrointestinal disorder
GABA	Gamma-aminobutyric acid
GI	Gastrointestinal
HIF-1α	Hypoxia-inducible factor 1 subunit alpha
IBS	Irritable bowel syndrome
IBS-C	IBS with constipation
IL	Interleukin
lncRNA	Long non-coding RNA
IJ	Intercellular junction
KO	Knockout
LTP	Long-term potentiation
miRNA	MicroRNA
mRNA	Messenger RNA
ncRNA	Non-coding RNA
NF-κB	Nuclear factor kappa-light-chain-enhancer of activated B cells
NHE3	Sodium/hydrogen exchanger isoform 3
NMDA	N-methyl-D-aspartate
NO	Nitric oxide
PAMPs	Pathogen-associated molecular patterns
PI3K	Phosphatidylinositol 3 kinase

PI-IBS	Post-infectious IBS
PI-IBS-C	PI-IBS with constipation
PI-IBS-D	PI-IBS with diarrhea
ROS	Reactive oxygen species
SHIP-1	Src homology 2 domain-containing inositol phosphatase 1
TGF-β	Transforming growth factor beta
TJ	Tight junction
TLR	Toll-like receptor
TNF-α	Tumor necrosis factor-alpha
TRPA1	Transient receptor potential ankyrin 1
TRPV1	Transient receptor potential vanilloid 1
vtRNA	Vault RNA

References

1. Videlock, E.J.; Chang, L. Latest Insights on the Pathogenesis of Irritable Bowel Syndrome. *Gastroenterol. Clin. N. Am.* **2021**, *50*, 505–522. [CrossRef] [PubMed]
2. Mayer, E.A.; Ryu, H.J.; Bhatt, R.R. The Neurobiology of Irritable Bowel Syndrome. *Mol. Psychiatry* **2023**, *28*, 1451–1465. [CrossRef] [PubMed]
3. Keefer, L.; Ballou, S.K.; Drossman, D.A.; Ringstrom, G.; Elsenbruch, S.; Ljótsson, B. A Rome Working Team Report on Brain-Gut Behavior Therapies for Disorders of Gut-Brain Interaction. *Gastroenterology* **2022**, *162*, 300–315. [CrossRef] [PubMed]
4. Zhou, Q.; Price, D.D.; Callam, C.S.; Woodruff, M.A.; Verne, G.N. Effects of the N-Methyl-D-Aspartate Receptor on Temporal Summation on Second Pain (Wind-up) in Irritable Bowel Syndrome. *J. Pain* **2011**, *12*, 297–303. [CrossRef] [PubMed]
5. Higgins, G.A.; Hong, S.; Wiley, J.W. The Role of Epigenomic Regulatory Pathways in the Gut-Brain Axis and Visceral Hyperalgesia. *Cell Mol. Neurobiol.* **2022**, *42*, 361–376. [CrossRef] [PubMed]
6. Grover, M.; Vanuytsel, T.; Chang, L. Intestinal Permeability in Disorders of Gut-Brain Interaction (DGBI): From Bench to Bedside. *Gastroenterology* **2024**, *168*, 480–495. [CrossRef] [PubMed]
7. Drossman, D.A.; Tack, J. Rome Foundation Clinical Diagnostic Criteria for Disorders of Gut-Brain Interaction. *Gastroenterology* **2022**, *162*, 675–679. [CrossRef] [PubMed]
8. Hockley, J.R.F.; Taylor, T.S.; Callejo, G.; Wilbrey, A.L.; Gutteridge, A.; Bach, K.; Winchester, W.J.; Bulmer, D.C.; McMurray, G.; Smith, E.S.J. Single-cell RNA-seq reveals seven classes of colonic sensory neuron. *Gut* **2019**, *68*, 633–644. [CrossRef] [PubMed]
9. Brierley, S.M.; Hibberd, T.J.; Spencer, N.J. Spinal afferent innervation of the colon and rectum. *Front. Cell Neurosci.* **2018**, *12*, 467. [CrossRef] [PubMed]
10. Edogawa, S.; Edwinson, A.L.; Peters, S.A.; Chikkamenahalli, L.L.; Sundt, W.; Graves, S.; Gurunathan, S.V.; Breen-Lyles, M.; Johnson, S.; Dyer, R.; et al. Serine proteases as luminal mediators of intestinal barrier dysfunction and symptom severity in IBS. *Gut* **2020**, *69*, 62–73. [CrossRef] [PubMed]
11. Mahurkar-Joshi, S.; Chang, L. Epigenetic Mechanisms in Irritable Bowel Syndrome. *Front. Psychiatry* **2020**, *11*, 805. [CrossRef] [PubMed]
12. Basbaum, A.I.; Bautista, D.M.; Scherrer, G.; Julius, D. Cellular and Molecular Mechanisms of Pain. *Cell* **2009**, *139*, 267–284. [CrossRef] [PubMed]
13. Zhou, Q.Q.; Price, D.D.; Verne, G.N. Reversal of Visceral and Somatic Hypersensitivity in a Subset of Hypersensitive Rats by Intracolonic Lidocaine. *Pain* **2008**, *139*, 218–224. [CrossRef] [PubMed]
14. Latremoliere, A.; Woolf, C.J. Central Sensitization: A Generator of Pain Hypersensitivity by Central Neural Plasticity. *J. Pain* **2009**, *10*, 895–926. [CrossRef] [PubMed]
15. Inquimbert, P.; Moll, M.; Latremoliere, A.; Tong, C.-K.; Whang, J.; Sheehan, G.F.; Smith, B.M.; Korb, E.; Athié, M.C.; Babaniyi, O.; et al. NMDA Receptor Activation Underlies the Loss of Spinal Dorsal Horn Neurons and the Transition to Persistent Pain after Peripheral Nerve Injury. *Cell Rep.* **2018**, *23*, 2678–2689. [CrossRef] [PubMed]
16. Verne, G.N.; Sen, A.; Price, D.D. Intrarectal lidocaine is an effective treatment for abdominal pain associated w diarrhea-predominant irritable bowel syndrome. *J. Pain* **2005**, *6*, 493–496. [CrossRef] [PubMed]
17. Young, R.L.; Kaelberer, M.M.; Cevikbas, V.; Berthoud, H.R.; Schneider, S.; Brierley, S.M. The Role of Enteroendocrine-Sensory Cell Communication in Visceral Nociception. *J. Neurochem.* **2023**, *164*, 748.
18. Ren, K.; Dubner, R. Molecular Mechanisms Underlying Nociception and Pain Plasticity: Bridging Peripheral and Central Mechanisms. *Nat. Rev. Neurosci.* **2024**, *25*, 189–205.

19. Woolf, C.J. Central Sensitization: Implications for the Diagnosis and Treatment of Pain. *Pain* **2011**, *152*, S2–S15. [CrossRef] [PubMed]
20. Gewandter, J.S.; Chaudari, J.; Iwan, K.B.; Kitt, R.; As-Sanie, S.; Bachmann, G.; Clemens, Q.; Lai, H.H.; Tu, F.; Verne, G.N.; et al. Research Design Characteristics of Published Pharmacologic Randomized Clinical Trials for Irritable Bowel Syndrome and Chronic Pelvic Pain Conditions: An ACTTION Systematic Review. *J. Pain* **2018**, *19*, 717–726. [CrossRef] [PubMed]
21. Atmani, K.; Wuestenberghs, F.; Baron, M.; Bouleté, I.; Guérin, C.; Bahlouli, W.; Vaudry, D.; de Rego, J.C.; Cornu, J.N.; Leroi, A.M.; et al. Bladder–colon chronic cross-sensitization involves neuro-glial pathways in male mice. *World J. Gastroenterol.* **2022**, *28*, 6935–6949. [CrossRef] [PubMed]
22. Zhou, Q.Q.; Verne, M.L.; Fields, J.Z.; Lefante, J.J.; Basra, S.; Salameh, H.; Verne, G.N. Randomised Placebo-Controlled Trial of Dietary Glutamine Supplements for Postinfectious Irritable Bowel Syndrome. *Gut* **2019**, *68*, 996–1002. [CrossRef] [PubMed]
23. Saha, K.; Zhou, Y.; Turner, J.R. Tight Junction Regulation, Intestinal Permeability, and Mucosal Immunity in Gastrointestinal Health and Disease. *Curr. Opin. Gastroenterol.* **2024**, *41*, 46–53. [CrossRef] [PubMed]
24. Kumar, A.; Priyamvada, S.; Ge, Y.; Jayawardena, D.; Singhal, M.; Anbazhagan, A.N.; Chatterjee, I.; Dayal, A.; Patel, M.; Zadeh, K.; et al. A Novel Role of SLC26A3 in Maintenance of Intestinal Epithelial Barrier Integrity. *Gastroenterology* **2020**, *160*, 1240–1255.e3. [CrossRef] [PubMed]
25. Calzadilla, N.; Qazi, A.; Sharma, A.; Mongan, K.; Comiskey, S.; Manne, J.; Youkhana, A.G.; Khanna, S.; Saksena, S.; Dudeja, P.K.; et al. Mucosal Metabolomic Signatures in Chronic Colitis: Novel Insights into the Pathophysiology of Inflammatory Bowel Disease. *Metabolites* **2023**, *13*, 873. [CrossRef] [PubMed]
26. Sharma, S.; Xiao, L.; Chung, H.K.; Chen, T.; Mallard, C.G.; Warner, B.; Yu, T.X.; Kwon, M.S.; Chae, S.; Raufman, J.P.; et al. Noncoding Vault RNA1-1 Impairs Intestinal Epithelial Renewal and Barrier Function by Interacting With CUG-Binding Protein 1. *Cell. Mol. Gastroenterol. Hepatol.* **2025**, *19*, 101410. [CrossRef] [PubMed]
27. Mahurkar-Joshi, S.; Rankin, C.R.; Videlock, E.J.; Soroosh, A.; Verma, A.; Khandadash, A.; Iliopoulos, D.; Pothoulakis, C.; Mayer, E.A.; Chang, L. The Colonic Mucosal MicroRNAs, MicroRNA-219a-5p, and MicroRNA-338-3p Are Downregulated in Irritable Bowel Syndrome and Are Associated With Barrier Function and MAPK Signaling. *Gastroenterology* **2021**, *160*, 2409–2422. [CrossRef] [PubMed]
28. Zhou, Q.Q.; Souba, W.W.; Croce, C.M.; Verne, G.N. MicroRNA-29a Regulates Intestinal Membrane Permeability in Patients with Irritable Bowel Syndrome. *Gut* **2010**, *59*, 775–784. [CrossRef] [PubMed]
29. Zhou, Q.Q.; Costinean, S.; Croce, C.M.; Brasier, A.R.; Merwat, S.; Larson, S.A.; Basra, S.; Verne, G.N. MicroRNA 29 Targets Nuclear Factor-KB-Repressing Factor and Claudin 1 to Increase Intestinal Permeability. *Gastroenterology* **2015**, *148*, 158–169.e8. [CrossRef] [PubMed]
30. Chung, H.K.; Xiao, L.; Han, N.; Chen, J.; Yao, V.; Cairns, C.M.; Raufman, B.; Rao, J.N.; Turner, D.J.; Kozar, R.; et al. Circular RNA Cdr1as Inhibits Proliferation and Delays Injury-Induced Regeneration of the Intestinal Epithelium. *JCI Insight* **2024**, *9*, e169716. [CrossRef] [PubMed]
31. Rawat, M.; Nighot, M.; Al-Sadi, R.; Gupta, Y.; Viszwapriya, D.; Yochum, G.; Koltun, W.; Ma, T.Y. IL1B Increases Intestinal Tight Junction Permeability by Up-Regulation of MIR200C-3p, Which Degrades Occludin MRNA. *Gastroenterology* **2020**, *159*, 1375–1389. [CrossRef] [PubMed]
32. Freire, V.S.; Burkhard, F.C.; Kessler, T.M.; Kuhn, A.; Draeger, A.; Monastyrskaya, K. MicroRNAs May Mediate the Down-Regulation of Neurokinin-1 Receptor in Chronic Bladder Pain Syndrome. *Am. J. Pathol.* **2010**, *176*, 288–303. [CrossRef] [PubMed]
33. He, Y.; Yang, C.; Kirkmire, C.M.; Wang, Z.J. Regulation of Opioid Tolerance by Let-7 Family MicroRNA Targeting the μ Opioid Receptor. *J. Neurosci.* **2010**, *30*, 10251–10258. [CrossRef] [PubMed]
34. Schank, J.R.; Heilig, M. Substance P and the Neurokinin-1 Receptor: The New Corticotropin-Releasing Factor. *Int. Rev. Neurobiol.* **2017**, *136*, 151–175. [PubMed]
35. Brierley, S.; Linden, D. Neuroplasticity and dysfunction after gastrointestinal inflammation. *Nat. Rev. Gastroenterol. Hepatol.* **2014**, *11*, 611–627. [CrossRef] [PubMed]
36. Hung, L.Y.; Alves, N.D.; Del Colle, A.; Talati, A.; Najjar, S.A.; Bouchard, V.; Gillet, V.; Tong, Y.; Huang, Z.; Browning, K.N.; et al. Intestinal epithelial serotonin as a novel target for treating disorders of gut–brain interaction and mood. *Gastroenterology* **2025**, *168*, 754–768. [CrossRef] [PubMed]
37. Liu, H.N.; Nakamura, M.; Kawashima, H. New role of serotonin as a biomarker of gut–brain interaction. *Life* **2023**, *14*, 1280. [CrossRef] [PubMed]
38. Julius, D. TRP Channels and Pain. *Annu. Rev. Cell Dev. Biol.* **2013**, *29*, 355–384. [CrossRef] [PubMed]
39. Zhou, Q.; Yang, L.; Larson, S.; Basra, S.; Merwat, S.; Tan, A.; Croce, C.; Nicholas Verne, G. Decreased MiR-199 Augments Visceral Pain in Patients with IBS through Translational Upregulation of TRPV1. *Gut* **2015**, *65*, 797–805. [CrossRef] [PubMed]

40. Osteen, J.D.; Sampson, K.; Iyer, V.; Julius, D.; Bosmans, F. Pharmacology of the Nav1.1 Domain IV Voltage Sensor Reveals Coupling between Inactivation Gating Processes. *Proc. Natl. Acad. Sci. USA* **2017**, *114*, 6836–6841. [CrossRef] [PubMed]
41. Osteen, J.D.; Herzig, V.; Gilchrist, J.; Emrick, J.J.; Zhang, C.; Wang, X.; Castro, J.; Garcia-Caraballo, S.; Grundy, L.; Rychkov, G.Y.; et al. Selective Spider Toxins Reveal a Role for the Nav1.1 Channel in Mechanical Pain. *Nature* **2016**, *534*, 494–499. [CrossRef] [PubMed]
42. Meloto, C.B.; Segall, S.K.; Smith, S.; Parisien, M.; Shabalina, S.A.; Rizzatti-Barbosa, C.M.; Gauthier, J.; Tsao, D.; Convertino, M.; Piltonen, M.H.; et al. COMT Gene Locus: New Functional Variants. *Pain* **2015**, *156*, 2072–2083. [CrossRef] [PubMed]
43. Vase, L.; Petersen, G.L.; Riley JL3rd Price, D.D. COMT Val158Met genotype modulates μ-opioid system activation during sustained pain and placebo analgesia. *Pain Rep.* **2019**, *4*, e780.
44. Zhou, Q.Q.; Yang, L.; Verne, M.L.; Zhang, B.B.; Fields, J.; Verne, G.N. Catechol-O-Methyltransferase Loss Drives Cell-Specific Nociceptive Signaling via the Enteric Catechol-O-Methyltransferase/MicroRNA-155/Tumor Necrosis Factor α Axis. *Gastroenterology* **2023**, *164*, 630–641.e34. [CrossRef] [PubMed]
45. Li, Y.C.; Zhang, F.C.; Xu, T.W.; Weng, R.X.; Zhang, H.H.; Chen, Q.Q.; Hu, S.; Gao, R.; Li, R.; Xu, G.Y. Advances in the pathological mechanisms and clinical treatments of visceral pain. *J. Neuroinflamm.* **2024**, *21*, 115. [CrossRef] [PubMed]
46. Singh, P.; Sayuk, G.S.; Rosenbaum, D.P.; Edelstein, S.; Kozuka, K.; Chang, L. An Overview of the Effects of Tenapanor on Visceral Hypersensitivity in the Treatment of Irritable Bowel Syndrome with Constipation. *Clin. Exp. Gastroenterol.* **2024**, *17*, 87–96. [CrossRef] [PubMed]
47. Ford, A.C.; Harris, L.A.; Lacy, B.E.; Quigley, E.M.M.; Moayyedi, P. Systematic review with meta-analysis: The efficacy of prebiotics, probiotics, synbiotics and antibiotics in irritable bowel syndrome. *Aliment. Pharmacol. Ther.* **2023**, *40*, 187–214. [CrossRef] [PubMed]

cells

MDPI

Review

BDNF Signaling and Pain Modulation

Mariacristina Mazzitelli [1], Takaki Kiritoshi [1], Peyton Presto [1], Zachary Hurtado [1], Nico Antenucci [1], Guangchen Ji [1] and Volker Neugebauer [1,2,3,*]

1 Department of Pharmacology and Neuroscience, School of Medicine, Texas Tech University Health Sciences Center, Lubbock, TX 79430, USA; mariacristina.mazzitelli@ttuhsc.edu (M.M.); takaki.kiritoshi@ttuhsc.edu (T.K.); peyton.presto@ttuhsc.edu (P.P.); zhurtado@ttuhsc.edu (Z.H.); nico.antenucci@ttuhsc.edu (N.A.); guangchen.ji@ttuhsc.edu (G.J.)

2 Center of Excellence for Translational Neuroscience and Therapeutics, Texas Tech University Health Sciences Center, Lubbock, TX 79430, USA

3 Garrison Institute on Aging, Texas Tech University Health Sciences Center, Lubbock, TX 79430, USA

* Correspondence: volker.neugebauer@ttuhsc.edu; Tel.: +1+806-743-3880; Fax: +1+806-732-2744

Abstract: Brain-derived neurotrophic factor (BDNF) is an important neuromodulator of nervous system functions and plays a key role in neuronal growth and survival, neurotransmission, and synaptic plasticity. The effects of BDNF are mainly mediated by the activation of tropomyosin receptor kinase B (TrkB), expressed in both the peripheral and central nervous system. BDNF has been implicated in several neuropsychiatric conditions such as schizophrenia and anxio-depressive disorders, as well as in pain states. This review summarizes the evidence for a critical role of BDNF throughout the pain system and describes contrasting findings of its pro- and anti-nociceptive effects. Different cellular sources of BDNF, its influence on neuroimmune signaling in pain conditions, and its effects in different cell types and regions are described. These and endogenous BDNF levels, downstream signaling mechanisms, route of administration, and approaches to manipulate BDNF functions could explain the bidirectional effects in pain plasticity and pain modulation. Finally, current knowledge gaps concerning BDNF signaling in pain are discussed, including sex- and pathway-specific differences.

Keywords: BDNF; TrkB; pain; neuroplasticity; neuroimmune signaling

1. Introduction

Since its discovery in 1982, brain-derived neurotrophic factor (BDNF) has been extensively studied for its critical involvement in the development and maintenance of the nervous system and its essential role in neuronal survival, growth, and differentiation and neuroplasticity [1,2]. Like other neutrophins, the initial BDNF precursor protein undergoes proteolytic cleavage to generate its mature form which exhibits biologically active properties [3,4]. The molecular processes governing its production and activation are tightly regulated and involve transcriptional, translational, and post-translational mechanisms [3]. BDNF exerts its influence through the activation of the high-affinity tropomyosin receptor kinase B (TrkB) and downstream signaling cascades to modulate neuronal function [5].

In clinical studies, alterations in BDNF levels in the nervous system have been linked to neurological and psychiatric disorders [6–8], including Alzheimer's disease [9–11], while only variations in serum BDNF have been reported in substance use disorders [12,13] and osteoarthritis patients [14].

In addition to its well-established functions in neuroplasticity and cognitive processes, BDNF has emerged as a key player in the complex realm of neuropathic pain [15]. Elevated BDNF levels have been observed in various pain conditions, contributing to increased neuronal excitability and synaptic plasticity in pain-processing circuits [16].

The objective of this review is to provide a unifying picture of BDNF's effects in pain modulation and to identify knowledge gaps and research directions. After reviewing the source and signaling mechanisms of BDNF, information about its role in peripheral and spinal nociception and in supraspinal pain processing will be presented. The role of BDNF in related diseases will also be discussed. This review should provide a scientifically grounded perspective on BDNF as a potential target for therapeutic interventions in chronic pain and related neurological and psychiatric disorders.

2. BDNF Signaling

2.1. BDNF Synthesis, Source, and Release

The BDNF gene, located on human chromosome 11, undergoes complex transcriptional regulation [17]. The transcription of BDNF is influenced by various transcription factors, with cyclic AMP (cAMP) response element-binding protein (CREB) being a key player [18]. BNDF gene expression is neuronal activity-dependent. The BDNF gene contains a unique structure with several $5'$ non-coding exons, also known as $5'$untranslated regions ($5'$-UTRs), and one $3'$ exon coding for the pre-pro-peptide [19]. The presence of multiple promoters adds an additional layer of complexity, allowing for tissue-specific and activity-dependent regulation of BDNF synthesis, as well as cellular and cognitive functions [19,20]. In rodents, there are nine functional promoters (I to IX) upstream of the nine non-coding exons, generating nine different mRNA transcripts [19]. In humans, there are also nine functional promoters but eleven exons [21]. Beyond transcriptional control, post-transcriptional processes contribute to the regulation of BDNF synthesis. RNA splicing variants give rise to different BDNF isoforms, each with distinct functional properties. Epigenetic modifications, including DNA methylation and histone acetylation, play a crucial role in shaping the chromatin landscape and, consequently, BDNF expression patterns [22]. BDNF synthesis begins with the formation of a pre-pro-peptide, pre-pro-BDNF, in the endoplasmic reticulum. Cleavage of the pre-domain results in the formation of pro-BDNF that will be transported to the Golgi apparatus for sorting and further processing in the trans-Golgi network [23,24]. Through post-translational processing, pro-BDNF is converted into the BDNF protein, which is cleaved into the mature BDNF (or BDNF) at 14 kDa [24]. BDNF is then processed through a pathway for packaging into large secretory vesicles [25]. Neuronal depolarization triggers the Ca^{2+} dependent release of BDNF-containing vesicles into the synaptic cleft, where it can act pre- or postsynaptically (see Section 2.2). This process ensures that BDNF release is tightly coupled to synaptic events, allowing it to modulate synaptic strength and plasticity in response to neuronal activity patterns [26].

BDNF gene transcription closely correlates with activity-induced Ca^{2+} increase [27], giving further evidence to activity-dependent transcription, localization, and subsequent release of BDNF. The location of BDNF in neurons can be axonal or dendritic, depending on cell type, activity, and connectivity. The different BDNF mRNA transcripts, classified in two main categories depending on the presence of either the short or long $3'$-UTR, seem to also govern the localization of BDNF in the neuron. For example, long $3'$-UTR transcripts prefer dendritic processes while short $3'$-UTRs are directed into the soma [20]. Dendritic localization of BDNF mRNA (summarized in [28]) was found in hippocampal and cortical cells and was associated with local (dendrites) translational processes contributing to plasticity and dendritic spinal remodeling [20], whereas axonal localization was found in other groups of neurons, such as cortical axons and mossy fibers projecting to hippocampal CA3 pyramidal cells [29]. Similarly, differences in the BDNF protein localization and isoforms were observed and showed temporal features. In early life, the pro-BDNF isoform seems to be more abundant than the mature BDNF, which is the predominant isoform in adulthood. Additionally, pro-BDNF is found throughout the hippocampus in juvenile animals, while its expression is restricted to the hippocampal mossy fibers of the dentate gyrus granule cells in adult brains [20]. Expression of BDNF is regulated by other neurotrophins, including nerve growth factor (NGF) and neurotrophin-3 (NT-3) through shared signaling pathways involving tropomyosin receptor kinase (Trk) receptors and p75 neurotrophin receptor [26].

BDNF expression in the central nervous system (CNS) glia cells is less clear and somewhat controversial. Many studies have reported that spinal [30,31], brain [32], and cultured [33,34] microglia are capable of expressing BDNF mRNA and protein. However, others have found very little BDNF or TrkB receptor expression in homeostatic or lipopolysaccharide (LPS)-activated microglia in the spinal cord or in brain regions including the somatomotor cortex and hippocampus [35]. Further groups reported that spinal microglia do not express significant levels of BDNF using transcriptomic analysis [36] or *BDNF-LacZ* reporter mice [37], and still others found that resting or ATP-activated microglia do not express BDNF transcriptionally or translationally in the mouse motor cortex [38]. This discrepancy indicates a significant knowledge gap in microglial-related BDNF signaling mechanisms throughout the neuraxis.

2.2. Targets and Downstream Signaling

TrkB, a prominent member of the neurotrophic tyrosine kinase receptor family, serves as a pivotal mediator in the signaling effects of BDNF. TrkB consists of an extracellular domain responsible for ligand binding, a dimeric transmembrane domain, and an intracellular domain hosting the tyrosine kinase activity. Three isoforms of TrkB have been identified: the full-length receptor glycoprotein, TrkB-FL, and two truncated forms, TrkB.T1 and TrkB-T-Shc. The latter two are obtained by alternative splicing processes and lack the tyrosine kinase domain at the C-terminus [39]. The binding of BDNF to TrkB-FL is the main mechanism of action for the neurotrophic effects of BDNF. The ligand-binding domain comprises distinct regions that interact with BDNF, fostering high-affinity and selective binding between the ligand and its receptor [26,40]. The binding of BDNF to TrkB results in conformational changes in the receptor, facilitating the formation of homodimers. This dimerization event triggers the autophosphorylation of specific tyrosine residues within the intracellular domain of TrkB. The two major autophosphorylation sites on TrkB are Tyr-515 and Tyr-816 [41]. Phosphorylation at Tyr-515 provides a docking site for proteins involved in the activation of the mitogen-activated protein kinases (MAPKs) like the extracellular signal-regulated kinase (ERK) and the phosphoinositide 3-kinase/protein kinase B (PI3K/AKT) pathway. The activation of this pathway regulates gene expression, contributing to neuronal survival and differentiation. On the other hand, phosphorylation at Tyr-816 mediates the activation of the phospholipase C-γ (PLCγ) pathway involving the engagement of type II calcium/calmodulin-dependent protein kinase (CaMKII) and resulting in the activation of CREB transcription factor, which plays an important role synaptic plasticity and neuronal functions [41]. Evidence also showed the formation of a complex between the pro-BDNF, the trafficking protein sortilin, and p75 neurotrophin receptor (p75NTR) with lower affinity for BDNF than TrkB [42]. This interaction is associated with apoptosis through the c-Jun N-terminal kinases (JNK) and p53 and caspase 3 pathways or with neuronal survival via the nuclear factor kappa B (NF-kB) signaling [43,44]. Additionally, the activation of p75NTR has been involved in the development of hippocampal long-term depression (LTD) [45].

TrkB is widely expressed in the peripheral and central nervous system, including the spinal cord, brain stem, hippocampus, cerebral cortex, and cerebellum, where it mediates the physiological effects of BDNF [43,46]. BDNF/TrkB signaling is essential for survival and growth, and transgenic mutations to the TrkB gene result in severe abnormalities in the nervous system and precocious death [47,48]. Within the cell, a pool of TrkB receptors is stored in synaptic-like vesicles. Its translocation to the membrane surface on dendritic spines and axon terminals is mediated by cAMP activity promoted by the Ca^{2+} influx induced by neuronal depolarization [49]. This process occurs rapidly and guides the sensitivity of the postsynaptic neuron to BDNF [50]. Additionally, it has been shown that TrkB receptor is expressed presynaptically in glutamatergic synapses, where it modulates the neurotransmitter release [51], and on the CA3 presynaptic axons to regulate long-term potentiation (LTP) [52] (see Section 2.3). While there is also evidence for bidirectional facilitatory and inhibitory interactions between BNDF/TrkB and endocannabinoid signaling, particularly 2-arachidonoylglycerol (2-AG), in the CNS, this remains to be determined for nociceptive processing and pain conditions [53].

2.3. Synaptic Plasticity

Synaptic plasticity is a lasting activity-dependent functional and/or structural change in neuronal connection strength and has been considered a critical component of learning and memory [54]. Accumulating evidence suggests an important contribution of BDNF-TrkB signaling to different forms of synaptic plasticity ranging from synaptogenesis to homeostatic plasticity [55]. Among them, LTP is one of the most studied forms of synaptic plasticity, and here, we will briefly review the contribution of BDNF to LTP.

The critical involvement of BDNF in LTP was demonstrated by earlier studies showing that the genetic deletion of BDNF impaired LTP at the CA3 Schaffer collateral-CA1 synapse in the hippocampus [56,57], and the impairment was rescued by incubation with recombinant BDNF [57] or viral expression of BDNF [58]. Because of conflicting results in the literature [52,58–61], the sites (pre- or postsynaptic) of BDNF release and activation of TrkB during LTP have been controversial, and their potentially distinct roles in LTP remain to be determined. A more recent study using region-specific deletion of BDNF or TrkB revealed a specific involvement of pre- and postsynaptic BDNF-TrkB signaling in different stages of LTP at CA3 Schaffer collateral-CA1 synapses [62]. Specifically, presynaptic BDNF contributes to the induction of LTP, while postsynaptic BDNF is required for its maintenance [62]. On the other hand, presynaptic TrkB receptors are required for LTP maintenance, while postsynaptic TrkB receptors are essential for both the induction and maintenance of LTP [62]. These findings suggest that BDNF release from presynaptic terminal induces initial potentiation, while BDNF release from postsynaptic sites prolongs this potentiation. In addition to pre- and postsynaptic neurons, microglia [63] and astrocytes [64,65] have been reported as other sources of BDNF during LTP. The source of BDNF contributing to LTP could depend on spinal or brain areas and on experimental protocols.

LTP is generally divided into early (E-LTP) and late phases (L-LTP) [66,67]. E-LTP lasts up to 2–3 h, requires modification and trafficking of proteins, and is independent of de novo protein synthesis [14,15]. In contrast, L-LTP requires gene expression and local protein synthesis, and lasts hours to days [66,67]. Although the molecular mechanisms of the role of BDNF in LTP are not fully understood, evidence suggests that BDNF can modulate glutamate receptors such as α-amino-3-hydroxy-5-methyl-4-isoxazolepropionic acid receptor (AMPAR) and N-methyl-D-aspartate receptor (NMDAR). During spike-timing-dependent LTP at the CA3 Schaffer collateral-CA1 synapse, postsynaptic release of BDNF induced the insertion of the new AMPAR-containing subunit GluA1 into postsynaptic membrane [68]. This process seems to involve protein kinase C (PKC) and CaMKII-mediated phosphorylation of newly synthesized GluA1s, followed by inositol 1,4,5-triphosphate (IP3) receptor (IP3R) and transient receptor potential canonical (TRPC)-mediated Ca^{2+} transient-dependent translocation of GluA1s to the postsynaptic membrane [69–71]. Importantly, BDNF has been demonstrated to induce GluA1 translation through a TRPC-CaMK kinase (CAMKK)–AKT–mammalian target of rapamycin (mTOR)-dependent pathway [70]. Additionally, BDNF has been shown to enhance interactions between AMPAR subunits GluA1 and GluA2 and their scaffolding proteins (synapse-associated protein of 97 kDa (SAP97) and glutamate receptor-interacting protein1 (GRIP1)) at synapses [72], suggesting that BDNF plays a key role in the long-term maintenance of the availability of AMPAR subunits and associated scaffolding proteins at synapses. NMDAR has also been shown to critically contribute to the action of BDNF in LTP. For example, one study reported that BDNF-dependent LTP in dentate granule cells (GCs) required the activation of NMDARs and Ca^{2+} channels [73]. Similarly to the AMPAR-mediated pathway, BDNF has been shown to upregulate NMDAR subunits NR1, NR2A, and NR2B in the plasma membrane, possibly through Ca^{2+}-dependent local synthesis and phosphorylation of the subunits [73–77]. Specifically for L-LTP, one study found that theta burst stimulation (TBS)-induced, but not HFS (four 100 Hz trains)-induced, L-LTP at the CA3 Schaffer collateral-CA1 synapse depended on BDNF-TrkB signaling-mediated modulation of subcellular distribution and nuclear translocation of the activated MAPK through cAMP-protein kinase A (PKA) signaling [78]. Interestingly, TrkB activation was not critical for the phosphorylation of MAPK in this particular form of LTP [78], suggesting a differential regulation of LTP through TrkB-independent MAPK activation and TrkB-dependent translocation. cAMP-PKA signaling has also been implicated in another type of LTP at the hilar mossy cell (MC)–GC synapse, where cAMP-PKA signaling was found to mediate LTP downstream of postsynaptic BDNF-TrkB signaling [79]. Phospholipase Cγ (PLCγ)-mediated phosphorylation of CREB and CaMKIV have also been suggested to act as key downstream targets of TrkB during both E-LTP and L-LTP at CA3 Schaffer collateral-CA1 synapse [80].

In addition to functional signaling mechanisms in LTP, BDNF has been demonstrated to induce an enlargement of CA1 dendritic spines during a spike-timing protocol-induced LTP [81], which aligns well with a study that showed that BDNF-induced mTOR-regulated reorganization of cytoskeleton mediated by the upregulation of RhoA protein, cofilin phosphorylation, and actin polymerization at CA1 dendritic spines [82]. These reports suggest that BDNF serves as a key regulator for structural synaptic consolidation underlying LTP.

Although most of the evidence described above comes from studies on hippocampal LTP, important contributions of BDNF-TrkB signaling to LTP have also been reported outside of the hippocampus, including key areas for pain processing such as the spinal dorsal horn [63,83,84], nucleus accumbens (NAc) [85], medial prefrontal cortex (mPFC) [86], and anterior cingulate cortex (ACC) [87]. These findings implicate BDNF in pain-related synaptic plasticity underlying pathological pain [88]. Accumulating evidence suggests the critical contribution of the BDNF signaling to pain mechanisms at the peripheral (Section 3), spinal (Section 4), and supraspinal (Section 5) levels (Figure 1), but the role of BDNF in non-LTP plasticity (Section 2.3) is relatively unknown compared to its role in neuroimmune signaling (Section 2.4).

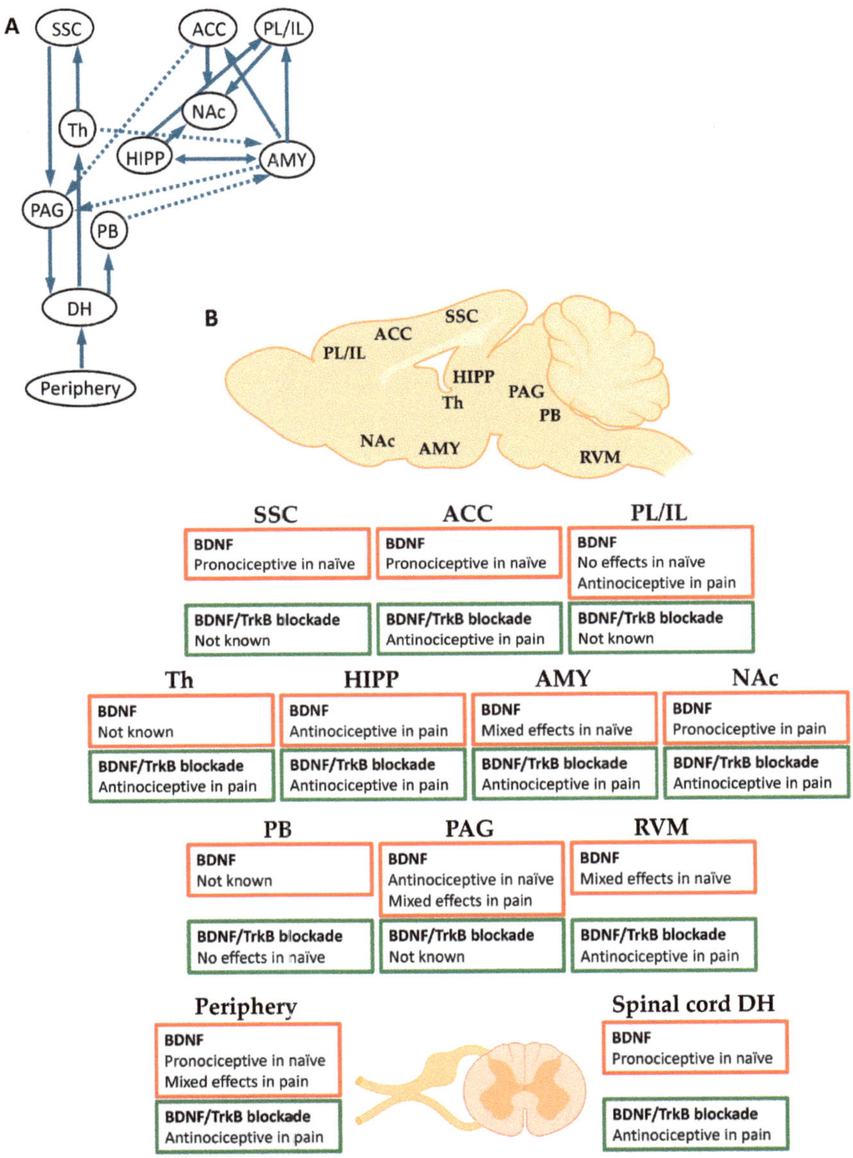

Figure 1. BDNF signaling in the pain system. (**A**) Elements of the pain system where BNDF signaling was explored. Dashed arrows indicate connections that were not explicitly tested. (**B**) Effects of BDNF-related manipulations tested in different brain areas, spinal cord, and periphery. Red boxes, BDNF; green boxes, BDNF/TrkB blockade. ACC, anterior cingulate cortex; AMY, amygdala; DH, dorsal horn; HIPP, hippocampus; NAc, nucleus accumbens; PAG, periaqueductal gray; PL/IL, pre/infra limbic cortex; PB, parabrachial nucleus; RVM, rostral ventromedial medulla; SSC, primary somatosensory cortex; Th, thalamus.

2.4. Neuroimmune Signaling

The field of pain research has overwhelmingly focused on the role of neurons in pain-related signaling mechanisms. However, a growing area of research is on the contribution of non-neuronal cells, such as astrocytes and microglia, due to their influence on and response to neuronal activity changes within nociceptive pathways [89,90]. In this sense, it is important to address the impact of BDNF in glia and other cells of relevance in pain modulation. Here, we describe the localization of BDNF in neuroimmune cell types, its mechanisms of action in these cells, and its role in neuroimmune signaling at baseline.

2.4.1. Expression of BDNF on Neuroimmune Cell Types

As discussed throughout this review, BDNF is widely distributed in many neuronal cell types throughout the peripheral and central nervous systems. Under physiological conditions, BDNF binds to the TrkB and other receptors and triggers distinct downstream pathways (see Section 2.2). This complex network of signaling cascades can influence the regulation of inflammatory cytokines in neuroimmune cells, further impacting endogenous inflammatory responses.

Peripheral Nervous System

Within the peripheral nervous system (PNS), BDNF has been shown to be secreted by activated macrophages and Schwann cells. Macrophages are versatile cells that play a crucial role in the innate immune system, originating from precursor cells in bone marrow and migrating to sites of injury to phagocytose pathogens, modulate inflammation, and signal tissue repair. They exist in several subtypes, resting in an inactivated (M0) state or polarized into pro-inflammatory (M1) or anti-inflammatory (M2) phenotypes in response to stimuli [91–93]. Immunohistochemistry and flow cytometry experiments in male rats revealed very low expression of BDNF protein in inactivated M0 macrophages, whereas M1 and M2 macrophages activated by myocardial infarction showed strong BDNF mRNA and protein expression [94]. Another study found that the application of BDNF to cultured macrophages from male mice stimulated the expression of both BDNF and its TrkB receptor, suggesting that BDNF/TrkB signaling plays an important role in the activation processes of macrophages through potential autocrine mechanisms [95]. However, macrophage expression of BDNF in pain states remains underexplored.

Schwann cells are the primary glial cell type of the PNS and play a critical role in the support and maintenance of nerve functions, chiefly through their ability to form myelin sheaths around peripheral nerves to provide electrical insulation to axons and increase conduction of action potentials [96]. Schwann cells are also involved in the regeneration and repair of damaged nerves through clearing debris and releasing mediators, including BDNF [97–100]. Rat Schwann cell cultures that were exposed to the passively secreted progesterone metabolite allopregnanolone showed significantly higher levels of BDNF mRNA expression and increased protein levels of the precursor proBDNF and mature BDNF [101]. Increased expression of BDNF has also been shown to promote the proliferation of Schwann cells [102,103], suggesting that, as with macrophages, BDNF may act on Schwann cells in an autocrine fashion.

Central Nervous System

Microglia are the resident immune cells of the CNS and play a key role in the maintenance of neural homeostasis and in the response to injury or disease. As highly dynamic cells, they possess many critical functions such as the phagocytosis of pathogens, the regulation of synaptic pruning, and the release of various cytokines and growth factors—including BDNF—that can either support or inhibit neural cell survival and differentiation [104,105]. Therefore, microglia can serve either neuroprotective or neurotoxic mechanisms [106].

Following peripheral nerve injury, intrathecal (i.th.) injection of BDNF was found to significantly upregulate BDNF protein expression and trigger M1 polarization of spinal dorsal horn microglia [107], suggesting a role for BDNF autocrine regulatory mechanisms in the CNS just like in the PNS. Peripheral nerve injury has also been shown to upregulate BDNF mRNA expression in somatosensory cortex (S1) microglia in male mice [108]. BDNF protein was also upregulated in the ACC and S1 in male rats with inflammatory pain [109]. Increased BDNF has additionally been shown to contribute to increased spinal microglia activation in different models of pain (see Section 4.1). The data suggest that BDNF expression in the spinal cord may be associated predominately with microglia's pro-inflammatory phenotype, though further studies are needed to characterize its expression in supraspinal regions.

Astrocytes are versatile CNS cells that play a critical role in supporting neuronal function and maintaining the brain's microenvironment; this includes providing structural and metabolic support, maintaining ion homeostasis, recycling neurotransmitters, forming the blood–brain barrier, and modulating synaptic activity [110–112]. One study proposed that cortical and hippocampal astrocytes may express BDNF during development but cease during adulthood, or small subsets of astrocytes may transiently express BDNF; however, astrocytes were found to predominately express the TrkB receptor in these regions [35]. Other in vitro studies have shown that brain (hippocampal and cortical) astrocytes can express BDNF both under normal conditions [27] and following injury [113,114]. Under pain conditions, astrocytic expression of BDNF has been demonstrated in the ACC and primary sensory cortex but was not compared to normal conditions [109] (see Section 5.1). Further exploration is needed to determine BDNF expression patterns in other regions throughout the neuraxis at baseline and in pain states.

Oligodendrocytes are the primary myelinating cell type in the CNS; this role allows them to supply vital nutrients to neurons, promote long-term axonal integrity, and coordinate the timing and strength of neuronal signaling [115–117]. A small portion of hippocampal oligodendrocytes were found to express BDNF or the TrkB receptor [35], though BDNF expression has been reported in cortical oligodendrocytes [118,119]. Most spinal oligodendrocytes were found to produce BDNF protein under normal conditions and upregulate its expression both 1 day and 1 week following spinal cord injury [120]. However, the role of oligodendrocytes and oligodendrocytic BDNF expression in pain conditions remains to be determined.

2.4.2. BDNF and Neuroimmune Signaling

Peripheral Nervous System

Application of BDNF was found to promote M2 polarization of mouse [121] and human [122] cultured macrophages through repression of the signal transducer and activator of transcription 3 (STAT3) pathway and inhibition of pro-inflammatory interleukin (IL)-1β, tumor necrosis factor (TNF)-α, and IL-6 expression. Furthermore, mice with diabetic coronary atherosclerosis had a downregulation of BDNF mRNA and an increased differentiation of M1 macrophages compared to control, whereas an overexpression of BDNF induced the differentiation of M2 macrophages [123]. Knockout of p75NTR significantly reduced the secretion of pro-inflammatory cytokines in both LPS-stimulated and unstimulated cultured macrophages, suggesting that p75NTR mediates the effects of BDNF in normal and inflammatory conditions [122]. In a mouse model of activity-induced muscle pain, muscle fatigue metabolites promoted macrophages to release IL-1 β, which then promoted the release of BDNF from primary dorsal root ganglion (DRG) neurons, though, interestingly, this finding was male-specific [124].

In Schwann cells, BDNF release has been shown to be dependent on P2X purinoceptor 4 (P2X4) activation. Overexpression of this receptor in male mouse Schwann cells accelerated nerve remyelination via BDNF release following a nerve crush injury [97]. Schwann cell BDNF secretion is also dependent on the T-type voltage-gated calcium channel [100]. Contrarily, the inhibition of the P2X4 prevents an increase in BDNF release [125]. The release of BDNF by Schwann cells was found to regulate PKCε in DRG neurons via TrkB activation in a paracrine manner [101], which has been shown to lead to the sensitization of primary afferent nociceptors [126]. Together, the data suggest that BDNF's role in the PNS may be beneficial for reducing inflammation and aiding in nerve repair, though pain-related BDNF neuroimmune signaling mechanisms remain to be determined.

Central Nervous System

Mechanisms controlling microglial BDNF signaling are complex and have been well-reviewed by others [26,127,128]. Many stimuli, such as extracellular nucleotides [30,129, 130] in spinal microglia and pro-inflammatory compounds [131] in cultured microglia, have been shown to stimulate BDNF secretion from microglia. Following peripheral nerve injury, BDNF was found to regulate spinal microglial autophagy through the AKT/mTOR pathway [108]. It has been proposed that nerve injury in male rats induces ATP release, activates microglial P2X4, and leads to spinal microglial BDNF secretion [132]. BDNF binding to the TrkB receptor on lamina I neurons may then interrupt chloride homeostasis and lead to increased intracellular chloride levels via a downregulation of the K^+/Cl^- cotransporter KCC2. These alterations may ultimately disinhibit GABA- and glycine-mediated synaptic transmission [132], leading to increased excitability and eventual action potential firing in lamina I neurons, which has been shown to promote the development of neuropathic pain [105]. BDNF has also been implicated in crosstalk between microglia and astrocytes in the spinal cord. Intrathecal injection of exogenous BDNF promoted spinal microglial (and astrocytic) activation and a subsequent increase in pro-inflammatory cytokine expression in a cyclophosphamide-induced cystitis model [133]. Similarly, increased signaling between the astrocytic colony-stimulating factor-1 (CSF1) and the microglial CSF1 receptor was found 6 h after injury in a chronic post ischemic pain model; this microglial activation led to a subsequent increase in the synthesis and secretion of BDNF, which heightened neuronal activity in the spinal dorsal horn [134]. Exogenous microglial activation caused similar effects in naïve rats, whereas inhibition of this astrocyte–microglia crosstalk suppressed BDNF upregulation and neuronal activity in ischemic rats [134]. These effects have predominately been explored in male animals. It is important to note that females do not exhibit the same upregulation of microglial P2X4 as males in neuropathic pain conditions [135]; therefore, potential sexually dimorphic mechanisms may occur with regard to microglial BDNF signaling in this region.

BDNF signaling mechanisms in astrocytes and oligodendrocytes have not been well characterized. In hippocampal or perirhinal cortex slices, proBDNF has been reported to be endocytosed by astrocytes in a p75-dependent manner [65,136]. The TrkB.T1 also mediates the storage of endocytosed proBDNF in astrocytes [137]. Following intracellular cleavage of neuronal pro-BDNF by astrocytes, mature BDNF was shown to be released and act on the TrkB receptor of adjacent neurons [65]. Astrocytic BDNF release may also be induced through the inhibition of their inwardly rectifying potassium (Kir) 4.1 channels [138]. Other factors that induce astrocytic release of recycled BDNF include glutamate release from the presynaptic terminal [139,140] and ATP through the P2X7 [141]. Oligodendrocyte modulation of neurotransmitter release was shown to utilize BDNF derived from oligodendrocytes and TrkB receptor signaling at presynaptic brainstem terminals that express the vesicular glutamate transporter VGluT1 [142]. Injection of a lentiviral vector expressing BDNF into the spinal cord was found to promote M2 polarization of local macrophages in a mouse spinal cord injury model [143]. Though further studies are needed, particularly among supraspinal regions, the neuroimmune-related signaling of BDNF in the CNS seems to confer detrimental effects. Most spinal BDNF actions appear to involve glia-to-neuron signaling pathways.

3. Peripheral Nociception

BDNF has consistently been linked to pronociceptive processes in the peripheral nervous system, but the regulatory mechanisms of peripheral BDNF levels in pain conditions remain a substantial knowledge gap.

3.1. Expression and Localization

Early immunostaining analyses found that BDNF was equally expressed in small-, medium-, and large-size DRG cells in rats under normal conditions [144,145], while a subsequent study using transgenic manipulation to express β-Gal under the control of the BDNF promoter (*BDNF*$^{LacZ/+}$ mice) demonstrated that BDNF is expressed only in a specific group of small-to-medium-sized nociceptors containing other neuropeptides that play important roles in nociceptive signaling [37,146], pointing to some inconsistencies that might be due to technical issues related to the use of non-specific antibodies. Electron microscopy immunohistochemical analyses revealed that approximately 55% of BDNF-immunoreactive neurons (L4–L5 DRG) showed positive signals for calcitonin gene-related peptide (CGRP), and many primary afferents in the laminae I–II of the spinal cord co-express CGRP and the low-affinity p75NTR, suggesting a parallel release of BDNF alongside neurotransmitters from specific nociceptive primary afferents in the spinal cord [147]. Importantly, BDNF did not seem to colocalize with the non-peptidergic marker isolectin B4 (IB4) [148]. A study using *BDNF-LacZ* reporter mice found that β-Gal positive signal, a marker of BDNF-expressing neurons, predominantly colocalized with NF200, a marker of myelinated fibers, suggesting that BDNF is contained in myelinated primary afferents and its expression is limited to the subclasses of nociceptors and pruritoceptors [37]. Notably, BDNF did not colocalize with glial markers in the DRG in a double transgenic *BDNF-Cre/floxed-tdTomato* mouse model [37]. Electron microscopy showed TrkB-FL receptor expression on primary afferent fibers of both mice and rats [144,149].

Peripheral BDNF expression and release has been implicated in various pain conditions, including neuropathic, inflammatory, visceral, bone, and musculoskeletal pain [150]. In neuropathic (sciatic nerve photochemical injury, Gazelius model) rats with allodynia, BDNF was upregulated in small- and medium-sized DRG cells as compared to the non-allodynic group that showed increased BDNF signals preferentially in large-sized neurons [145]. Significant BDNF immunoreactivity upregulation was found in lumbar (L4–L5) DRGs and in the axonal terminals in the L4/L5 dorsal horn in neuropathic (chronic constriction injury, CCI; and spinal nerve ligation, SNL) [151–153] and NGF-induced [154] or Complete Freund's Adjuvant (CFA) inflammatory [155] pain models, as well as in cervical DRGs in the bilateral cervical facet joint distraction pain model [156], pointing to the involvement of the endogenous BDNF in the initiation of different types of pain. BDNF and TrkB positive signals were colocalized in DRGs of diabetic rats (model established by the combination a specific diet with a systemic streptozotocin (STZ) injection) [157]. Similarly, the protein and mRNA levels of BDNF and TrkB were increased in DRGs of STZ-induced diabetic animals compared to the control group, and chronic (6 weeks) exogenous administration of BDNF or a BDNF-sequestering fusion protein (TrkB–Fc) failed to revert these effects [158]. CFA inflammatory pain or TNF-α intraplantar injection induced the upregulation of BDNF and TrkB receptor signals in rat DRGs compared to the contralateral side, while primary (L1–L6) DRG cultures chronically (24 or 48 h) treated with TNF-α showed increased the mRNA and protein levels of BDNF and TrkB receptor, as well as enhanced the release of BDNF [159]. After the induction of inflammatory pain (formalin and CFA), proBDNF protein levels were found to be upregulated in local tissue of mice, while BDNF was downregulated [160], suggesting an impairment in BDNF cleavage processes in the inflamed area.

Several studies have suggested that peripheral BDNF may be linked to different signaling pathways in pain conditions. For example, BDNF immunoreactivity or mRNA levels were associated with P2X4 and P2X7 [125,152], NGF and TrkA [153], p75NTR [161,162] and mTOR [162], and ERK1/2 and tumor necrosis factor receptor 1 (TNFR1) signals [159], while changes in BDNF total protein levels were related to Huntington-associated protein 1 (HAP1)-induced reductions in L-type calcium channel (Cav1.2) signaling [163] in pain. Finally, BDNF release by peripheral immune cells, particularly Schwann cells, was linked to TrkB-mediated PKCε signaling [101].

3.2. Cellular Functions

Only a few studies have addressed the peripheral neuronal actions of BDNF in pain processing. In a saphenous nerve–skin preparation obtained from normal rats, acute BDNF application induced sensitization to the heat stimulation of the C-fibers [164], suggesting its crucial role in peripheral heat responses. Moreover, BDNF significantly increased the release of CGRP and substance P in (L1–L6) DRG cultures pretreated chronically (48 h) with TNF-α compared to their untreated (non-TNF-α-treated) counterpart, although it failed to potentiate the release of the two mediators in the TNF-α group when applied for 30–60 min [159]. In contrast, chronic (6 weeks) i.th. administration of BDNF in STZ-induced diabetic rats effectively reversed the changes in acutely cultured DRG neurons induced by the neuropathic pain model, which included depolarized resting membrane potential (RMP), reduced rheobase, and enhanced action potential frequency [158], pointing to antinociceptive BDNF properties in pain conditions, which may be mediated by compensatory mechanisms induced by the prolonged treatment. These effects were not observed in the presence of TrkB–Fc in neuropathic animals or in the control (non-neuropathic) group [158]. Conditional *Bdnf* knockout mice obtained by the preferential deletion of BDNF from peripheral sensory neurons (Avil-CreERT2 mice) did not show changes in the responses of spinal L3–L5 dorsal horn wide-dynamic-range neurons evoked by mechanical or thermal stimulation under normal conditions [165]. Although little information is currently available about the action of peripheral BDNF on pain processing, these electrophysiological results point to the facilitatory effects of BDNF in the development of pain-related responses (Table 1).

3.3. Behavioral Studies

In the postnatal life period, BDNF plays essential functions for the survival of peptidergic and non-peptidergic nociceptors of the spinal L1 and L4 segments [166]. Mice carrying mutations of the *Bdnf* gene (homozygous mutant, $Bdnf^{-/-}$) lost approximately half of all nociceptive neurons during the first 2 weeks of life, and $Bdnf^{+/-}$ heterozygous animals, which were used instead of the $Bdnf^{-/-}$ homozygous mutant animals because of their poor life expectancy, exhibited decreased nociceptive behavior in the hot plate test compared to their wild-type counterpart under basal conditions [166]. Conditional *Bdnf* knockout (*Avil-CreERT2* to selectively delete BDNF from primary sensory neurons) mice did not have altered baseline nociceptive responses or motor functions [37,165]; there was no difference between male and female mice, whereas the conditional *Bdnf* knockout males, but not females, showed increased thermal withdrawal latencies in the tail immersion test [37]. No significant differences were found between the knockout and wild-type groups in pruritogen-induced scratching responses or in pain-like behaviors in spared nerve injury (SNI)- or paclitaxel-induced neuropathies or the CFA inflammatory model [37]. *Avil-CreERT2* male mice showed reduced nociceptive responses in the second phase of the formalin test [37,165], while (knockout) females displayed diminished histamine-induced scratching [37], pointing to complex and sexually dimorphic BDNF functions in peripheral pain processing. Moreover, *Avil-CreERT2* male and female mice showed a reversal of mechanical hypersensitivity at the chronic stage of the SNL model of neuropathic pain, while no differences were found in the partial sciatic nerve ligation (pSNL) model [165]. In the same study, the prolonged hyperalgesia in the priming model achieved by the intraplantar injection of prostaglandin E2 (PGE2) following an intraplantar administration of carrageenan was lost in BNDF knockout mice [165], suggesting that BDNF effects on chronic pain may depend on the type of injury. When BDNF was injected into the plantar surface of the rat paw, it resulted in hyperalgesic responses compared to the vehicle-treated paw in the same animals [164] and induced mechanical allodynia in naïve mice [97]. Similarly, the application of BDNF onto intact L5 DRGs by osmotic pumps promoted a long-lasting (7 days) mechanical allodynia in uninjured rats [160], pointing to a strong contribution of BDNF to the development of peripheral pain responses. Interestingly, exogenous BDNF had no effects in the sciatic nerve crush injury mouse model [97], while blockade of BDNF signaling by the injection of an antibody against BDNF on L5 DRGs reversed mechanical allodynia in a L5 spinal nerve lesion rat model [160]. Importantly, intraplantar application of an adenovirus vector encoding proBDNF gene (Ad-proBDNF), but not (mature) BDNF, promoted nociceptive responses (licking and flitching) in combination with a low (0.5%) dose of formalin and decreased mechanical withdrawal thresholds in normal mice [167]. Pretreatment with a polyclonal anti-human proBDNF antibody (poly-Ab-proBDNF) inhibited both phases of the formalin- and visceral pain (induced by the systemic injection of acetic acid)-induced pain responses [167], suggesting a critical role of proBDNF in peripheral nociception. Intra-articular injection of BDNF into the rat knee had no effects in naïve animals but exacerbated weight-bearing impairments and mechanical allodynia in an osteoarthritis pain model induced by monoiodoacetate injection; intra-articular TrkB-Fctreatment partially reversed weight-bearing asymmetries and mechanical allodynia in surgically or monoiodoacetate induced osteoarthritis pain models [168]. Little has been studied with regard to the role of neuroimmune signaling mechanisms involving BDNF on pain-related behaviors in the periphery. Selective inhibition of BDNF signaling in macrophages via the neurotrophin inhibitor Y1036 prevented hypersensitivity induced by the application of nucleus pulposus to the sciatic nerve in male and female mice [169].

Collectively, these reports suggest that BDNF from peripheral afferents has facilitatory effects in the development of pain-related behaviors under normal conditions, although some inconsistent results were obtained with exogenous BDNF applications into the knee joint, perhaps pointing to important differences in routes of administration and tissue-specific actions with respect to the appropriate behavioral outcome measures. Transgenic manipulations did not yield conclusive information, while blockade of peripheral BDNF and/or TrkB signaling can decrease pain-like behaviors, particularly in models of prolonged or chronic pain, although important differences in BDNF effects were reported in different pain models (Table 2).

4. Spinal Nociception

The pronociceptive role of BDNF in the spinal cord has been extensively studied at the molecular, cellular, and behavioral levels. In fact, most of the information about the contribution of BDNF to pain processing originates from research on spinal mechanisms.

4.1. Expression and Localization

There is some controversy about the cellular source of BDNF in the spinal cord. Despite the evidence for microglia as a main source of BDNF in the spinal cord [30], immunostaining analyses showed that only a small subset of cells immunoreactive for the microglia marker Iba-1 co-expressed BDNF in dorsal horn laminae I-II of naïve post-natal [144] or of chronic post-ischemic pain (CPIP) [134] rats. Additionally, in double transgenic *BDNF-Cre/floxed-tdTomato* mice, BDNF seemed to be selectively expressed by neurons and not by glial cells in lamina I-V of the spinal cord [37]. The colocalization of GFP-expressing microglia with BDNF immunostaining in the spinal dorsal horns of transgenic CX3CR1$^{+/GFP}$ mice after partial nerve ligation (PNL) [130] was not confirmed by subsequent studies. On the other hand, in primary microglia cultures, morphine-induced BDNF release was linked to ATP-mediated P2X4R activation [129,170], which was mediated by SNARE (soluble N-ethylmaleimide-sensitive factor attachment protein receptor) and p38/MAPK (mitogen-activated protein kinase) mechanisms [129]. Increased BDNF has additionally been shown to contribute to increased spinal microglia activation in neuropathic (streptozotocin-induced neuropathic pain and SNI) [171,172] and inflammatory (experimental autoimmune prostatitis) [173] pain models, and higher BDNF levels in microglia of the spinal nucleus of the tri-geminal nerve were associated with increased trigeminal allodynia in male rats [174].

Increased BDNF and TrkB levels were found in different pain conditions. In a model of inflammatory pain induced by systemic administration of NGF, BDNF immunoreactivity in the superficial layers of spinal dorsal horns was significantly enhanced compared to control in neonatal rats [154]. BDNF immunoreactivity in the ipsilateral (to the injury) superficial layers of the spinal cord was significantly enhanced after pSNL surgery compared to the contralateral side [175]. Similarly, an upregulation of BDNF mRNA and protein levels was detected in the superficial layers of spinal cord in rats with bilateral cervical facet joint distraction compared to a sham control group [156], while enhanced BDNF protein expression was found in the spinal L4–L5 segments ipsilaterally to the side of mammary gland carcinoma cells implantation in a model of bone cancer pain, and this upregulation was mediated by the activation of the proteinase-activated receptors 2 (PAR2)/-NFκB pathway [176].

Several studies focused on the presence of the TrkB receptor and its functional full length (TrkB-FL) isoform as potential targets and mediators of BDNF signaling. Electron microscopy analyses showed that TrkB-FL receptors were expressed postsynaptically on the dendrites and somata of second order neurons in the mouse and rat spinal dorsal horns [144,149], which led to the concept that BDNF released from afferent nerve terminals could act on pre- and postsynaptic elements to modulate nociceptive messages. However, only a small portion of TrkB-FL immunoreactive dendrites was found to form synapses with BDNF immunoreactive axons, though this was explained with technical limitations [149]. Additionally, enhanced BDNF and TrkB mRNA and protein levels were detected in the spinal cord of diabetic rats (STZ model) [157]; they were also increased in rats after C-fiber stimulation and in the CFA model [155] and in mice with neuropathic pain in the experimental autoimmune encephalomyelitis (EAE) model of multiple sclerosis [177]. Phosphorylation of TrkB is a critical mechanism for BDNF actions. The decrease in spinal phosphorylated TrkB protein levels by the blockade of TrkB autophosphorylation with an antagonist (1NM-PP1, a small protein derivate of the protein kinase inhibitor protein phosphatase 1) in *TrkB*^{F616A} knock-in mice blocked the mechanical hypersensitivity in the capsaicin model of inflammatory pain [178].

The spinal cord seems to be a key structure for BDNF actions. There is good evidence for the presence of BDNF in spinal neurons. Still, it is commonly accepted that BDNF is released from microglia to act postsynaptically on neurons in the spinal cord (see Section 4.2), although the actual evidence for microglia-derived BDNF is not quite strong, with opposite findings (see Section 2.4.1), and therefore remains a knowledge gap.

4.2. Cellular Functions

BDNF signaling in the spinal cord plays a role in pain-related central sensitization and the regulation of neurotransmitters and neuromodulators.

In spinal cord slices from normal rats, BDNF application depolarized lamina I dorsal horn neurons and converted GABA-A receptor-mediated hyperpolarizing responses into depolarizing responses in a third of the recorded cells [30]. Importantly, in slices obtained from rats with peripheral nerve injury (PNI), treatment with a function-blocking antibody against the TrkB receptor (anti-TrkB) restored the hyperpolarizing GABA-induced effects that were lost in the untreated group, suggesting that endogenous BDNF is required for the pain-induced changes; here, BDNF was shown to be released from microglia and to act via TrkB [30].

BDNF application also significantly increased the responses of spinal neurons induced by NMDA or C-fiber stimulation in spinal cord slices, while a facilitatory trend was observed for A-fiber stimulation [154,179]. These effects were reversed by pretreatment with a BDNF-sequestering antibody (TrkB-IgG) [154]. Moreover, BDNF induced Ca^{2+} oscillations in lamina II neurons and resulted in an increase in the frequency, but not amplitude or decay time, of miniature excitatory synaptic currents (EPSCs) in dorsal horn neurons, suggesting the involvement of presynaptic mechanisms [144]. These effects were blocked by the application of a TrkB antagonist (K252a) or anti-trkB antibody (IgG1—clone 47), by the co-administration of AMPA and NMDA receptor antagonists (NBQX and D-AP5, respectively), and by a substance P-NK1 receptor blocker (L-732-138), suggesting that BDNF has facilitatory effects on glutamatergic and peptidergic transmission through TrkB activation [144]. Pretreatment with BDNF resulted in the failure of acutely applied capsaicin to induce Ca^{2+} oscillation in lamina II neurons, which could be explained by occlusion, consistent with BDNF engaging the release of neurotransmitters from nociceptive terminals [144]. In spinal cord slices from rats treated systemically with NGF to induce an inflammatory pain state, TrkB-IgG superfusion decreased the enhanced responses of spinal neurons to C-, but not A-, fiber stimulation, suggesting that BDNF signaling mediates spinal nociceptive processing in inflammatory pain [154]. Likewise, the i.th. injection of the TrkB-Fcchimera (to sequester BDNF) decreased excitatory synaptic responses (EPSCs) of lamina II neurons to the electrical DRG stimulation in a bone cancer-induced rat pain model [176]. Importantly, BDNF-mediated facilitatory effects were associated with the activation of the PLC/PKC pathway [175,179] that is linked to TrkB signaling (see Section 2.2).

It should be noted that early evidence from preclinical studies pointed to antinociceptive properties of exogenous BDNF at the spinal level under normal conditions, mainly mediated by an increased release of GABA from spinal interneurons. In an isolated dorsal horn preparation, BDNF bath application decreased the electrically or capsaicin-induced release of substance P from sensory neurons through concerted mechanisms involving GABA-B and TrkB signaling [180]. Importantly, in the same experimental setup, naloxone application failed to block BDNF-related effects, suggesting that the opioid system was not required in this process. Additionally, exogenous BDNF facilitated the release of GABA caused by K^+ depolarization through TrkB receptors under normal conditions [180] and restored GABA levels that were depleted in a neuropathic pain model (7 days after SNL surgery) [181].

4.3. Behavioral Studies

A large body of evidence supports the involvement of BDNF in the development of pain at the level of the spinal cord (Table 2). A single i.th. injection of BDNF or extended delivery using a BDNF-transducing recombinant adenovirus (adBDNF) caused tactile allodynia and thermal hyperalgesia in naïve animals [30,175,182], and these effects were reversed by the pretreatment with antisense oligonucleotide against TrkB-FL mRNA that downregulated the expression of the receptor [182]. Heat hyperalgesia was also produced in normal rats by the i.th. injection of a high-affinity TrkB ligand, neuroptrophin-4/5 (NT-4/5), providing further evidence for TrkB-mediated behavioral effects [182]. Disruption of BDNF-TrkB signaling by the spinal application of antisense oligonucleotide against BDNF mRNA or anti-BDNF antibody or a BDNF-sequestering fusion protein (TrkB–Fc) reversed the enhanced mechanical and thermal responses in neuropathic (PNI, pSNL, or SNL) [30,153,175,183], bone cancer [176], bilateral cervical facet joint distraction [156], and carrageenan-induced inflammatory [182] pain models. In adult rats, i.th. injection of TrkB-IgG ameliorated nociceptive behaviors in both phases of the formalin test when animals were previously primed with systemic NGF injection compared to a saline pretreated control group and improved thermal withdrawal responses in a carrageenan-induced inflammatory pain model [154], suggesting that endogenous BDNF is involved in inflammatory pain.

There is evidence for the release of BDNF from microglial cells as a mechanism of pain at the spinal level. ATP-challenged microglia have been shown to cause tactile allodynia that was associated with the release of BDNF, demonstrating for the first time the association between BDNF and microglia in the modulation of pain behavior [30]. In fact, ATP-stimulated microglia previously treated with anti-TrkB or TrkB–Fc or transfected with BDNF siRNA failed to evoke changes in the mechanical withdrawal thresholds when delivered in the spinal cord of rats [30]. In support of these findings, P2X4 receptor-mediated release of BDNF by microglia activation with ATP was shown [129], and transgenic mice lacking P2X4 receptors ($P2X4^{-/-}$) did not develop mechanical allodynia in PNL and SNI models of neuropathic pain via impaired BDNF release mechanisms in the spinal dorsal horns [130]. In the SNI model, the neuronal KCC2 downregulation induced by microglial BDNF release in the spinal dorsal horn led to dynamic mechanical allodynia; inhibiting microglia (and subsequent BDNF secretion) suppressed the induction of mechanical allodynia in male mice [184]. Conditional knockout of BDNF from microglia also prevented pain hypersensitivity in male mice with peripheral nerve injury [108]. BDNF-related crosstalk between glial cells has also been reported to influence pain behavior. Exogenous activation of spinal microglia with a CSF1 receptor agonist increased BDNF secretion and promoted mechanical allodynia in naïve rats, whereas inhibition of astrocytic-microglial CSF1-CSF1 receptor signaling with PLX-3397 (a CSF1 receptor antagonist) prevented BDNF release and relieved mechanical allodynia and thermal hyperalgesia in rats with ischemic pain [134]. Inhibiting spinal astrocyte activation with the BDNF/TrkB inhibitor ANA12 alleviated mechanical allodynia in a mouse partial crush injury model [185]. Together, the data suggest a pro-nociceptive role of BDNF within the spinal cord, and interventions that modulate BDNF signaling by targeting microglia and astrocytes show promise as potential therapeutic strategies for neuropathic pain conditions. It should be noted, however, that the colocalization of BDNF and microglial (e.g., Iba-1) markers in immunostaining has never been shown (see Section 4.1). This seems to be a controversial matter, perhaps due to technical limitations. Advanced technologies, such as single-cell transcriptomic analyses, offer the opportunity to resolve this important knowledge gap.

Importantly, a sexual dimorphism of BDNF signaling in microglia was reported with respect to pain processing. I.th. administration of an NGF/BDNF inhibitor (Y1036) or the BDNF-sequestering fusion protein TrkB-Fc and tamoxifen-induced Cre-loxP-mediated deletion of the Bdnf gene in CX3CR1-positive cells blocked SNI-induced mechanical allodynia in male, but not female, mice [186].

It should be noted that early reports point to antinociceptive mechanisms of spinal BDNF (see the last paragraph in Section 4.2). In these studies, i.th. injection of BDNF ameliorated thermal, but not mechanical, withdrawal thresholds in the injured paws of SNL rats (7 days after surgery) without affecting the responses of the contralateral paw, and this effect was blocked by a GABA-B receptor antagonist (CGP55445), suggesting that BDNF effects involved GABA-B signaling [181]. Surprisingly, similar effects were observed in normal conditions [180].

5. Pain Processing and Modulation in the Brain

5.1. Expression and Localization

BDNF signaling has been investigated at different supraspinal levels in the brainstem and brain in several models of pain.

In the brainstem, enhanced BDNF protein and immunomarker levels were found in the periaqueductal gray (PAG) in an inflammatory (CFA) pain condition [187]. Moreover, upregulation of TrkB (full-length but not truncated) and TrkB phosphorylation were detected in the rostral ventromedial medulla (RVM) of CFA rats [187], suggesting that the BDNF neurons projecting from PAG to RVM are activated in the inflammatory pain model. The PAG-RVM system is a critical component of the descending pain modulatory system [188]. Importantly, TrkB was shown to be expressed in the RVM serotoninergic neurons that project to the spinal cord as part of the descending pain control pathway [189]. In vitro studies performed in RVM slices revealed that the facilitatory effects of BDNF were associated with the tyrosine phosphorylation of the NMDA NR2A subunit through the IP3, PKC, and Src signaling pathway [187]. Higher BDNF protein levels in the RVM were detected in a model of pain induced by the combination of plantar incision with presurgical (24 h) paradoxical sleep deprivation as compared to the injury group [190], suggesting the critical involvement of BDNF signaling in nociceptive responses aggravated by sleep impairments. In the VTA, increased BDNF protein level and release were observed in CCI mice.

Within the limbic system, increased BDNF protein levels were detected in the central nucleus of amygdala (CeA) of CFA animals, which were associated with reduced activity of the transcriptional repressor histone dimethyltransferase G9a, supporting the idea of BDNF facilitatory effects in the amygdala [191]. Similarly, BDNF mRNA and protein expression was upregulated in the medial thalamus (MT) of central poststroke pain (CPSP) rats [192]. In a pain condition induced by chronic intermittent stress (CIS) followed by the induction of thermal injury (burn), increased BDNF, TrkB, and phosphorylated TrkB protein levels were observed in the hypothalamus, but not mPFC, as compared to the un-stressed group [193]. On the other hand, in the thalamic paraventricular nucleus, pro-BDNF and BDNF were downregulated in chronic restraint stress (CRS) mice [194]. In the reward system, BDNF protein and release were found to be upregulated NAc of neuropathic (CCI) mice [195].

In the cortex, BDNF mRNA and protein levels were upregulated in the ACC of rats in models of bone cancer [196], inflammatory (CFA) [109] and neuropathic (SNI) [197, 198] pain. Importantly, BDNF immunostaining was increased in the contralateral (to the injury) S1 hindlimb portion of rats with CFA-induced inflammatory pain and was detected in GFAP-, Iba-1-, and NeuN-expressing cells, suggesting that BDNF is expressed in astrocytes, microglia, and neurons [109]. Accordingly, fluorescence in situ hybridization (FISH) experiments detected a higher percentage of BDNF mRNA and number of BDNF mRNA puncta in microglia in the contralateral (to the injury) S1 of neuropathic mice (SNI model) as compared to a sham group [108]. Interestingly, BDNF did not show changes at the protein or mRNA levels in PFC of SNI rats [197], and decreased BDNF levels were found in the IL but not prelimbic (PL) cortex of CFA rats [199], suggesting region-specific changes in BDNF in the mPFC. These bidirectional changes in cortical BDNF expression in pain conditions were also detected at the receptor level. TrkB mRNA and protein expression was enhanced in the ACC, but not PFC, in neuropathic (SNI) pain [197]. Reduced levels of BDNF were also observed in the hippocampal dentate gyrus (DG) [200] and CA1, as well as in the IL cortex [201], in an inflammatory (CFA) pain condition. Similarly, BDNF was found to be downregulated, while TrkB was upregulated, in hippocampal tissue in a mouse model of thalamic hemorrhage-induced CPSP [202].

In contrast to the periphery and spinal cord, supraspinal changes in BDNF in pain seem to be dependent on the targeted (sub-)region. Overall, a shift towards increased BDNF expression is observed in the brainstem and limbic regions in pain conditions. Within the cortex, the picture is less clear, and mixed and regional differences in pain-related expression change have been detected.

5.2. Cellular Functions

The cellular effects of BDNF signaling in supraspinal structures related to pain are not well understood but have become the focus of recent studies. Despite differential changes in the level of BDNF in different brain and brainstem structures (see Section 5.1), BDNF has been shown to primarily increase neuronal activity in these areas in pain models. For example, pain-induced upregulation of BDNF was associated with increased neuronal activity or evoked synaptic responses in the same areas such as S1 (SNI) [108], MT (CPSP) [192], nucleus raphe magnus (NRM) (CFA) [203,204], and trigeminal nucleus caudalis (TNC) (trigeminal allodynia induced by inflammatory soup) [174]. Importantly, these pain-induced abnormal neuronal activities were decreased by the application of TrkB antagonists, TrkB-IgG fusion protein, a BDNF scavenger, or depletion of BDNF [108,174,192,203,204]. On the other hand, inflammatory (CFA) pain-induced downregulation of BDNF in the ventral hippocampal CA1 to infralimbic cortex (vCA1-IL) pathway was accompanied by lower spontaneous neuronal firing and gamma power in the IL, decreased vCA1 to IL information flow, weakened phase-amplitude coupling (PAC) between vCA1theta phase and IL gamma amplitudes in in vivo electrophysiological experiments [199,201]. Infusion or overexpression of BDNF increased neuronal activity in the IL and normalized disrupted vCA1-IL connectivity in the pain model [199,201]. Similar to the beneficial effects of BDNF-induced neuronal activations in the vCA1-IL pathway under pain conditions, morphine-induced analgesia was mediated by BDNF-TrkB signaling and increased neuronal activity (c-Fos) in the amygdala (CeA) and BNST, which was decreased by the deletion of BDNF in the PB [205].

Several mechanisms have been linked to BDNF signaling in the brain. Upregulation of BDNF in inflammatory (CFA) and chronic (PNI or CPSP) pain conditions resulted in a TrkB-mediated downregulation or increased phosphorylation of KCC2 to disrupt Cl^- homeostasis and increase neuronal excitability [192,204,206]. Furthermore, multiple lines of evidence suggest that BDNF-TrkB signaling engages glutamatergic neurotransmission in pain conditions, including the upregulation of NMDA receptor 2B subunit (NR2B) [196, 198], phosphorylation of AMPA receptor GluA1 subunit [203], and excitatory amino acid transporter (EAAT3) [174], to mediate increased excitatory synaptic transmissions and excitability or activation in pain. Downregulation of BDNF in a chronic neuropathic pain (CCI) model impaired the maintenance, but not induction, of hippocampal LTP, which was ameliorated by increasing BDNF in the hippocampus [207]. In the NRM, the exogenous application of BDNF increased frequency of mIPSCs and the accumulation of GAD65 in synaptic terminals under normal conditions [208], suggesting that BDNF-TrkB signaling can also modulate the release of inhibitory neurotransmitters. The effects of exogenous BDNF were lost in the CFA pain model due to increased endogenous BDNF [208]. Additionally, recent studies showed TrkB-dependent phosphorylation of ERK and CREB in inflammatory (CFA) and bone cancer pain models [109,196]. Activation of ERK-CREB signaling by BDNF could modulate gene expression and lead to long-lasting structural neuroplasticity underlying chronic pain conditions [209]. In line with this notion, systemic depletion of BDNF decreased neuropathic pain-induced dendritic spine remodeling in S1 [108], whereas rescue strategies to upregulate BDNF in the hippocampus (CA1) mitigated the reduction in dendritic spine and PSD-95 in neuropathic (CCI) pain [207] and promoted neurogenesis in an inflammatory (CFA) model of pain [200]. These findings collectively suggest that region-specific bidirectional changes in BDNF-TrkB signaling in supraspinal structures in pain conditions correlate with neuronal activity changes and neuroplasticity that can be mitigated by normalizing BDNF-TrkB signaling (Table 1).

5.3. Behavioral Studies

Pro- and anti-nociceptive effects have been reported in different brainstem regions. In the RVM, the injection of BDNF in naïve animals induced thermal hyperalgesia and mechanical allodynia [187,189] through NMDA receptors [187]. Conversely, suppression of BDNF-TrkB signaling in the RVM attenuated thermal hyperalgesia in an inflammatory pain model (CFA) [187] and paradoxical sleep deprivation-induced mechanical hypersensitivity in an incision pain model [190]. Interestingly, the injection of BDNF into the RVM in serotonin (5-HT)-depleted animals was antinociceptive, rather than pronociceptive [189], suggesting an important role of the 5-HT system in the pronociceptive effect of BDNF signaling in the RVM. The concentration of BDNF appears to be a key factor in its bidirectional effects on descending facilitation versus inhibition, because higher doses of exogenous BDNF in the RVM were antinociceptive, whereas lower doses were pronociceptive in naïve animals [187]. Consistent with the antinociceptive role of BDNF signaling in the RVM, blockade of TrkB in the RVM reversed histone deacetylase (HDAC) inhibitor-induced analgesic effects in an inflammatory pain model (CFA) [208]. BDNF signaling in the VTA may have beneficial effects in pain conditions, because overexpression of BDNF improved spatial memory formation in a neuropathic pain model (CCI), while BDNF knockdown blocked spatial memory improvement induced by the chemogenetic activation of DG-projecting VTA dopaminergic neurons [210].

In other supraspinal structures, the role of BDNF in pain modulation remains unknown or controversial. For example, infusion of BDNF into the midbrain near the (PAG) and dorsal raphe nuclei induced antinociceptive effects in naïve and formalin-injected animals [211–213]. In contrast, injection of BDNF into the PAG reversed the analgesic effect of transcranial direct current stimulation (tDCS) in a knee osteoarthritis model [214], pointing to a pronociceptive effect of BDNF in the PAG. In the central nucleus of the amygdala (CeA), the infusion of BDNF in naïve animals induced thermal hyperalgesia, promoted morphine reward, and rescued impaired morphine-induced CPP in animals with knockdown of a transcriptional regulator methyl CpG-binding protein 2 (MeCP2) [191]. Conversely, blockade of BDNF signaling in the CeA inhibited thermal hyperalgesia and morphine-induced CPP in an inflammatory pain model (CFA) [191]. On the other hand, intra-CeA injection of a BDNF scavenger (TrkB-Fc) decreased morphine-induced analgesia in naïve animals [205], suggesting perhaps an antinociceptive, rather than pronociceptive, role of endogenous BDNF signaling in the CeA BDNF. A potential source of BDNF in the CeA may be the parabrachial (PB) input because localized deletion of BDNF in the PB also decreased morphine-induced analgesia without affecting basal nociceptive responses and anxiety-like behaviors [205], indicating an important contribution of BDNF signaling in the PB-CeA pathway to opiate analgesia. In the thalamus, overexpression of BDNF in the parafascicular nucleus alleviated anxiety- and depression-like behaviors as well as hyperalgesia, while BDNF knockdown induced the opposite results in a chronic restraint stress (CRS) model [194]. In contrast, a BDNF scavenger (TrkB-Fc) or a TrkB antagonist (CTX-B) into the MT reversed hypersensitivity, but not thermal allodynia, in a CPSP model [192], indicating subregion-specific differential roles of BDNF signaling in thalamic pain processing.

In subcortical regions, pronociceptive effects of BDNF signaling have been consistently reported in the NAc. Injections of BDNF into the NAc shell increased thermal hyperalgesia without modulating depression-like behaviors in the CUMS model [215], while intra-NAc injections of a BDNF scavenger (TrkB-Fc) and a TrkB antagonist (ANA-12) decreased thermal hyperalgesia in a neuropathic pain model (CCI) [195] and optogenetically induced hypersensitivity in naïve animals [216], respectively. Injections of BDNF into the NAc reversed the antinociceptive effects of pharmacological inhibition, achieved with the injection of a GABA-B receptor antagonist (baclofen) or a I_h blocker (DK-AH269), of ventral tegmental area (VTA) neurons in a neuropathic pain model (CCI) [195]. Conversely, thermal nociceptive responses induced by the intra-VTA injection of morphine were prevented by the injection of a BDNF scavenger into the NAc shell without affecting depression-like behaviors in the CUMS model [215]. Furthermore, selective knockdown of BDNF in NAc-projecting VTA neurons reversed thermal hyperalgesia in a neuropathic pain model (CCI) [195]. These findings suggest that the VTA-NAc pathway is an important site for the pronociceptive action of BDNF signaling in the NAc. In contrast to the VTA-NAc pathway, BDNF signaling in the VTA-mPFC pathway seems to contribute to depression-like, but not nociceptive, behaviors [215].

Region-specific pro- and anti-nociceptive effects of BDNF signaling have been reported in the cortex. BDNF injections or viral vector-mediated upregulation of BDNF in the ACC produced cold hypersensitivity [109] and conditioned place avoidance (CPA) [196,197], as well as clonidine-induced conditioned place preference (CPP) [198], in naïve animals. Furthermore, exogenous BDNF into the ACC reversed the spinal clonidine-induced pain relief in a neuropathic pain model (SNI) [198]. Conversely, intra-ACC injections of a TrkB antagonist (Tat-CTX-B) blocked pain-related behaviors in inflammatory (CFA) [109], neuropathic (SNI) [197,198], and bone cancer [196] pain models. Several lines of evidence suggest a critical involvement of NR2B and the ERK-CREB signaling pathway in these pronociceptive effects of BDNF signaling within the ACC [196–198]. It is important to note that these effects were not observed with intra-PFC injections of BDNF or the TrkB antagonist [196–198]. Interestingly, intra-ACC injections of anti-proBDNF, but not anti-BDNF, antibodies mitigated chronic unpredictable mild stress (CUMS)-induced anxiety- and depression-like behaviors [217], suggesting a specific contribution of proBDNF signaling in the ACC to these emotional–affective behaviors. Similar pronociceptive effects of BDNF have been reported in the S1 cortex [108,109] and appear to involve microglial BDNF based on local depletion of microglial BDNF in transgenic mice [108]. In contrast to these pronociceptive effects, BDNF infusion into the IL subregion of the mPFC alleviated thermal hyperalgesia and mechanical allodynia in an inflammatory pain model (CFA) [199] and failed to promote thermal hypersensitivity in CUMS mice [215], suggesting region-specific effects of BDNF and differential roles of mPFC subregions in pain modulation in line with previous work [218]. Inputs from the hippocampus may play a key role in this antinociceptive effect, since overexpression of BDNF in the vCA1-IL pathway alleviated spontaneous pain, thermal hyperalgesia, mechanical allodynia, and anxiety-like behaviors in an inflammatory pain model (CFA) [201]. Overexpression of BDNF in the dentate gyrus (DG) also produced analgesic and anxiolytic effects and attenuated cognitive impairment in an inflammatory pain model (CFA) [200], suggesting that the IL and hippocampus are important brain areas for BDNF-induced beneficial effects under pain conditions, which is consistent with the merging role of disrupted hippocampal–mPFC (PL) connectivity in pain hypersensitivity and cognitive deficits [219].

In summary, activation of BDNF/TrkB signaling by exogenous BDNF has mixed effects in the brainstem but pronociceptive properties in subcortical areas, such as NAc, under normal conditions. Within the mPFC, the picture is less clear, with BDNF in the ACC having facilitatory effects while inhibitory effects were observed when IL or hippocampal inputs to the IL were targeted (and also the hippocampus itself). Similarly, blockade of BDNF/TrkB signaling had antinociceptive properties in the brainstem and subcortical regions, but mixed effects were observed within the mPFC, pointing to potentially region-specific roles of BDNF signaling in different brain regions in pain modulation (Table 2).

6. Other Brain Disorders

The involvement of BDNF in the brain in pain states remains a relatively understudied domain and a clear picture has yet to emerge. However, BDNF-related signaling pathways in the brain play important roles in other neurological and psychiatric conditions, which may be relevant to pain conditions as they are frequently comorbid with a wide spectrum of disorders such as anxiety and depression [220–222]. Therefore, information about BDNF signaling in these disorders may help inform about its role in chronic pain and the neurobiological interplay between chronic pain and its neuropsychiatric components.

6.1. Depression

Abundant research has implicated BDNF in the pathophysiology of depression and in the mechanisms of action of antidepressants, and BDNF-related signaling has become a therapeutic target for depression disorders [223–226]. In the rat learned helplessness model of depression, infusion of BDNF into the midbrain [227] or bilaterally into the hippocampal dentate gyrus [228] produced antidepressive effects. In the latter study, blockade of BDNF-TrkB signaling with the broad spectrum Trk inhibitor K252a prevented these effects [228]. In mice, decreased levels of BDNF in the hippocampus and PFC were linked to depression-like behavior in an inflammation-induced mouse model of depression [229], and conditional knockout of BDNF in the forebrain attenuated the actions of the antidepressant desipramine [230]. Similarly, corticosterone-induced depression in mice was associated with hyperactive neuronal autophagy in the dentate gyrus, which triggered increased lysosomal degradation of BDNF in neurons; inhibition of this autophagy with selective short hairpin RNA (shRNA) reversed the decreased neuronal BDNF expression and led to increased antidepressive effects [231].

However, BDNF in the brain seems to have region-specific effects on depression-like states. In contrast to the decreased levels of BDNF in forebrain areas, increased BDNF in the ventral tegmental area (VTA)–NAc pathway has been linked to the onset of depression [232,

233]. Infusions of BDNF into the VTA resulted in increased depression-like behavior, whereas local deletion of the gene encoding BDNF in VTA neurons (projecting to the NAc) had an antidepressant-like effect in mice in the social defeat stress model [234]. Increased levels of BDNF in the NAc were also associated with a depression-like phenotype in mice [229]. Together, the data suggest that while BDNF may exert antidepressive actions in forebrain regions such as the hippocampus and PFC, it may contribute to a depression-like phenotype in the mesolimbic pathway centered on the VTA and NAc [235]. Understanding the differential effects of BDNF in various brain regions with different roles in depressive states could further illuminate its complex functions in pain perception and modulation.

6.2. Schizophrenia

Clinical studies have consistently shown that schizophrenic patients have lower BDNF levels in the hippocampus [7], PFC [236], and parietal cortex [6] compared to normal individuals. At the preclinical level, rats with neonatal ibotenic acid lesions of the ventral hippocampus (a neurodevelopmental animal model of schizophrenia) showed decreased BDNF mRNA levels in the hippocampus and PFC [237,238]. Rats in the methylazoxymethanol acetate (MAM) model of schizophrenia had significantly decreased BDNF in the hippocampus compared to controls, which was associated with cognitive deficits in the acquisition and retention phases of the Morris water maze [239]. Mice with a forebrain-specific knockout of the TrkB receptor showed hyperlocomotion, cognitive impairments, and other stereotypical behaviors associated with animal models of schizophrenia [240]. More recently, BDNF-haploinsufficient mice with lower levels of BDNF in the amygdala, dorsal hippocampus, NAc, and PFC showed deficits in attentional set shifting, increased startle magnitudes, and prepulse inhibition deficits, which are behavioral phenotypes associated with schizophrenia models; interestingly, these endophenotypes were rescued by environmental enrichment [241]. Intracerebroventricular administration of BDNF decreased schizophrenic-like behaviors (startle response and disrupted prepulse inhibition) in the DBA/2J mouse strain, which presented several behavioral features relevant to schizophrenia [242]. The specific TrkB agonist 7,8-dihydroxyflavone promoted hippocampal synaptic plasticity and reversed cognitive deficits in a MK-801-induced rat model of schizophrenia [243]. Together, the data suggest that restoration of BDNF-mediated signaling in these brain regions could restore cognitive functioning, which is also an important dimension in the experience of pain.

6.3. Neurodegeneration

Neurodegenerative diseases encompass a wide range of neurological disorders, including Alzheimer's disease, Parkinson's disease, and Huntington's disease. While there are no pharmacological treatments currently available to alter the pathophysiology or provide a cure, beneficial effects of BDNF on cognitive functioning have been established [244–246]. It has been proposed that the decreased level of BDNF in neurodegenerative diseases leads to dysregulated GABAergic transmission via altered GABA release and transport in astrocytes and neurons, as well as through a decreased transcription of the GABA-A receptor [246]. Pharmacologically (aminopropyl carbazole) induced increases in hippocampal BDNF levels ameliorated cognitive function in a mouse model of Alzheimer's disease [247]. In animal models of Parkinson's disease, BDNF administration or increasing BDNF levels through gene transduction via viral delivery have been shown to enhance the survival of dopaminergic neurons and protect dopaminergic transmission to the striatum (reviewed in [248]). Inactivation of BDNF in the mouse forebrain led to a Huntington's disease-like behavioral phenotype [249], and inactivation of one BDNF allele in a Huntington's disease mouse model led to an earlier and worse behavioral and motor phenotype with severe striatal neuron loss [250]. Overexpression of BDNF in the forebrain in a mouse model of Huntington's disease improved the behavioral and motor phenotype and reduced neuropathological signs in the striatum [251,252]. Therefore, BDNF-mediated signaling within the brain may represent a promising therapeutic target to restore or protect cognitive function in pain and other conditions.

7. Conclusions

Although BDNF signaling in pain has been extensively studied throughout the pain neuroaxis (Figure 1), several knowledge gaps remain. In addition to the neuronal release of BDNF, accumulating evidence suggests a critical role of BDNF from the neuroimmune system as a pain mechanism, but despite that, the picture remains unclear. Peripherally, BDNF seems to be released by neuronal afferents into the spinal cord. Spinal BDNF promotes facilitation, but the cellular source of BDNF remains unclear. Several studies claimed that it is microglial BDNF acting on spinal neurons that serves pronociceptive functions, but this has not been demonstrated directly. Undoubtedly, neuronal BDNF plays a critical role in pain facilitation by acting on several cell types, including neuroimmune elements, and therefore engages neuron-to-glia interactions. Mixed and bidirectional effects of BDNF signaling were observed in the brainstem and brain in different models of pain, which may point to (sub-)region specific differences in BDNF function as well as subregion-specific pain mechanisms at the supraspinal level. Additionally, the cellular sources of BDNF in the brainstem and brain have not been fully identified and may contribute to the differential roles of BDNF in pain modulation. Finally, BDNF deficits and beneficial effects of BDNF have been reported in several neuropsychiatric diseases. The contribution of BDNF, its source(s), and signaling pathway(s) to pain mechanisms across the nervous system remain to be determined, especially with respect to neuroimmune signaling and sex differences.

Table 1. BDNF effects on cellular functions.

Intervention	Region and Assay	Species	Pain Model	Effect	Reference
Periphery					
BDNF	DRG culture (patch-clamp)	Rat	STZ-induced neuropathy	↓neuronal properties	[158]
	DRG culture	Rat	TNF-α treatment	↑substance P and CGRP release	[159]
Avil-CreERT2 (condition BDNF knockout from primary sensory neurons)	Spinal WDR neurons (in vivo electrophysiology)	Mouse	Naïve	No effects	[165]
Spinal cord					
BDNF	Isolated hemisected spinal cord	Rat	Naïve	↑NMDA-induced, C- and A-fiber evoked responses	[154]
BDNF	Lamina II neurons in slice (patch-clamp)	Rat	Naïve	↑C-fiber evoked responses (EPSCs)	[179]
BDNF	Lamina II neurons in slice (Ca^{2+} imaging)	Rat	Naïve	↑Ca^{2+} oscillations	[144]
			Capsaicin challenge	↓Ca^{2+} oscillations	
	Lamina II neurons in slice (patch-clamp)		Naïve	↑EPSC frequency; No effects on EPSC decay or amplitude	
TrkB-IgG	Lamina II neurons in slice (patch-clamp)	Rat	NGF-induced inflammation	↓C-fiber evoked responses	[154]
TrkB-Fc chimera	Lamina II neurons in slice (patch-clamp)	Rat	Bone cancer-induced pain	↓DRG evoked EPSCs	[176]
BDNF	Isolated dorsal horn with dorsal root attached	Rat	Naïve	↓electrical- or capsaicin-induced substance P release	[180]
				↑K^+-mediated GABA release	
			SNL	↑GABA release	[181]
BDNF	Lamina I neurons in slice (patch-clamp)	Rat	Naïve	↑GABA-mediated Ca^{2+} responses; depolarized E_{anion}	[30]
anti-TrkB			PNI	hyperpolarized E_{anion}	

Table 1. *Cont.*

Intervention	Region and Assay	Species	Pain Model	Effect	Reference
Brain and brainstem					
TrkB-Fc	MT (in vivo electrophysiology)	Rat	CPSP	↓SNS-electrically evoked neuronal response	[192]
BDNF	NRM (patch-clamp)	Rat	Naïve	↑frequency and amplitude of AMPA mEPSCs	[203]
TrkB-IgG			CFA	↓AMPA EPSCs	
BDNF	NRM (patch-clamp)	Rat	Naïve	↓mIPSC frequency	[208]
				Depolarizing shift in EPSC and ↑excitability in MOR-expressing neurons	[204]
TrkB-IgG			CFA	Hyperpolarizing shift in EPSC and ↓excitability in MOR-expressing neurons	
BDNF				No effect on mIPSC frequency	[208]
pAAV2-hSyn-Cre-GFP, (AAV2-Retro) + pAAV2-CAG-DIO-BDNF-mCherry-3*flag(vCA1-IL pathway-specific overexpression of BDN)	IL (in vivo electrophysiology)	Rat	CFA	↑spontaneous neuronal firing, power spectral density in low gamma band, gPDC	[201]
p156sinRRLpptCAG-BDNF (BDNF lentiviral vector)	S1 (hindlimb part)	Rat	CFA	↑LTP	[109]
Cx3cr1$^{CreER/+}$;*Bdnf*$^{fl/fl}$ (systemic depletion of microglial BDNF)	Layer 5 S1 (in vivo two-photon imaging)	Mouse	SNI	↓spontaneous and mechanically induced Ca^{2+} activity	[108]
EE-induced BDNF increase	Hippocampus	Mouse	CCI	↑LTP maintenance (fEPSP)	[207]

CCI = chronic constriction injury; CGRP = calcitonin gene-related peptide; CFA = Complete Freund's Adjuvant; CPSP = central poststroke pain; DRG = dorsal root ganglion; EE = environmental enrichment; fEPSP = field excitatory postsynaptic potentials; gPDC = generalized partial directed coherence; EPSCs = excitatory postsynaptic currents; IL = infralimbic cortex; LTP = long-term potentiation; mIPSC = miniature inhibitory postsynaptic currents; MT = medial thalamus; NGF = nerve growth factor; NRM = nucleus raphe magnus; PNI = peripheral nerve injury; S1 = somatosensory cortex; SM1 = somatosensory cortex; SNL = spinal nerve ligation; SNI = spared nerve injury; SNS = sciatic nerve stimulation; STZ = streptozotocin; WDR neurons = wide-dynamic-range neurons.

Table 2. BDNF effects on pain-like behaviors.

Intervention	Region/Assay	Species	Pain Model	Effect	Reference
Periphery					
Avil-CreERT2 (condition BDNF knockout from primary sensory neurons)	Nocifensive behaviors	Mouse	Formalin test	↓second phase	[37,165]
	Mechanical allodynia		SNI- or paclitaxel-induced neuropathy	No effects	[37]
	Mechanical and thermal hypersensitivity		CFA inflammatory pain		
	Mechanical allodynia		pSNL		[165]
	Mechanical hypersensitivity		SNL		
			Hyperalgesic priming	↓	
BDNF	Mechanical allodynia	Mouse	Normal	↑	[97]
			SNC	No effects	
	Thermal hypersensitivity	Rat	Normal	↑	[164]
	Weight-bearing deficits and mechanical allodynia	Rat	Normal	No effects	[168]
			MIA	↓	
			MNX		
Anti-BDNF antibody	Mechanical allodynia	Rat	Normal	↑	[160]
			L5 spinal nerve lesion model	↓	
Ad-proBDNF (adenovirus vector-encoding proBDNF gene)	Mechanical allodynia	Mouse	Normal	↑	[167]
	Flitching and licking (second phase)		Formalin (0.5%)		

Table 2. *Cont.*

Intervention	Region/Assay		Species	Pain Model	Effect	Reference
		Spinal cord				
Anti- BDNF antibody	Thermal hyperalgesia		Rat	SNL	↓	[153]
BDNF						[181]
				Normal		[180]
	Mechanical allodynia and thermal hyperalgesia		Mouse		↑	[175] [182]
adBDNF	Mechanical allodynia		Rat			[30]
neuroptrophin-4/5						
Antisense oligonucleotide against BDNF mRNA	Thermal hyperalgesia		Mouse	Carrageenan inflammatory pain		[182]
	Mechanical allodynia			pSNL		[183] [175]
TrK-Fc	Mechanical allodynia and thermal hyperalgesia		Rat	Bone cancer pain	↓	[176]
				Bilateral cervical facet joint distraction		[156]
				PNI		
ATP-challenged microglia with anti-TrkB, TrkB-Fc BDNF siRNA	Mechanical allodynia		Rat	Normal	No effect	[30]
TrKB-IgG	Nocifensive behaviors		Rat	NGF primed in formalin test	↓	[154]
Y1036, TrkB-Fc, *Cx3cr1*^{CreER} × *loxP-Bdnf* (tamoxifen-induced Cre-loxP–mediated deletion of the Bdnf gene in CX3CR1-positive cells)	Mechanical allodynia		Mouse	SNL	↓	[186]
		Brain and brainstem				
BDNF, p156sinRRLpptCAG-BDNF (BDNF lentiviral vector)	ACC	Cold hypersensitivity, CPA, clonidine- induced CPP	Rat	Naïve	↑	[109,196–198]
Tat-CTX-B				CFA, SNI, bone cancer pain	↓	
BDNF	IL	Thermal hyperalgesia and mechanical allodynia	Rat	Naïve	No effects	[199]
				CFA	↓	
pAAV2-hSyn-Cre-GFP, (AAV2-Retro) + pAAV2-CAG-DIO-BDNF-mCherry-3*flag (vCA1-IL pathway-specific overexpression of BDNF)	vCA1-IL	Spontaneous nociceptive behaviors, thermal hyperalgesia, mechanical allodynia, anxiety-like behaviors	Rat		↓	[201]
pAAV-CMV-MCS-EGFP-3Flag (BDNF-specific overexpression)	ventral DG	Thermal hyperalgesia, mechanical allodynia, anxiety-like behaviors	Mouse			[200]
BDNF	NAc	Thermal hyperalgesia	Mouse	CUMS	↑	[215]
TrkB-Fc				Morphine-induced	↓	
				CCI		[195]
ANA-12				Optogenetically induced hypersensitivity		[216]
BDNF	CeA	Thermal hyperalgesia	Mouse	Naïve	↑	[191]
TrkB-IgG				CFA	↓	
TrkB-Fc	CeA	Morphine-induced analgesia	Mouse	Naïve	↓	[205]
oe-BDNF lentivirus (BDNF overexpression)	Parafascicular nucleus of thalamus	Anxiety-like behaviors and mechanical allodynia	Mouse	CRS	↓	[194]
sh-BDNF lentivirus (BDNF knockdown)					↑	
TrkB-Fc, CTX-B	MT	Mechanical allodynia	Rat	CPSP	↓	[192]
		Thermal hyperalgesia			No effects	
K252a	NRM	Mechanical allodynia	Rat	CFA	↓	[208]
BDNF	Midbrain (PAG-DRN)	Thermal hyperalgesia	Rat	Normal	↓	[211,212]
		Nociceptive responses		Formalin		[213]
BDNF	PAG	tDCS-induced analgesic effects	Rat	MIA	↓	[214]
AAV-eGFP-Cre virus	PB	Thermal hyperalgesia and mechanical allodynia	Floxed-BDNF mouse	Normal	No effects	[205]
BDNF	RVM	Thermal hyper-algesia and mechanical allodynia	Rat	Naïve	↑	[187,189]
		Thermal hyperalgesia		CFA	↓	[187]
				RVM 5-HT-depleted animals		[189]
rabbit anti-BDNF antibody	RVM	PSD-induced cumulative pain scores and mechanical allodynia	Rat	Incisional pain	↓	[190]

CFA = Complete Freund's Adjuvant; CPA = conditioned placed avoidance; CPP = conditioned place preference; CPSP, central poststroke pain; CRS = chronic restraint stress; CUMS = chronic unpredictable mild stress; DG = dentate gyrus; DRN = dorsal raphe nucleus; MIA, monoiodoacetate; MNX = transection of the medial meniscal; MT = medial thalamus; NGF = nerve growth factor; PNI = peripheral nerve injury; NRM = nucleus raphe magnus; PAG = periaqueductal gray; PSD = paradoxical sleep deprivation; pSNL = partial sciatic nerve ligation; SNC = sciatic nerve crush; SNL = spinal nerve ligation; tDCS = transcranial direct current stimulation.

Author Contributions: Conceptualization, M.M. and V.N.; methodology, M.M., T.K., P.P., Z.H., N.A., G.J. and V.N.; investigation, M.M., T.K., P.P., Z.H., N.A. and G.J.; writing—original draft preparation, M.M.; writing—review and editing, M.M., T.K., P.P., Z.H., N.A., G.J, and V.N.; visualization, M.M. and V.N.; project administration, V.N.; funding acquisition, V.N. All authors have read and agreed to the published version of the manuscript.

Funding: This work was supported by National Institutes of Health (NIH) grants R01 NS038261 and R01 NS118731 (VN).

Institutional Review Board Statement: Not applicable.

Informed Consent Statement: Not applicable.

Data Availability Statement: Not applicable.

Conflicts of Interest: The authors declare no conflicts of interest.

References

1. Barde, Y.A.; Edgar, D.; Thoenen, H. Purification of a new neurotrophic factor from mammalian brain. *EMBO J.* **1982**, *1*, 549–553. [CrossRef]
2. Miranda, M.; Morici, J.F.; Zanoni, M.B.; Bekinschtein, P. Brain-Derived Neurotrophic Factor: A Key Molecule for Memory in the Healthy and the Pathological Brain. *Front. Cell Neurosci.* **2019**, *13*, 363. [CrossRef]
3. Mowla, S.J.; Farhadi, H.F.; Pareek, S.; Atwal, J.K.; Morris, S.J.; Seidah, N.G.; Murphy, R.A. Biosynthesis and post-translational processing of the precursor to brain-derived neurotrophic factor. *J. Biol. Chem.* **2001**, *276*, 12660–12666. [CrossRef] [PubMed]
4. Chao, M.V.; Bothwell, M. Neurotrophins: To cleave or not to cleave. *Neuron* **2002**, *33*, 9–12. [CrossRef] [PubMed]
5. Minichiello, L. TrkB signalling pathways in LTP and learning. *Nat. Rev. Neurosci.* **2009**, *10*, 850–860. [CrossRef] [PubMed]
6. Yang, B.; Ren, Q.; Zhang, J.C.; Chen, Q.X.; Hashimoto, K. Altered expression of BDNF, BDNF pro-peptide and their precursor proBDNF in brain and liver tissues from psychiatric disorders: Rethinking the brain-liver axis. *Transl. Psychiatry* **2017**, *7*, e1128. [CrossRef]
7. Thompson Ray, M.; Weickert, C.S.; Wyatt, E.; Webster, M.J. Decreased BDNF, trkB-TK+ and GAD67 mRNA expression in the hippocampus of individuals with schizophrenia and mood disorders. *J. Psychiatry Neurosci.* **2011**, *36*, 195–203. [CrossRef]
8. Castren, E.; Rantamaki, T. The role of BDNF and its receptors in depression and antidepressant drug action: Reactivation of developmental plasticity. *Dev. Neurobiol.* **2010**, *70*, 289–297. [CrossRef]
9. Garzon, D.; Yu, G.; Fahnestock, M. A new brain-derived neurotrophic factor transcript and decrease in brain-derived neurotrophic factor transcripts 1, 2 and 3 in Alzheimer's disease parietal cortex. *J. Neurochem.* **2002**, *82*, 1058–1064. [CrossRef]
10. Phillips, H.S.; Hains, J.M.; Armanini, M.; Laramee, G.R.; Johnson, S.A.; Winslow, J.W. BDNF mRNA is decreased in the hippocampus of individuals with Alzheimer's disease. *Neuron* **1991**, *7*, 695–702. [CrossRef]
11. Hock, C.; Heese, K.; Hulette, C.; Rosenberg, C.; Otten, U. Region-specific neurotrophin imbalances in Alzheimer disease: Decreased levels of brain-derived neurotrophic factor and increased levels of nerve growth factor in hippocampus and cortical areas. *Arch. Neurol.* **2000**, *57*, 846–851. [CrossRef] [PubMed]
12. Angelucci, F.; Ricci, V.; Pomponi, M.; Conte, G.; Mathe, A.A.; Attilio Tonali, P.; Bria, P. Chronic heroin and cocaine abuse is associated with decreased serum concentrations of the nerve growth factor and brain-derived neurotrophic factor. *J. Psychopharmacol.* **2007**, *21*, 820–825. [CrossRef]
13. Kim, D.J.; Roh, S.; Kim, Y.; Yoon, S.J.; Lee, H.K.; Han, C.S.; Kim, Y.K. High concentrations of plasma brain-derived neurotrophic factor in methamphetamine users. *Neurosci. Lett.* **2005**, *388*, 112–115. [CrossRef]
14. Simao, A.P.; Mendonca, V.A.; de Oliveira Almeida, T.M.; Santos, S.A.; Gomes, W.F.; Coimbra, C.C.; Lacerda, A.C. Involvement of BDNF in knee osteoarthritis: The relationship with inflammation and clinical parameters. *Rheumatol. Int.* **2014**, *34*, 1153–1157. [CrossRef] [PubMed]
15. Geng, S.J.; Liao, F.F.; Dang, W.H.; Ding, X.; Liu, X.D.; Cai, J.; Han, J.S.; Wan, Y.; Xing, G.G. Contribution of the spinal cord BDNF to the development of neuropathic pain by activation of the NR2B-containing NMDA receptors in rats with spinal nerve ligation. *Exp. Neurol.* **2010**, *222*, 256–266. [CrossRef]
16. Hildebrand, M.E.; Xu, J.; Dedek, A.; Li, Y.; Sengar, A.S.; Beggs, S.; Lombroso, P.J.; Salter, M.W. Potentiation of Synaptic GluN2B NMDAR Currents by Fyn Kinase Is Gated through BDNF-Mediated Disinhibition in Spinal Pain Processing. *Cell Rep.* **2016**, *17*, 2753–2765. [CrossRef]
17. Binder, D.K.; Scharfman, H.E. Brain-derived neurotrophic factor. *Growth Factors* **2004**, *22*, 123–131. [CrossRef]

18. Esvald, E.E.; Tuvikene, J.; Sirp, A.; Patil, S.; Bramham, C.R.; Timmusk, T. CREB Family Transcription Factors Are Major Mediators of BDNF Transcriptional Autoregulation in Cortical Neurons. *J. Neurosci.* **2020**, *40*, 1405–1426. [CrossRef] [PubMed]
19. You, H.; Lu, B. Diverse Functions of Multiple Bdnf Transcripts Driven by Distinct Bdnf Promoters. *Biomolecules* **2023**, *13*, 655. [CrossRef]
20. Greenberg, M.E.; Xu, B.; Lu, B.; Hempstead, B.L. New insights in the biology of BDNF synthesis and release: Implications in CNS function. *J. Neurosci.* **2009**, *29*, 12764–12767. [CrossRef]
21. Cattaneo, A.; Cattane, N.; Begni, V.; Pariante, C.M.; Riva, M.A. The human BDNF gene: Peripheral gene expression and protein levels as biomarkers for psychiatric disorders. *Transl. Psychiatry* **2016**, *6*, e958. [CrossRef]
22. Lu, B. BDNF and activity-dependent synaptic modulation. *Learn. Mem.* **2003**, *10*, 86–98. [CrossRef] [PubMed]
23. Lessmann, V.; Brigadski, T. Mechanisms, locations, and kinetics of synaptic BDNF secretion: An update. *Neurosci. Res.* **2009**, *65*, 11–22. [CrossRef] [PubMed]
24. Brigadski, T.; Lessmann, V. The physiology of regulated BDNF release. *Cell Tissue Res.* **2020**, *382*, 15–45. [CrossRef] [PubMed]
25. Dieni, S.; Matsumoto, T.; Dekkers, M.; Rauskolb, S.; Ionescu, M.S.; Deogracias, R.; Gundelfinger, E.D.; Kojima, M.; Nestel, S.; Frotscher, M.; et al. BDNF and its pro-peptide are stored in presynaptic dense core vesicles in brain neurons. *J. Cell Biol.* **2012**, *196*, 775–788. [CrossRef]
26. Arevalo, J.C.; Deogracias, R. Mechanisms Controlling the Expression and Secretion of BDNF. *Biomolecules* **2023**, *13*, 789. [CrossRef]
27. Zafra, F.; Lindholm, D.; Castren, E.; Hartikka, J.; Thoenen, H. Regulation of brain-derived neurotrophic factor and nerve growth factor mRNA in primary cultures of hippocampal neurons and astrocytes. *J. Neurosci.* **1992**, *12*, 4793–4799. [CrossRef]
28. Tongiorgi, E. Activity-dependent expression of brain-derived neurotrophic factor in dendrites: Facts and open questions. *Neurosci. Res.* **2008**, *61*, 335–346. [CrossRef]
29. Danzer, S.C.; McNamara, J.O. Localization of brain-derived neurotrophic factor to distinct terminals of mossy fiber axons implies regulation of both excitation and feedforward inhibition of CA3 pyramidal cells. *J. Neurosci.* **2004**, *24*, 11346–11355. [CrossRef]
30. Coull, J.A.; Beggs, S.; Boudreau, D.; Boivin, D.; Tsuda, M.; Inoue, K.; Gravel, C.; Salter, M.W.; De Koninck, Y. BDNF from microglia causes the shift in neuronal anion gradient underlying neuropathic pain. *Nature* **2005**, *438*, 1017–1021. [CrossRef]
31. Parkhurst, C.N.; Yang, G.; Ninan, I.; Savas, J.N.; Yates, J.R., 3rd; Lafaille, J.J.; Hempstead, B.L.; Littman, D.R.; Gan, W.B. Microglia promote learning-dependent synapse formation through brain-derived neurotrophic factor. *Cell* **2013**, *155*, 1596–1609. [CrossRef]
32. Elkabes, S.; DiCicco-Bloom, E.M.; Black, I.B. Brain microglia/macrophages express neurotrophins that selectively regulate microglial proliferation and function. *J. Neurosci.* **1996**, *16*, 2508–2521. [CrossRef] [PubMed]
33. Nakajima, K.; Honda, S.; Tohyama, Y.; Imai, Y.; Kohsaka, S.; Kurihara, T. Neurotrophin secretion from cultured microglia. *J. Neurosci. Res.* **2001**, *65*, 322–331. [CrossRef]
34. Yao, W.; Cao, Q.; Luo, S.; He, L.; Yang, C.; Chen, J.; Qi, Q.; Hashimoto, K.; Zhang, J.C. Microglial ERK-NRBP1-CREB-BDNF signaling in sustained antidepressant actions of (R)-ketamine. *Mol. Psychiatry* **2022**, *27*, 1618–1629. [CrossRef] [PubMed]
35. Niu, C.; Yue, X.; An, J.J.; Bass, R.; Xu, H.; Xu, B. Genetic Dissection of BDNF and TrkB Expression in Glial Cells. *Biomolecules* **2024**, *14*, 91. [CrossRef] [PubMed]
36. Denk, F.; Crow, M.; Didangelos, A.; Lopes, D.M.; McMahon, S.B. Persistent Alterations in Microglial Enhancers in a Model of Chronic Pain. *Cell Rep.* **2016**, *15*, 1771–1781. [CrossRef]
37. Dembo, T.; Braz, J.M.; Hamel, K.A.; Kuhn, J.A.; Basbaum, A.I. Primary Afferent-Derived BDNF Contributes Minimally to the Processing of Pain and Itch. *eNeuro* **2018**, *5*. [CrossRef]
38. Honey, D.; Wosnitzka, E.; Klann, E.; Weinhard, L. Analysis of microglial BDNF function and expression in the motor cortex. *Front. Cell Neurosci.* **2022**, *16*, 961276. [CrossRef]
39. De Santi, L.; Annunziata, P.; Sessa, E.; Bramanti, P. Brain-derived neurotrophic factor and TrkB receptor in experimental autoimmune encephalomyelitis and multiple sclerosis. *J. Neurol. Sci.* **2009**, *287*, 17–26. [CrossRef]
40. Bathina, S.; Das, U.N. Brain-derived neurotrophic factor and its clinical implications. *Arch. Med. Sci.* **2015**, *11*, 1164–1178. [CrossRef]
41. Enkavi, G.; Girych, M.; Moliner, R.; Vattulainen, I.; Castren, E. TrkB transmembrane domain: Bridging structural understanding with therapeutic strategy. *Trends Biochem. Sci.* **2024**, *49*, 445–456. [CrossRef]
42. Teng, H.K.; Teng, K.K.; Lee, R.; Wright, S.; Tevar, S.; Almeida, R.D.; Kermani, P.; Torkin, R.; Chen, Z.Y.; Lee, F.S.; et al. ProBDNF induces neuronal apoptosis via activation of a receptor complex of p75NTR and sortilin. *J. Neurosci.* **2005**, *25*, 5455–5463. [CrossRef] [PubMed]
43. Schiro, G.; Iacono, S.; Ragonese, P.; Aridon, P.; Salemi, G.; Balistreri, C.R. A Brief Overview on BDNF-Trk Pathway in the Nervous System: A Potential Biomarker or Possible Target in Treatment of Multiple Sclerosis? *Front. Neurol.* **2022**, *13*, 917527. [CrossRef]

44. Turkistani, A.; Al-Kuraishy, H.M.; Al-Gareeb, A.I.; Albuhadily, A.K.; Elhussieny, O.; Al-Farga, A.; Aqlan, F.; Saad, H.M.; Batiha, G.E. The functional and molecular roles of p75 neurotrophin receptor (p75(NTR)) in epilepsy. *J. Cent. Nerv. Syst. Dis.* **2024**, *16*, 11795735241247810. [CrossRef] [PubMed]
45. Woo, N.H.; Teng, H.K.; Siao, C.J.; Chiaruttini, C.; Pang, P.T.; Milner, T.A.; Hempstead, B.L.; Lu, B. Activation of p75NTR by proBDNF facilitates hippocampal long-term depression. *Nat. Neurosci.* **2005**, *8*, 1069–1077. [CrossRef]
46. Gupta, V.K.; You, Y.; Gupta, V.B.; Klistorner, A.; Graham, S.L. TrkB receptor signalling: Implications in neurodegenerative, psychiatric and proliferative disorders. *Int. J. Mol. Sci.* **2013**, *14*, 10122–10142. [CrossRef]
47. Klein, R.; Smeyne, R.J.; Wurst, W.; Long, L.K.; Auerbach, B.A.; Joyner, A.L.; Barbacid, M. Targeted disruption of the trkB neurotrophin receptor gene results in nervous system lesions and neonatal death. *Cell* **1993**, *75*, 113–122. [PubMed]
48. Rohrer, B.; Korenbrot, J.I.; LaVail, M.M.; Reichardt, L.F.; Xu, B. Role of neurotrophin receptor TrkB in the maturation of rod photoreceptors and establishment of synaptic transmission to the inner retina. *J. Neurosci.* **1999**, *19*, 8919–8930. [CrossRef]
49. Andreska, T.; Luningschror, P.; Sendtner, M. Regulation of TrkB cell surface expression-a mechanism for modulation of neuronal responsiveness to brain-derived neurotrophic factor. *Cell Tissue Res.* **2020**, *382*, 5–14. [CrossRef]
50. Zhao, L.; Sheng, A.L.; Huang, S.H.; Yin, Y.X.; Chen, B.; Li, X.Z.; Zhang, Y.; Chen, Z.Y. Mechanism underlying activity-dependent insertion of TrkB into the neuronal surface. *J. Cell Sci.* **2009**, *122*, 3123–3136. [CrossRef]
51. Li, Y.X.; Zhang, Y.; Lester, H.A.; Schuman, E.M.; Davidson, N. Enhancement of neurotransmitter release induced by brain-derived neurotrophic factor in cultured hippocampal neurons. *J. Neurosci.* **1998**, *18*, 10231–10240. [CrossRef]
52. Xu, B.; Gottschalk, W.; Chow, A.; Wilson, R.I.; Schnell, E.; Zang, K.; Wang, D.; Nicoll, R.A.; Lu, B.; Reichardt, L.F. The role of brain-derived neurotrophic factor receptors in the mature hippocampus: Modulation of long-term potentiation through a presynaptic mechanism involving TrkB. *J. Neurosci.* **2000**, *20*, 6888–6897. [CrossRef]
53. Luongo, L.; Maione, S.; Di Marzo, V. Endocannabinoids and neuropathic pain: Focus on neuron-glia and endocannabinoid-neurotrophin interactions. *Eur. J. Neurosci.* **2014**, *39*, 401–408. [CrossRef] [PubMed]
54. Magee, J.C.; Grienberger, C. Synaptic Plasticity Forms and Functions. *Annu. Rev. Neurosci.* **2020**, *43*, 95–117. [CrossRef] [PubMed]
55. Wang, C.S.; Kavalali, E.T.; Monteggia, L.M. BDNF signaling in context: From synaptic regulation to psychiatric disorders. *Cell* **2022**, *185*, 62–76. [CrossRef] [PubMed]
56. Korte, M.; Carroll, P.; Wolf, E.; Brem, G.; Thoenen, H.; Bonhoeffer, T. Hippocampal long-term potentiation is impaired in mice lacking brain-derived neurotrophic factor. *Proc. Natl. Acad. Sci. USA* **1995**, *92*, 8856–8860. [CrossRef]
57. Patterson, S.L.; Abel, T.; Deuel, T.A.; Martin, K.C.; Rose, J.C.; Kandel, E.R. Recombinant BDNF rescues deficits in basal synaptic transmission and hippocampal LTP in BDNF knockout mice. *Neuron* **1996**, *16*, 1137–1145. [CrossRef]
58. Korte, M.; Griesbeck, O.; Gravel, C.; Carroll, P.; Staiger, V.; Thoenen, H.; Bonhoeffer, T. Virus-mediated gene transfer into hippocampal CA1 region restores long-term potentiation in brain-derived neurotrophic factor mutant mice. *Proc. Natl. Acad. Sci. USA* **1996**, *93*, 12547–12552. [CrossRef]
59. An, J.J.; Gharami, K.; Liao, G.Y.; Woo, N.H.; Lau, A.G.; Vanevski, F.; Torre, E.R.; Jones, K.R.; Feng, Y.; Lu, B.; et al. Distinct role of long 3′ UTR BDNF mRNA in spine morphology and synaptic plasticity in hippocampal neurons. *Cell* **2008**, *134*, 175–187. [CrossRef]
60. Gartner, A.; Polnau, D.G.; Staiger, V.; Sciarretta, C.; Minichiello, L.; Thoenen, H.; Bonhoeffer, T.; Korte, M. Hippocampal long-term potentiation is supported by presynaptic and postsynaptic tyrosine receptor kinase B-mediated phospholipase Cgamma signaling. *J. Neurosci.* **2006**, *26*, 3496–3504. [CrossRef]
61. Zakharenko, S.S.; Patterson, S.L.; Dragatsis, I.; Zeitlin, S.O.; Siegelbaum, S.A.; Kandel, E.R.; Morozov, A. Presynaptic BDNF required for a presynaptic but not postsynaptic component of LTP at hippocampal CA1-CA3 synapses. *Neuron* **2003**, *39*, 975–990. [CrossRef] [PubMed]
62. Lin, P.Y.; Kavalali, E.T.; Monteggia, L.M. Genetic Dissection of Presynaptic and Postsynaptic BDNF-TrkB Signaling in Synaptic Efficacy of CA3-CA1 Synapses. *Cell Rep.* **2018**, *24*, 1550–1561. [CrossRef] [PubMed]
63. Zhou, L.J.; Peng, J.; Xu, Y.N.; Zeng, W.J.; Zhang, J.; Wei, X.; Mai, C.L.; Lin, Z.J.; Liu, Y.; Murugan, M.; et al. Microglia Are Indispensable for Synaptic Plasticity in the Spinal Dorsal Horn and Chronic Pain. *Cell Rep.* **2019**, *27*, 3844–3859.e6. [CrossRef]
64. Liu, J.H.; Zhang, M.; Wang, Q.; Wu, D.Y.; Jie, W.; Hu, N.Y.; Lan, J.Z.; Zeng, K.; Li, S.J.; Li, X.W.; et al. Distinct roles of astroglia and neurons in synaptic plasticity and memory. *Mol. Psychiatry* **2022**, *27*, 873–885. [CrossRef]
65. Vignoli, B.; Battistini, G.; Melani, R.; Blum, R.; Santi, S.; Berardi, N.; Canossa, M. Peri-Synaptic Glia Recycles Brain-Derived Neurotrophic Factor for LTP Stabilization and Memory Retention. *Neuron* **2016**, *92*, 873–887. [CrossRef] [PubMed]
66. Baltaci, S.B.; Mogulkoc, R.; Baltaci, A.K. Molecular Mechanisms of Early and Late LTP. *Neurochem. Res.* **2019**, *44*, 281–296. [CrossRef]
67. Sweatt, J.D. Toward a molecular explanation for long-term potentiation. *Learn. Mem.* **1999**, *6*, 399–416. [CrossRef]

68. Edelmann, E.; Cepeda-Prado, E.; Franck, M.; Lichtenecker, P.; Brigadski, T.; Lessmann, V. Theta Burst Firing Recruits BDNF Release and Signaling in Postsynaptic CA1 Neurons in Spike-Timing-Dependent LTP. *Neuron* **2015**, *86*, 1041–1054. [CrossRef]
69. Caldeira, M.V.; Melo, C.V.; Pereira, D.B.; Carvalho, R.; Correia, S.S.; Backos, D.S.; Carvalho, A.L.; Esteban, J.A.; Duarte, C.B. Brain-derived neurotrophic factor regulates the expression and synaptic delivery of alpha-amino-3-hydroxy-5-methyl-4-isoxazole propionic acid receptor subunits in hippocampal neurons. *J. Biol. Chem.* **2007**, *282*, 12619–12628. [CrossRef]
70. Fortin, D.A.; Srivastava, T.; Dwarakanath, D.; Pierre, P.; Nygaard, S.; Derkach, V.A.; Soderling, T.R. Brain-derived neurotrophic factor activation of CaM-kinase kinase via transient receptor potential canonical channels induces the translation and synaptic incorporation of GluA1-containing calcium-permeable AMPA receptors. *J. Neurosci.* **2012**, *32*, 8127–8137. [CrossRef]
71. Nakata, H.; Nakamura, S. Brain-derived neurotrophic factor regulates AMPA receptor trafficking to post-synaptic densities via IP3R and TRPC calcium signaling. *FEBS Lett.* **2007**, *581*, 2047–2054. [CrossRef]
72. Jourdi, H.; Kabbaj, M. Acute BDNF treatment upregulates GluR1-SAP97 and GluR2-GRIP1 interactions: Implications for sustained AMPA receptor expression. *PLoS ONE* **2013**, *8*, e57124. [CrossRef] [PubMed]
73. Kovalchuk, Y.; Hanse, E.; Kafitz, K.W.; Konnerth, A. Postsynaptic Induction of BDNF-Mediated Long-Term Potentiation. *Science* **2002**, *295*, 1729–1734. [CrossRef]
74. Caldeira, M.V.; Melo, C.V.; Pereira, D.B.; Carvalho, R.F.; Carvalho, A.L.; Duarte, C.B. BDNF regulates the expression and traffic of NMDA receptors in cultured hippocampal neurons. *Mol. Cell Neurosci.* **2007**, *35*, 208–219. [CrossRef]
75. Schratt, G.M.; Nigh, E.A.; Chen, W.G.; Hu, L.; Greenberg, M.E. BDNF regulates the translation of a select group of mRNAs by a mammalian target of rapamycin-phosphatidylinositol 3-kinase-dependent pathway during neuronal development. *J. Neurosci.* **2004**, *24*, 7366–7377. [CrossRef] [PubMed]
76. Suen, P.C.; Wu, K.; Levine, E.S.; Mount, H.T.; Xu, J.L.; Lin, S.Y.; Black, I.B. Brain-derived neurotrophic factor rapidly enhances phosphorylation of the postsynaptic N-methyl-D-aspartate receptor subunit 1. *Proc. Natl. Acad. Sci. USA* **1997**, *94*, 8191–8195. [CrossRef] [PubMed]
77. Lin, S.Y.; Wu, K.; Levine, E.S.; Mount, H.T.; Suen, P.C.; Black, I.B. BDNF acutely increases tyrosine phosphorylation of the NMDA receptor subunit 2B in cortical and hippocampal postsynaptic densities. *Brain Res. Mol. Brain Res.* **1998**, *55*, 20–27. [CrossRef]
78. Patterson, S.L.; Pittenger, C.; Morozov, A.; Martin, K.C.; Scanlin, H.; Drake, C.; Kandel, E.R. Some forms of cAMP-mediated long-lasting potentiation are associated with release of BDNF and nuclear translocation of phospho-MAP kinase. *Neuron* **2001**, *32*, 123–140. [CrossRef]
79. Hashimotodani, Y.; Nasrallah, K.; Jensen, K.R.; Chavez, A.E.; Carrera, D.; Castillo, P.E. LTP at Hilar Mossy Cell-Dentate Granule Cell Synapses Modulates Dentate Gyrus Output by Increasing Excitation/Inhibition Balance. *Neuron* **2017**, *95*, 928–943.e3. [CrossRef]
80. Minichiello, L.; Calella, A.M.; Medina, D.L.; Bonhoeffer, T.; Klein, R.; Korte, M. Mechanism of TrkB-mediated hippocampal long-term potentiation. *Neuron* **2002**, *36*, 121–137. [CrossRef]
81. Tanaka, J.; Horiike, Y.; Matsuzaki, M.; Miyazaki, T.; Ellis-Davies, G.C.; Kasai, H. Protein synthesis and neurotrophin-dependent structural plasticity of single dendritic spines. *Science* **2008**, *319*, 1683–1687. [CrossRef] [PubMed]
82. Briz, V.; Zhu, G.; Wang, Y.; Liu, Y.; Avetisyan, M.; Bi, X.; Baudry, M. Activity-dependent rapid local RhoA synthesis is required for hippocampal synaptic plasticity. *J. Neurosci.* **2015**, *35*, 2269–2282. [CrossRef]
83. Ding, X.; Cai, J.; Li, S.; Liu, X.D.; Wan, Y.; Xing, G.G. BDNF contributes to the development of neuropathic pain by induction of spinal long-term potentiation via SHP2 associated GluN2B-containing NMDA receptors activation in rats with spinal nerve ligation. *Neurobiol. Dis.* **2015**, *73*, 428–451. [CrossRef] [PubMed]
84. Zhou, L.J.; Ren, W.J.; Zhong, Y.; Yang, T.; Wei, X.H.; Xin, W.J.; Liu, C.C.; Zhou, L.H.; Li, Y.Y.; Liu, X.G. Limited BDNF contributes to the failure of injury to skin afferents to produce a neuropathic pain condition. *Pain* **2010**, *148*, 148–157. [CrossRef] [PubMed]
85. White, A.O.; Kramar, E.A.; Lopez, A.J.; Kwapis, J.L.; Doan, J.; Saldana, D.; Davatolhagh, M.F.; Alaghband, Y.; Blurton-Jones, M.; Matheos, D.P.; et al. BDNF rescues BAF53b-dependent synaptic plasticity and cocaine-associated memory in the nucleus accumbens. *Nat. Commun.* **2016**, *7*, 11725. [CrossRef]
86. Tanqueiro, S.R.; Mouro, F.M.; Ferreira, C.B.; Freitas, C.F.; Fonseca-Gomes, J.; Simoes do Couto, F.; Sebastiao, A.M.; Dawson, N.; Diogenes, M.J. Sustained NMDA receptor hypofunction impairs brain-derived neurotropic factor signalling in the PFC, but not in the hippocampus, and disturbs PFC-dependent cognition in mice. *J. Psychopharmacol.* **2021**, *35*, 730–743. [CrossRef]
87. Miao, H.H.; Miao, Z.; Pan, J.G.; Li, X.H.; Zhuo, M. Brain-derived neurotrophic factor produced long-term synaptic enhancement in the anterior cingulate cortex of adult mice. *Mol. Brain* **2021**, *14*, 140. [CrossRef]
88. Luo, C.; Kuner, T.; Kuner, R. Synaptic plasticity in pathological pain. *Trends Neurosci.* **2014**, *37*, 343–355. [CrossRef]
89. Ji, R.R.; Chamessian, A.; Zhang, Y.Q. Pain regulation by non-neuronal cells and inflammation. *Science* **2016**, *354*, 572–577. [CrossRef]

90. Ji, R.R.; Nackley, A.; Huh, Y.; Terrando, N.; Maixner, W. Neuroinflammation and Central Sensitization in Chronic and Widespread Pain. *Anesthesiology* **2018**, *129*, 343–366. [CrossRef]
91. Funes, S.C.; Rios, M.; Escobar-Vera, J.; Kalergis, A.M. Implications of macrophage polarization in autoimmunity. *Immunology* **2018**, *154*, 186–195. [CrossRef] [PubMed]
92. Wang, C.; Ma, C.; Gong, L.; Guo, Y.; Fu, K.; Zhang, Y.; Zhou, H.; Li, Y. Macrophage Polarization and Its Role in Liver Disease. *Front. Immunol.* **2021**, *12*, 803037. [CrossRef]
93. Shapouri-Moghaddam, A.; Mohammadian, S.; Vazini, H.; Taghadosi, M.; Esmaeili, S.A.; Mardani, F.; Seifi, B.; Mohammadi, A.; Afshari, J.T.; Sahebkar, A. Macrophage plasticity, polarization, and function in health and disease. *J. Cell Physiol.* **2018**, *233*, 6425–6440. [CrossRef]
94. Hong, J.H.; Park, H.M.; Byun, K.H.; Lee, B.H.; Kang, W.C.; Jeong, G.B. BDNF expression of macrophages and angiogenesis after myocardial infarction. *Int. J. Cardiol.* **2014**, *176*, 1405–1408. [CrossRef]
95. Asami, T.; Ito, T.; Fukumitsu, H.; Nomoto, H.; Furukawa, Y.; Furukawa, S. Autocrine activation of cultured macrophages by brain-derived neurotrophic factor. *Biochem. Biophys. Res. Commun.* **2006**, *344*, 941–947. [CrossRef] [PubMed]
96. Salzer, J.L. Schwann cell myelination. *Cold Spring Harb. Perspect. Biol.* **2015**, *7*, a020529. [CrossRef] [PubMed]
97. Su, W.F.; Wu, F.; Jin, Z.H.; Gu, Y.; Chen, Y.T.; Fei, Y.; Chen, H.; Wang, Y.X.; Xing, L.Y.; Zhao, Y.Y.; et al. Overexpression of P2X4 receptor in Schwann cells promotes motor and sensory functional recovery and remyelination via BDNF secretion after nerve injury. *Glia* **2019**, *67*, 78–90. [CrossRef]
98. Jessen, K.R.; Mirsky, R. The repair Schwann cell and its function in regenerating nerves. *J. Physiol.* **2016**, *594*, 3521–3531. [CrossRef]
99. Wilhelm, J.C.; Xu, M.; Cucoranu, D.; Chmielewski, S.; Holmes, T.; Lau, K.S.; Bassell, G.J.; English, A.W. Cooperative roles of BDNF expression in neurons and Schwann cells are modulated by exercise to facilitate nerve regeneration. *J. Neurosci.* **2012**, *32*, 5002–5009. [CrossRef]
100. Luo, B.; Huang, J.; Lu, L.; Hu, X.; Luo, Z.; Li, M. Electrically induced brain-derived neurotrophic factor release from Schwann cells. *J. Neurosci. Res.* **2014**, *92*, 893–903. [CrossRef]
101. Bonalume, V.; Caffino, L.; Castelnovo, L.F.; Faroni, A.; Giavarini, F.; Liu, S.; Caruso, D.; Schmelz, M.; Fumagalli, F.; Carr, R.W.; et al. Schwann Cell Autocrine and Paracrine Regulatory Mechanisms, Mediated by Allopregnanolone and BDNF, Modulate PKCepsilon in Peripheral Sensory Neurons. *Cells* **2020**, *9*, 1874. [CrossRef]
102. Castelnovo, L.F.; Thomas, P. Membrane progesterone receptor alpha (mPRalpha/PAQR7) promotes migration, proliferation and BDNF release in human Schwann cell-like differentiated adipose stem cells. *Mol. Cell Endocrinol.* **2021**, *531*, 111298. [CrossRef] [PubMed]
103. Bierlein De la Rosa, M.; Sharma, A.D.; Mallapragada, S.K.; Sakaguchi, D.S. Transdifferentiation of brain-derived neurotrophic factor (BDNF)-secreting mesenchymal stem cells significantly enhance BDNF secretion and Schwann cell marker proteins. *J. Biosci. Bioeng.* **2017**, *124*, 572–582. [CrossRef]
104. Nayak, D.; Roth, T.L.; McGavern, D.B. Microglia development and function. *Annu. Rev. Immunol.* **2014**, *32*, 367–402. [CrossRef]
105. Colonna, M.; Butovsky, O. Microglia Function in the Central Nervous System During Health and Neurodegeneration. *Annu. Rev. Immunol.* **2017**, *35*, 441–468. [CrossRef]
106. Kwon, H.S.; Koh, S.H. Neuroinflammation in neurodegenerative disorders: The roles of microglia and astrocytes. *Transl. Neurodegener.* **2020**, *9*, 42. [CrossRef] [PubMed]
107. Bai, J.; Geng, B.; Wang, X.; Wang, S.; Yi, Q.; Tang, Y.; Xia, Y. Exercise Facilitates the M1-to-M2 Polarization of Microglia by Enhancing Autophagy via the BDNF/AKT/mTOR Pathway in Neuropathic Pain. *Pain. Physician* **2022**, *25*, E1137–E1151.
108. Huang, L.; Jin, J.; Chen, K.; You, S.; Zhang, H.; Sideris, A.; Norcini, M.; Recio-Pinto, E.; Wang, J.; Gan, W.B.; et al. BDNF produced by cerebral microglia promotes cortical plasticity and pain hypersensitivity after peripheral nerve injury. *PLoS Biol.* **2021**, *19*, e3001337. [CrossRef]
109. Thibault, K.; Lin, W.K.; Rancillac, A.; Fan, M.; Snollaerts, T.; Sordoillet, V.; Hamon, M.; Smith, G.M.; Lenkei, Z.; Pezet, S. BDNF-dependent plasticity induced by peripheral inflammation in the primary sensory and the cingulate cortex triggers cold allodynia and reveals a major role for endogenous BDNF as a tuner of the affective aspect of pain. *J. Neurosci.* **2014**, *34*, 14739–14751. [CrossRef]
110. Lawal, O.; Ulloa Severino, F.P.; Eroglu, C. The role of astrocyte structural plasticity in regulating neural circuit function and behavior. *Glia* **2022**, *70*, 1467–1483. [CrossRef]
111. Endo, F.; Kasai, A.; Soto, J.S.; Yu, X.; Qu, Z.; Hashimoto, H.; Gradinaru, V.; Kawaguchi, R.; Khakh, B.S. Molecular basis of astrocyte diversity and morphology across the CNS in health and disease. *Science* **2022**, *378*, eadc9020. [CrossRef] [PubMed]
112. Sofroniew, M.V. Astrocyte Reactivity: Subtypes, States, and Functions in CNS Innate Immunity. *Trends Immunol.* **2020**, *41*, 758–770. [CrossRef] [PubMed]

113. Schwartz, J.P.; Sheng, J.G.; Mitsuo, K.; Shirabe, S.; Nishiyama, N. Trophic factor production by reactive astrocytes in injured brain. *Ann. N. Y Acad. Sci.* **1993**, *679*, 226–234. [CrossRef] [PubMed]
114. Schwartz, J.P.; Nishiyama, N. Neurotrophic factor gene expression in astrocytes during development and following injury. *Brain Res. Bull.* **1994**, *35*, 403–407. [CrossRef]
115. Elbaz, B.; Popko, B. Molecular Control of Oligodendrocyte Development. *Trends Neurosci.* **2019**, *42*, 263–277. [CrossRef]
116. Butt, A.M.; Papanikolaou, M.; Rivera, A. Physiology of Oligodendroglia. *Adv. Exp. Med. Biol.* **2019**, *1175*, 117–128. [CrossRef]
117. Kuhn, S.; Gritti, L.; Crooks, D.; Dombrowski, Y. Oligodendrocytes in Development, Myelin Generation and Beyond. *Cells* **2019**, *8*, 1424. [CrossRef]
118. Dai, X.; Lercher, L.D.; Clinton, P.M.; Du, Y.; Livingston, D.L.; Vieira, C.; Yang, L.; Shen, M.M.; Dreyfus, C.F. The trophic role of oligodendrocytes in the basal forebrain. *J. Neurosci.* **2003**, *23*, 5846–5853. [CrossRef]
119. Bagayogo, I.P.; Dreyfus, C.F. Regulated release of BDNF by cortical oligodendrocytes is mediated through metabotropic glutamate receptors and the PLC pathway. *ASN Neuro* **2009**, *1*, AN20090006. [CrossRef]
120. Dougherty, K.D.; Dreyfus, C.F.; Black, I.B. Brain-derived neurotrophic factor in astrocytes, oligodendrocytes, and microglia/macrophages after spinal cord injury. *Neurobiol. Dis.* **2000**, *7*, 574–585. [CrossRef]
121. Bi, C.; Fu, Y.; Li, B. Brain-derived neurotrophic factor alleviates diabetes mellitus-accelerated atherosclerosis by promoting M2 polarization of macrophages through repressing the STAT3 pathway. *Cell Signal* **2020**, *70*, 109569. [CrossRef] [PubMed]
122. Yu, H.C.; Huang, H.B.; Huang Tseng, H.Y.; Lu, M.C. Brain-Derived Neurotrophic Factor Suppressed Proinflammatory Cytokines Secretion and Enhanced MicroRNA(miR)-3168 Expression in Macrophages. *Int. J. Mol. Sci.* **2022**, 23. [CrossRef]
123. Bi, C.; Fu, Y.; Zhang, Z.; Li, B. Prostaglandin E2 confers protection against diabetic coronary atherosclerosis by stimulating M2 macrophage polarization via the activation of the CREB/BDNF/TrkB signaling pathway. *FASEB J.* **2020**, *34*, 7360–7371. [CrossRef]
124. Hayashi, K.; Lesnak, J.B.; Plumb, A.N.; Janowski, A.J.; Smith, A.F.; Hill, J.K.; Sluka, K.A. Brain-derived neurotrophic factor contributes to activity-induced muscle pain in male but not female mice. *bioRxiv* **2023**, *120*, 471–487. [CrossRef] [PubMed]
125. Lalisse, S.; Hua, J.; Lenoir, M.; Linck, N.; Rassendren, F.; Ulmann, L. Sensory neuronal P2RX4 receptors controls BDNF signaling in inflammatory pain. *Sci. Rep.* **2018**, *8*, 964. [CrossRef] [PubMed]
126. Velazquez, K.T.; Mohammad, H.; Sweitzer, S.M. Protein kinase C in pain: Involvement of multiple isoforms. *Pharmacol. Res.* **2007**, *55*, 578–589. [CrossRef]
127. Dadkhah, M.; Baziar, M.; Rezaei, N. The regulatory role of BDNF in neuroimmune axis function and neuroinflammation induced by chronic stress: A new therapeutic strategies for neurodegenerative disorders. *Cytokine* **2024**, *174*, 156477. [CrossRef]
128. Sun, C.; Deng, J.; Ma, Y.; Meng, F.; Cui, X.; Li, M.; Li, J.; Li, J.; Yin, P.; Kong, L.; et al. The dual role of microglia in neuropathic pain after spinal cord injury: Detrimental and protective effects. *Exp. Neurol.* **2023**, *370*, 114570. [CrossRef]
129. Trang, T.; Beggs, S.; Wan, X.; Salter, M.W. P2X4-receptor-mediated synthesis and release of brain-derived neurotrophic factor in microglia is dependent on calcium and p38-mitogen-activated protein kinase activation. *J. Neurosci.* **2009**, *29*, 3518–3528. [CrossRef]
130. Ulmann, L.; Hatcher, J.P.; Hughes, J.P.; Chaumont, S.; Green, P.J.; Conquet, F.; Buell, G.N.; Reeve, A.J.; Chessell, I.P.; Rassendren, F. Up-regulation of P2X4 receptors in spinal microglia after peripheral nerve injury mediates BDNF release and neuropathic pain. *J. Neurosci.* **2008**, *28*, 11263–11268. [CrossRef]
131. Nakajima, K.; Tohyama, Y.; Kohsaka, S.; Kurihara, T. Ceramide activates microglia to enhance the production/secretion of brain-derived neurotrophic factor (BDNF) without induction of deleterious factors in vitro. *J. Neurochem.* **2002**, *80*, 697–705. [CrossRef]
132. Ferrini, F.; De Koninck, Y. Microglia control neuronal network excitability via BDNF signalling. *Neural Plast.* **2013**, *2013*, 429815. [CrossRef]
133. Ding, H.; Chen, J.; Su, M.; Lin, Z.; Zhan, H.; Yang, F.; Li, W.; Xie, J.; Huang, Y.; Liu, X.; et al. BDNF promotes activation of astrocytes and microglia contributing to neuroinflammation and mechanical allodynia in cyclophosphamide-induced cystitis. *J. Neuroinflamm.* **2020**, *17*, 19. [CrossRef] [PubMed]
134. Tang, Y.; Liu, L.; Xu, D.; Zhang, W.; Zhang, Y.; Zhou, J.; Huang, W. Interaction between astrocytic colony stimulating factor and its receptor on microglia mediates central sensitization and behavioral hypersensitivity in chronic post ischemic pain model. *Brain Behav. Immun.* **2018**, *68*, 248–260. [CrossRef] [PubMed]
135. Mapplebeck, J.C.S.; Beggs, S.; Salter, M.W. Sex differences in pain: A tale of two immune cells. *Pain* **2016**, *157* (Suppl. S1), S2–S6. [CrossRef]
136. Bergami, M.; Santi, S.; Formaggio, E.; Cagnoli, C.; Verderio, C.; Blum, R.; Berninger, B.; Matteoli, M.; Canossa, M. Uptake and recycling of pro-BDNF for transmitter-induced secretion by cortical astrocytes. *J. Cell Biol.* **2008**, *183*, 213–221. [CrossRef] [PubMed]

137. Alderson, R.F.; Curtis, R.; Alterman, A.L.; Lindsay, R.M.; DiStefano, P.S. Truncated TrkB mediates the endocytosis and release of BDNF and neurotrophin-4/5 by rat astrocytes and schwann cells in vitro. *Brain Res.* **2000**, *871*, 210–222. [CrossRef]
138. Kinboshi, M.; Mukai, T.; Nagao, Y.; Matsuba, Y.; Tsuji, Y.; Tanaka, S.; Tokudome, K.; Shimizu, S.; Ito, H.; Ikeda, A.; et al. Inhibition of Inwardly Rectifying Potassium (Kir) 4.1 Channels Facilitates Brain-Derived Neurotrophic Factor (BDNF) Expression in Astrocytes. *Front. Mol. Neurosci.* **2017**, *10*, 408. [CrossRef]
139. Jean, Y.Y.; Lercher, L.D.; Dreyfus, C.F. Glutamate elicits release of BDNF from basal forebrain astrocytes in a process dependent on metabotropic receptors and the PLC pathway. *Neuron Glia Biol.* **2008**, *4*, 35–42. [CrossRef]
140. Santi, S.; Cappello, S.; Riccio, M.; Bergami, M.; Aicardi, G.; Schenk, U.; Matteoli, M.; Canossa, M. Hippocampal neurons recycle BDNF for activity-dependent secretion and LTP maintenance. *EMBO J.* **2006**, *25*, 4372–4380. [CrossRef]
141. Stenovec, M.; Lasic, E.; Bozic, M.; Bobnar, S.T.; Stout, R.F., Jr.; Grubisic, V.; Parpura, V.; Zorec, R. Ketamine Inhibits ATP-Evoked Exocytotic Release of Brain-Derived Neurotrophic Factor from Vesicles in Cultured Rat Astrocytes. *Mol. Neurobiol.* **2016**, *53*, 6882–6896. [CrossRef] [PubMed]
142. Jang, M.; Gould, E.; Xu, J.; Kim, E.J.; Kim, J.H. Oligodendrocytes regulate presynaptic properties and neurotransmission through BDNF signaling in the mouse brainstem. *eLife* **2019**, *8*, e42156. [CrossRef]
143. Ji, X.C.; Dang, Y.Y.; Gao, H.Y.; Wang, Z.T.; Gao, M.; Yang, Y.; Zhang, H.T.; Xu, R.X. Local Injection of Lenti-BDNF at the Lesion Site Promotes M2 Macrophage Polarization and Inhibits Inflammatory Response After Spinal Cord Injury in Mice. *Cell Mol. Neurobiol.* **2015**, *35*, 881–890. [CrossRef] [PubMed]
144. Merighi, A.; Bardoni, R.; Salio, C.; Lossi, L.; Ferrini, F.; Prandini, M.; Zonta, M.; Gustincich, S.; Carmignoto, G. Presynaptic functional trkB receptors mediate the release of excitatory neurotransmitters from primary afferent terminals in lamina II (substantia gelatinosa) of postnatal rat spinal cord. *Dev. Neurobiol.* **2008**, *68*, 457–475. [CrossRef] [PubMed]
145. Tender, G.C.; Li, Y.Y.; Cui, J.G. Brain-derived neurotrophic factor redistribution in the dorsal root ganglia correlates with neuropathic pain inhibition after resiniferatoxin treatment. *Spine J.* **2010**, *10*, 715–720. [CrossRef]
146. Lopez-Perez, A.E.; Nurgali, K.; Abalo, R. Painful neurotrophins and their role in visceral pain. *Behav. Pharmacol.* **2018**, *29*, 120–139. [CrossRef]
147. Luo, X.G.; Rush, R.A.; Zhou, X.F. Ultrastructural localization of brain-derived neurotrophic factor in rat primary sensory neurons. *Neurosci. Res.* **2001**, *39*, 377–384. [CrossRef]
148. Salio, C.; Ferrini, F. BDNF and GDNF expression in discrete populations of nociceptors. *Ann. Anat.* **2016**, *207*, 55–61. [CrossRef]
149. Salio, C.; Lossi, L.; Ferrini, F.; Merighi, A. Ultrastructural evidence for a pre- and postsynaptic localization of full-length trkB receptors in substantia gelatinosa (lamina II) of rat and mouse spinal cord. *Eur. J. Neurosci.* **2005**, *22*, 1951–1966. [CrossRef]
150. Cappoli, N.; Tabolacci, E.; Aceto, P.; Dello Russo, C. The emerging role of the BDNF-TrkB signaling pathway in the modulation of pain perception. *J. Neuroimmunol.* **2020**, *349*, 577406. [CrossRef]
151. Ha, S.O.; Kim, J.K.; Hong, H.S.; Kim, D.S.; Cho, H.J. Expression of brain-derived neurotrophic factor in rat dorsal root ganglia, spinal cord and gracile nuclei in experimental models of neuropathic pain. *Neuroscience* **2001**, *107*, 301–309. [CrossRef] [PubMed]
152. Wu, Y.; Shen, Z.; Xu, H.; Zhang, K.; Guo, M.; Wang, F.; Li, J. BDNF Participates in Chronic Constriction Injury-Induced Neuropathic Pain via Transcriptionally Activating P2X(7) in Primary Sensory Neurons. *Mol. Neurobiol.* **2021**, *58*, 4226–4236. [CrossRef]
153. Fukuoka, T.; Kondo, E.; Dai, Y.; Hashimoto, N.; Noguchi, K. Brain-derived neurotrophic factor increases in the uninjured dorsal root ganglion neurons in selective spinal nerve ligation model. *J. Neurosci.* **2001**, *21*, 4891–4900. [CrossRef] [PubMed]
154. Kerr, B.J.; Bradbury, E.J.; Bennett, D.L.; Trivedi, P.M.; Dassan, P.; French, J.; Shelton, D.B.; McMahon, S.B.; Thompson, S.W. Brain-derived neurotrophic factor modulates nociceptive sensory inputs and NMDA-evoked responses in the rat spinal cord. *J. Neurosci.* **1999**, *19*, 5138–5148. [CrossRef] [PubMed]
155. Mannion, R.J.; Costigan, M.; Decosterd, I.; Amaya, F.; Ma, Q.P.; Holstege, J.C.; Ji, R.R.; Acheson, A.; Lindsay, R.M.; Wilkinson, G.A.; et al. Neurotrophins: Peripherally and centrally acting modulators of tactile stimulus-induced inflammatory pain hypersensitivity. *Proc. Natl. Acad. Sci. USA* **1999**, *96*, 9385–9390. [CrossRef]
156. Kras, J.V.; Weisshaar, C.L.; Quindlen, J.; Winkelstein, B.A. Brain-derived neurotrophic factor is upregulated in the cervical dorsal root ganglia and spinal cord and contributes to the maintenance of pain from facet joint injury in the rat. *J. Neurosci. Res.* **2013**, *91*, 1312–1321. [CrossRef]
157. Ge, H.; Guan, S.; Shen, Y.; Sun, M.; Hao, Y.; He, L.; Liu, L.; Yin, C.; Huang, R.; Xiong, W.; et al. Dihydromyricetin affects BDNF levels in the nervous system in rats with comorbid diabetic neuropathic pain and depression. *Sci. Rep.* **2019**, *9*, 14619. [CrossRef]
158. Li, L.; Yu, T.; Yu, L.; Li, H.; Liu, Y.; Wang, D. Exogenous brain-derived neurotrophic factor relieves pain symptoms of diabetic rats by reducing excitability of dorsal root ganglion neurons. *Int. J. Neurosci.* **2016**, *126*, 749–758. [CrossRef]
159. Lin, Y.T.; Ro, L.S.; Wang, H.L.; Chen, J.C. Up-regulation of dorsal root ganglia BDNF and trkB receptor in inflammatory pain: An in vivo and in vitro study. *J. Neuroinflamm.* **2011**, *8*, 126. [CrossRef]

160. Zhou, X.F.; Deng, Y.S.; Xian, C.J.; Zhong, J.H. Neurotrophins from dorsal root ganglia trigger allodynia after spinal nerve injury in rats. *Eur. J. Neurosci.* **2000**, *12*, 100–105. [CrossRef]
161. Obata, K.; Katsura, H.; Sakurai, J.; Kobayashi, K.; Yamanaka, H.; Dai, Y.; Fukuoka, T.; Noguchi, K. Suppression of the p75 neurotrophin receptor in uninjured sensory neurons reduces neuropathic pain after nerve injury. *J. Neurosci.* **2006**, *26*, 11974–11986. [CrossRef] [PubMed]
162. Meng, X.W.; Jin, X.H.; Wei, X.; Wang, L.N.; Yang, J.P.; Ji, F.H. Erratum: Low-affinity neurotrophin receptor p75 of brain-derived neurotrophic factor contributes to cancer-induced bone pain by upregulating mTOR signaling. *Exp. Ther. Med.* **2020**, *19*, 2804. [CrossRef]
163. Pan, J.; Zhao, Y.; Sang, R.; Yang, R.; Bao, J.; Wu, Y.; Fei, Y.; Wu, J.; Chen, G. Huntington-associated protein 1 inhibition contributes to neuropathic pain by suppressing Cav1.2 activity and attenuating inflammation. *Pain* **2023**, *164*, e286–e302. [CrossRef] [PubMed]
164. Shu, X.Q.; Llinas, A.; Mendell, L.M. Effects of trkB and trkC neurotrophin receptor agonists on thermal nociception: A behavioral and electrophysiological study. *Pain* **1999**, *80*, 463–470. [CrossRef]
165. Sikandar, S.; Minett, M.S.; Millet, Q.; Santana-Varela, S.; Lau, J.; Wood, J.N.; Zhao, J. Brain-derived neurotrophic factor derived from sensory neurons plays a critical role in chronic pain. *Brain* **2018**, *141*, 1028–1039. [CrossRef] [PubMed]
166. Valdes-Sanchez, T.; Kirstein, M.; Perez-Villalba, A.; Vega, J.A.; Farinas, I. BDNF is essentially required for the early postnatal survival of nociceptors. *Dev. Biol.* **2010**, *339*, 465–476. [CrossRef]
167. Luo, C.; Zhong, X.L.; Zhou, F.H.; Li, J.Y.; Zhou, P.; Xu, J.M.; Song, B.; Li, C.Q.; Zhou, X.F.; Dai, R.P. Peripheral Brain Derived Neurotrophic Factor Precursor Regulates Pain as an Inflammatory Mediator. *Sci. Rep.* **2016**, *6*, 27171. [CrossRef]
168. Gowler, P.R.W.; Li, L.; Woodhams, S.G.; Bennett, A.J.; Suzuki, R.; Walsh, D.A.; Chapman, V. Peripheral brain-derived neurotrophic factor contributes to chronic osteoarthritis joint pain. *Pain.* **2020**, *161*, 61–73. [CrossRef]
169. Tu, Y.; Muley, M.M.; Beggs, S.; Salter, M.W. Microglia-independent peripheral neuropathic pain in male and female mice. *Pain* **2022**, *163*, e1129–e1144. [CrossRef]
170. Ferrini, F.; Trang, T.; Mattioli, T.A.; Laffray, S.; Del'Guidice, T.; Lorenzo, L.E.; Castonguay, A.; Doyon, N.; Zhang, W.; Godin, A.G.; et al. Morphine hyperalgesia gated through microglia-mediated disruption of neuronal Cl(-) homeostasis. *Nat. Neurosci.* **2013**, *16*, 183–192. [CrossRef]
171. Ismail, C.A.N.; Suppian, R.; Ab Aziz, C.B.; Long, I. Ifenprodil Reduced Expression of Activated Microglia, BDNF and DREAM Proteins in the Spinal Cord Following Formalin Injection During the Early Stage of Painful Diabetic Neuropathy in Rats. *J. Mol. Neurosci.* **2021**, *71*, 379–393. [CrossRef] [PubMed]
172. Zhou, T.T.; Wu, J.R.; Chen, Z.Y.; Liu, Z.X.; Miao, B. Effects of dexmedetomidine on P2X4Rs, p38-MAPK and BDNF in spinal microglia in rats with spared nerve injury. *Brain Res.* **2014**, *1568*, 21–30. [CrossRef]
173. Wong, L.; Done, J.D.; Schaeffer, A.J.; Thumbikat, P. Experimental autoimmune prostatitis induces microglial activation in the spinal cord. *Prostate* **2015**, *75*, 50–59. [CrossRef]
174. Liu, C.; Zhang, Y.; Liu, Q.; Jiang, L.; Li, M.; Wang, S.; Long, T.; He, W.; Kong, X.; Qin, G.; et al. P2X4-receptor participates in EAAT3 regulation via BDNF-TrkB signaling in a model of trigeminal allodynia. *Mol. Pain.* **2018**, *14*, 1744806918795930. [CrossRef]
175. Yajima, Y.; Narita, M.; Usui, A.; Kaneko, C.; Miyatake, M.; Narita, M.; Yamaguchi, T.; Tamaki, H.; Wachi, H.; Seyama, Y.; et al. Direct evidence for the involvement of brain-derived neurotrophic factor in the development of a neuropathic pain-like state in mice. *J. Neurochem.* **2005**, *93*, 584–594. [CrossRef]
176. Bao, Y.; Hou, W.; Liu, R.; Gao, Y.; Kong, X.; Yang, L.; Shi, Z.; Li, W.; Zheng, H.; Jiang, S.; et al. PAR2-mediated upregulation of BDNF contributes to central sensitization in bone cancer pain. *Mol. Pain.* **2014**, *10*, 28. [CrossRef] [PubMed]
177. Khan, N.; Gordon, R.; Woodruff, T.M.; Smith, M.T. Antiallodynic effects of alpha lipoic acid in an optimized RR-EAE mouse model of MS-neuropathic pain are accompanied by attenuation of upregulated BDNF-TrkB-ERK signaling in the dorsal horn of the spinal cord. *Pharmacol. Res. Perspect.* **2015**, *3*, e00137. [CrossRef]
178. Wang, X.; Ratnam, J.; Zou, B.; England, P.M.; Basbaum, A.I. TrkB signaling is required for both the induction and maintenance of tissue and nerve injury-induced persistent pain. *J. Neurosci.* **2009**, *29*, 5508–5515. [CrossRef] [PubMed]
179. Garraway, S.M.; Petruska, J.C.; Mendell, L.M. BDNF sensitizes the response of lamina II neurons to high threshold primary afferent inputs. *Eur. J. Neurosci.* **2003**, *18*, 2467–2476. [CrossRef]
180. Pezet, S.; Cunningham, J.; Patel, J.; Grist, J.; Gavazzi, I.; Lever, I.J.; Malcangio, M. BDNF modulates sensory neuron synaptic activity by a facilitation of GABA transmission in the dorsal horn. *Mol. Cell Neurosci.* **2002**, *21*, 51–62. [CrossRef]
181. Lever, I.; Cunningham, J.; Grist, J.; Yip, P.K.; Malcangio, M. Release of BDNF and GABA in the dorsal horn of neuropathic rats. *Eur. J. Neurosci.* **2003**, *18*, 1169–1174. [CrossRef]
182. Groth, R.; Aanonsen, L. Spinal brain-derived neurotrophic factor (BDNF) produces hyperalgesia in normal mice while antisense directed against either BDNF or trkB, prevent inflammation-induced hyperalgesia. *Pain* **2002**, *100*, 171–181. [CrossRef]

183. Yajima, Y.; Narita, M.; Narita, M.; Matsumoto, N.; Suzuki, T. Involvement of a spinal brain-derived neurotrophic factor/full-length TrkB pathway in the development of nerve injury-induced thermal hyperalgesia in mice. *Brain Res.* **2002**, *958*, 338–346. [CrossRef]
184. Hu, Z.; Yu, X.; Chen, P.; Jin, K.; Zhou, J.; Wang, G.; Yu, J.; Wu, T.; Wang, Y.; Lin, F.; et al. BDNF-TrkB signaling pathway-mediated microglial activation induces neuronal KCC2 downregulation contributing to dynamic allodynia following spared nerve injury. *Mol. Pain.* **2023**, *19*, 17448069231185439. [CrossRef]
185. Phan, T.T.; Jayathilake, N.J.; Lee, K.P.; Park, J.M. BDNF/TrkB Signaling Inhibition Suppresses Astrogliosis and Alleviates Mechanical Allodynia in a Partial Crush Injury Model. *Exp. Neurobiol.* **2023**, *32*, 343–353. [CrossRef] [PubMed]
186. Sorge, R.E.; Mapplebeck, J.C.; Rosen, S.; Beggs, S.; Taves, S.; Alexander, J.K.; Martin, L.J.; Austin, J.S.; Sotocinal, S.G.; Chen, D.; et al. Different immune cells mediate mechanical pain hypersensitivity in male and female mice. *Nat. Neurosci.* **2015**, *18*, 1081–1083. [CrossRef]
187. Guo, W.; Robbins, M.T.; Wei, F.; Zou, S.; Dubner, R.; Ren, K. Supraspinal brain-derived neurotrophic factor signaling: A novel mechanism for descending pain facilitation. *J. Neurosci.* **2006**, *26*, 126–137. [CrossRef]
188. Heinricher, M.M.; Tavares, I.; Leith, J.L.; Lumb, B.M. Descending control of nociception: Specificity, recruitment and plasticity. *Brain Res. Rev.* **2009**, *60*, 214–225. [CrossRef]
189. Wei, F.; Dubner, R.; Zou, S.; Ren, K.; Bai, G.; Wei, D.; Guo, W. Molecular depletion of descending serotonin unmasks its novel facilitatory role in the development of persistent pain. *J. Neurosci.* **2010**, *30*, 8624–8636. [CrossRef] [PubMed]
190. Xue, J.; Li, H.; Xu, Z.; Ma, D.; Guo, R.; Yang, K.; Wang, Y. Paradoxical Sleep Deprivation Aggravates and Prolongs Incision-Induced Pain Hypersensitivity via BDNF Signaling-Mediated Descending Facilitation in Rats. *Neurochem. Res.* **2018**, *43*, 2353–2361. [CrossRef] [PubMed]
191. Zhang, Z.; Tao, W.; Hou, Y.Y.; Wang, W.; Kenny, P.J.; Pan, Z.Z. MeCP2 repression of G9a in regulation of pain and morphine reward. *J. Neurosci.* **2014**, *34*, 9076–9087. [CrossRef] [PubMed]
192. Shih, H.C.; Kuan, Y.H.; Shyu, B.C. Targeting brain-derived neurotrophic factor in the medial thalamus for the treatment of central poststroke pain in a rodent model. *Pain.* **2017**, *158*, 1302–1313. [CrossRef]
193. Sosanya, N.M.; Garza, T.H.; Stacey, W.; Crimmins, S.L.; Christy, R.J.; Cheppudira, B.P. Involvement of brain-derived neurotrophic factor (BDNF) in chronic intermittent stress-induced enhanced mechanical allodynia in a rat model of burn pain. *BMC Neurosci.* **2019**, *20*, 17. [CrossRef]
194. Liu, X.; Hou, Z.; Han, M.; Chen, K.; Wang, Y.; Qing, J.; Yang, F. Salvianolic acid B alleviates comorbid pain in depression induced by chronic restraint stress through inhibiting GABAergic neuron excitation via an ERK-CREB-BDNF axis-dependent mechanism. *J. Psychiatr. Res.* **2022**, *151*, 205–216. [CrossRef]
195. Zhang, H.; Qian, Y.L.; Li, C.; Liu, D.; Wang, L.; Wang, X.Y.; Liu, M.J.; Liu, H.; Zhang, S.; Guo, X.Y.; et al. Brain-Derived Neurotrophic Factor in the Mesolimbic Reward Circuitry Mediates Nociception in Chronic Neuropathic Pain. *Biol. Psychiatry* **2017**, *82*, 608–618. [CrossRef]
196. Li, J.; Wang, X.; Wang, H.; Wang, R.; Guo, Y.; Xu, L.; Zhang, G.; Wu, J.; Wang, G. The BDNF-TrkB signaling pathway in the rostral anterior cingulate cortex is involved in the development of pain aversion in rats with bone cancer via NR2B and ERK-CREB signaling. *Brain Res. Bull.* **2022**, *185*, 18–27. [CrossRef] [PubMed]
197. Wang, X.; Zhang, L.; Zhan, Y.; Li, D.; Zhang, Y.; Wang, G.; Zhang, M. Contribution of BDNF/TrkB signalling in the rACC to the development of pain-related aversion via activation of ERK in rats with spared nerve injury. *Brain Res.* **2017**, *1671*, 111–120. [CrossRef]
198. Zhang, L.; Wang, G.; Ma, J.; Liu, C.; Liu, X.; Zhan, Y.; Zhang, M. Brain-derived neurotrophic factor (BDNF) in the rostral anterior cingulate cortex (rACC) contributes to neuropathic spontaneous pain-related aversion via NR2B receptors. *Brain Res. Bull.* **2016**, *127*, 56–65. [CrossRef]
199. Yue, L.; Ma, L.Y.; Cui, S.; Liu, F.Y.; Yi, M.; Wan, Y. Brain-derived neurotrophic factor in the infralimbic cortex alleviates inflammatory pain. *Neurosci. Lett.* **2017**, *655*, 7–13. [CrossRef]
200. Zheng, J.; Jiang, Y.Y.; Xu, L.C.; Ma, L.Y.; Liu, F.Y.; Cui, S.; Cai, J.; Liao, F.F.; Wan, Y.; Yi, M. Adult Hippocampal Neurogenesis along the Dorsoventral Axis Contributes Differentially to Environmental Enrichment Combined with Voluntary Exercise in Alleviating Chronic Inflammatory Pain in Mice. *J. Neurosci.* **2017**, *37*, 4145–4157. [CrossRef]
201. Ma, L.; Yue, L.; Zhang, Y.; Wang, Y.; Han, B.; Cui, S.; Liu, F.Y.; Wan, Y.; Yi, M. Spontaneous Pain Disrupts Ventral Hippocampal CA1-Infralimbic Cortex Connectivity and Modulates Pain Progression in Rats with Peripheral Inflammation. *Cell Rep.* **2019**, *29*, 1579–1593.e6. [CrossRef]
202. Infantino, R.; Schiano, C.; Luongo, L.; Paino, S.; Mansueto, G.; Boccella, S.; Guida, F.; Ricciardi, F.; Iannotta, M.; Belardo, C.; et al. MED1/BDNF/TrkB pathway is involved in thalamic hemorrhage-induced pain and depression by regulating microglia. *Neurobiol. Dis.* **2022**, *164*, 105611. [CrossRef] [PubMed]

203. Tao, W.; Chen, Q.; Zhou, W.; Wang, Y.; Wang, L.; Zhang, Z. Persistent inflammation-induced up-regulation of brain-derived neurotrophic factor (BDNF) promotes synaptic delivery of alpha-amino-3-hydroxy-5-methyl-4-isoxazolepropionic acid receptor GluA1 subunits in descending pain modulatory circuits. *J. Biol. Chem.* **2014**, *289*, 22196–22204. [CrossRef] [PubMed]
204. Zhang, Z.; Wang, X.; Wang, W.; Lu, Y.G.; Pan, Z.Z. Brain-derived neurotrophic factor-mediated downregulation of brainstem K^+-Cl^- cotransporter and cell-type-specific GABA impairment for activation of descending pain facilitation. *Mol. Pharmacol.* **2013**, *84*, 511–520. [CrossRef]
205. Sarhan, M.; Pawlowski, S.A.; Barthas, F.; Yalcin, I.; Kaufling, J.; Dardente, H.; Zachariou, V.; Dileone, R.J.; Barrot, M.; Veinante, P. BDNF parabrachio-amygdaloid pathway in morphine-induced analgesia. *Int. J. Neuropsychopharmacol.* **2013**, *16*, 1649–1660. [CrossRef]
206. Taylor, A.M.; Castonguay, A.; Taylor, A.J.; Murphy, N.P.; Ghogha, A.; Cook, C.; Xue, L.; Olmstead, M.C.; De Koninck, Y.; Evans, C.J.; et al. Microglia disrupt mesolimbic reward circuitry in chronic pain. *J. Neurosci.* **2015**, *35*, 8442–8450. [CrossRef] [PubMed]
207. Wang, X.M.; Pan, W.; Xu, N.; Zhou, Z.Q.; Zhang, G.F.; Shen, J.C. Environmental enrichment improves long-term memory impairment and aberrant synaptic plasticity by BDNF/TrkB signaling in nerve-injured mice. *Neurosci. Lett.* **2019**, *694*, 93–98. [CrossRef]
208. Tao, W.; Chen, Q.; Wang, L.; Zhou, W.; Wang, Y.; Zhang, Z. Brainstem brain-derived neurotrophic factor signaling is required for histone deacetylase inhibitor-induced pain relief. *Mol. Pharmacol.* **2015**, *87*, 1035–1041. [CrossRef]
209. Kuner, R.; Flor, H. Structural plasticity and reorganisation in chronic pain. *Nat. Rev. Neurosci.* **2016**, *18*, 20–30. [CrossRef]
210. Xia, S.H.; Hu, S.W.; Ge, D.G.; Liu, D.; Wang, D.; Zhang, S.; Zhang, Q.; Yuan, L.; Li, Y.Q.; Yang, J.X.; et al. Chronic Pain Impairs Memory Formation via Disruption of Neurogenesis Mediated by Mesohippocampal Brain-Derived Neurotrophic Factor Signaling. *Biol. Psychiatry* **2020**, *88*, 597–610. [CrossRef]
211. Frank, L.; Wiegand, S.J.; Siuciak, J.A.; Lindsay, R.M.; Rudge, J.S. Effects of BDNF infusion on the regulation of TrkB protein and message in adult rat brain. *Exp. Neurol.* **1997**, *145*, 62–70. [CrossRef] [PubMed]
212. Siuciak, J.A.; Altar, C.A.; Wiegand, S.J.; Lindsay, R.M. Antinociceptive effect of brain-derived neurotrophic factor and neurotrophin-3. *Brain Res.* **1994**, *633*, 326–330. [CrossRef] [PubMed]
213. Siuciak, J.A.; Wong, V.; Pearsall, D.; Wiegand, S.J.; Lindsay, R.M. BDNF produces analgesia in the formalin test and modifies neuropeptide levels in rat brain and spinal cord areas associated with nociception. *Eur. J. Neurosci.* **1995**, *7*, 663–670. [CrossRef] [PubMed]
214. Ye, Y.; Yan, X.; Wang, L.; Xu, J.; Li, T. Transcranial direct current stimulation attenuates chronic pain in knee osteoarthritis by modulating BDNF/TrkB signaling in the descending pain modulation system. *Neurosci. Lett.* **2023**, *810*, 137320. [CrossRef]
215. Liu, D.; Tang, Q.Q.; Yin, C.; Song, Y.; Liu, Y.; Yang, J.X.; Liu, H.; Zhang, Y.M.; Wu, S.Y.; Song, Y.; et al. Brain-derived neurotrophic factor-mediated projection-specific regulation of depressive-like and nociceptive behaviors in the mesolimbic reward circuitry. *Pain* **2018**, *159*, 175. [CrossRef]
216. Ma, Y.; Zhao, W.; Chen, D.; Zhou, D.; Gao, Y.; Bian, Y.; Xu, Y.; Xia, S.H.; Fang, T.; Yang, J.X.; et al. Disinhibition of Mesolimbic Dopamine Circuit by the Lateral Hypothalamus Regulates Pain Sensation. *J. Neurosci.* **2023**, *43*, 4525–4540. [CrossRef]
217. Yang, C.R.; Bai, Y.Y.; Ruan, C.S.; Zhou, F.H.; Li, F.; Li, C.Q.; Zhou, X.F. Injection of Anti-proBDNF in Anterior Cingulate Cortex (ACC) Reverses Chronic Stress-Induced Adverse Mood Behaviors in Mice. *Neurotox. Res.* **2017**, *31*, 298–308. [CrossRef] [PubMed]
218. Thompson, J.M.; Neugebauer, V. Cortico-limbic pain mechanisms. *Neurosci. Lett.* **2019**, *702*, 15–23. [CrossRef]
219. Neugebauer, V.; Kiritoshi, T. Corticolimbic plasticity in pain: Hippocampus joins the party. *Pain* **2023**, *165*, 965–967. [CrossRef]
220. Moriarty, O.; McGuire, B.E.; Finn, D.P. The effect of pain on cognitive function: A review of clinical and preclinical research. *Prog. Neurobiol.* **2011**, *93*, 385–404. [CrossRef]
221. Vachon-Presseau, E.; Centeno, M.V.; Ren, W.; Berger, S.E.; Tetreault, P.; Ghantous, M.; Baria, A.; Farmer, M.; Baliki, M.N.; Schnitzer, T.J.; et al. The Emotional Brain as a Predictor and Amplifier of Chronic Pain. *J. Dent. Res.* **2016**, *95*, 605–612. [CrossRef] [PubMed]
222. Veinante, P.; Yalcin, I.; Barrot, M. The amygdala between sensation and affect: A role in pain. *J. Mol. Psychiatry* **2013**, *1*, 9. [CrossRef]
223. Altar, C.A. Neurotrophins and depression. *Trends Pharmacol. Sci.* **1999**, *20*, 59–61. [CrossRef] [PubMed]
224. Hashimoto, K.; Shimizu, E.; Iyo, M. Critical role of brain-derived neurotrophic factor in mood disorders. *Brain Res. Brain Res. Rev.* **2004**, *45*, 104–114. [CrossRef]
225. Martinowich, K.; Manji, H.; Lu, B. New insights into BDNF function in depression and anxiety. *Nat. Neurosci.* **2007**, *10*, 1089–1093. [CrossRef]
226. Rana, T.; Behl, T.; Sehgal, A.; Srivastava, P.; Bungau, S. Unfolding the Role of BDNF as a Biomarker for Treatment of Depression. *J. Mol. Neurosci.* **2021**, *71*, 2008–2021. [CrossRef] [PubMed]
227. Siuciak, J.A.; Lewis, D.R.; Wiegand, S.J.; Lindsay, R.M. Antidepressant-like effect of brain-derived neurotrophic factor (BDNF). *Pharmacol. Biochem. Behav.* **1997**, *56*, 131–137. [CrossRef]

228. Shirayama, Y.; Chen, A.C.; Nakagawa, S.; Russell, D.S.; Duman, R.S. Brain-derived neurotrophic factor produces antidepressant effects in behavioral models of depression. *J. Neurosci.* **2002**, *22*, 3251–3261. [CrossRef]
229. Zhang, J.C.; Wu, J.; Fujita, Y.; Yao, W.; Ren, Q.; Yang, C.; Li, S.X.; Shirayama, Y.; Hashimoto, K. Antidepressant effects of TrkB ligands on depression-like behavior and dendritic changes in mice after inflammation. *Int. J. Neuropsychopharmacol.* **2014**, *18*, pyu077. [CrossRef]
230. Monteggia, L.M.; Luikart, B.; Barrot, M.; Theobold, D.; Malkovska, I.; Nef, S.; Parada, L.F.; Nestler, E.J. Brain-derived neurotrophic factor conditional knockouts show gender differences in depression-related behaviors. *Biol. Psychiatry* **2007**, *61*, 187–197. [CrossRef]
231. Zhang, K.; Wang, F.; Zhai, M.; He, M.; Hu, Y.; Feng, L.; Li, Y.; Yang, J.; Wu, C. Hyperactive neuronal autophagy depletes BDNF and impairs adult hippocampal neurogenesis in a corticosterone-induced mouse model of depression. *Theranostics* **2023**, *13*, 1059–1075. [CrossRef]
232. Eisch, A.J.; Bolanos, C.A.; de Wit, J.; Simonak, R.D.; Pudiak, C.M.; Barrot, M.; Verhaagen, J.; Nestler, E.J. Brain-derived neurotrophic factor in the ventral midbrain-nucleus accumbens pathway: A role in depression. *Biol. Psychiatry* **2003**, *54*, 994–1005. [CrossRef]
233. Nestler, E.J.; Carlezon, W.A., Jr. The mesolimbic dopamine reward circuit in depression. *Biol. Psychiatry* **2006**, *59*, 1151–1159. [CrossRef]
234. Berton, O.; McClung, C.A.; Dileone, R.J.; Krishnan, V.; Renthal, W.; Russo, S.J.; Graham, D.; Tsankova, N.M.; Bolanos, C.A.; Rios, M.; et al. Essential role of BDNF in the mesolimbic dopamine pathway in social defeat stress. *Science* **2006**, *311*, 864–868. [CrossRef]
235. Zhang, J.C.; Yao, W.; Hashimoto, K. Brain-derived Neurotrophic Factor (BDNF)-TrkB Signaling in Inflammation-related Depression and Potential Therapeutic Targets. *Curr. Neuropharmacol.* **2016**, *14*, 721–731. [CrossRef]
236. Weickert, C.S.; Hyde, T.M.; Lipska, B.K.; Herman, M.M.; Weinberger, D.R.; Kleinman, J.E. Reduced brain-derived neurotrophic factor in prefrontal cortex of patients with schizophrenia. *Mol. Psychiatry* **2003**, *8*, 592–610. [CrossRef]
237. Ashe, P.C.; Chlan-Fourney, J.; Juorio, A.V.; Li, X.M. Brain-derived neurotrophic factor (BDNF) mRNA in rats with neonatal ibotenic acid lesions of the ventral hippocampus. *Brain Res.* **2002**, *956*, 126–135. [CrossRef]
238. Lipska, B.K.; Khaing, Z.Z.; Weickert, C.S.; Weinberger, D.R. BDNF mRNA expression in rat hippocampus and prefrontal cortex: Effects of neonatal ventral hippocampal damage and antipsychotic drugs. *Eur. J. Neurosci.* **2001**, *14*, 135–144. [CrossRef] [PubMed]
239. Fiore, M.; Korf, J.; Antonelli, A.; Talamini, L.; Aloe, L. Long-lasting effects of prenatal MAM treatment on water maze performance in rats: Associations with altered brain development and neurotrophin levels. *Neurotoxicol Teratol.* **2002**, *24*, 179–191. [CrossRef]
240. Zorner, B.; Wolfer, D.P.; Brandis, D.; Kretz, O.; Zacher, C.; Madani, R.; Grunwald, I.; Lipp, H.P.; Klein, R.; Henn, F.A.; et al. Forebrain-specific trkB-receptor knockout mice: Behaviorally more hyperactive than "depressive". *Biol. Psychiatry* **2003**, *54*, 972–982. [CrossRef] [PubMed]
241. Harb, M.; Jagusch, J.; Durairaja, A.; Endres, T.; Lessmann, V.; Fendt, M. BDNF haploinsufficiency induces behavioral endophenotypes of schizophrenia in male mice that are rescued by enriched environment. *Transl. Psychiatry* **2021**, *11*, 233. [CrossRef]
242. Naumenko, V.S.; Bazovkina, D.V.; Morozova, M.V.; Popova, N.K. Effects of brain-derived and glial cell line-derived neurotrophic factors on startle response and disrupted prepulse inhibition in mice of DBA/2J inbred strain. *Neurosci. Lett.* **2013**, *550*, 115–118. [CrossRef] [PubMed]
243. Yang, Y.J.; Li, Y.K.; Wang, W.; Wan, J.G.; Yu, B.; Wang, M.Z.; Hu, B. Small-molecule TrkB agonist 7,8-dihydroxyflavone reverses cognitive and synaptic plasticity deficits in a rat model of schizophrenia. *Pharmacol. Biochem. Behav.* **2014**, *122*, 30–36. [CrossRef]
244. Allen, S.J.; Watson, J.J.; Shoemark, D.K.; Barua, N.U.; Patel, N.K. GDNF, NGF and BDNF as therapeutic options for neurodegeneration. *Pharmacol. Ther.* **2013**, *138*, 155–175. [CrossRef]
245. Colucci-D'Amato, L.; Speranza, L.; Volpicelli, F. Neurotrophic Factor BDNF, Physiological Functions and Therapeutic Potential in Depression, Neurodegeneration and Brain Cancer. *Int. J. Mol. Sci.* **2020**, *21*, 7777. [CrossRef]
246. Kim, J.; Lee, S.; Kang, S.; Kim, S.H.; Kim, J.C.; Yang, M.; Moon, C. Brain-derived neurotropic factor and GABAergic transmission in neurodegeneration and neuroregeneration. *Neural Regen. Res.* **2017**, *12*, 1733–1741. [CrossRef]
247. Choi, S.H.; Bylykbashi, E.; Chatila, Z.K.; Lee, S.W.; Pulli, B.; Clemenson, G.D.; Kim, E.; Rompala, A.; Oram, M.K.; Asselin, C.; et al. Combined adult neurogenesis and BDNF mimic exercise effects on cognition in an Alzheimer's mouse model. *Science* **2018**, *361*, eaan8821. [CrossRef]
248. Palasz, E.; Wysocka, A.; Gasiorowska, A.; Chalimoniuk, M.; Niewiadomski, W.; Niewiadomska, G. BDNF as a Promising Therapeutic Agent in Parkinson's Disease. *Int. J. Mol. Sci.* **2020**, *21*, 1170. [CrossRef]
249. Baquet, Z.C.; Gorski, J.A.; Jones, K.R. Early striatal dendrite deficits followed by neuron loss with advanced age in the absence of anterograde cortical brain-derived neurotrophic factor. *J. Neurosci.* **2004**, *24*, 4250–4258. [CrossRef] [PubMed]

250. Canals, J.M.; Pineda, J.R.; Torres-Peraza, J.F.; Bosch, M.; Martin-Ibanez, R.; Munoz, M.T.; Mengod, G.; Ernfors, P.; Alberch, J. Brain-derived neurotrophic factor regulates the onset and severity of motor dysfunction associated with enkephalinergic neuronal degeneration in Huntington's disease. *J. Neurosci.* **2004**, *24*, 7727–7739. [CrossRef]
251. Gharami, K.; Xie, Y.; An, J.J.; Tonegawa, S.; Xu, B. Brain-derived neurotrophic factor over-expression in the forebrain ameliorates Huntington's disease phenotypes in mice. *J. Neurochem.* **2008**, *105*, 369–379. [CrossRef] [PubMed]
252. Xie, Y.; Hayden, M.R.; Xu, B. BDNF overexpression in the forebrain rescues Huntington's disease phenotypes in YAC128 mice. *J. Neurosci.* **2010**, *30*, 14708–14718. [CrossRef] [PubMed]

MDPI AG
Grosspeteranlage 5
4052 Basel
Switzerland
Tel.: +41 61 683 77 34

Cells Editorial Office
E-mail: cells@mdpi.com
www.mdpi.com/journal/cells

www.ingramcontent.com/pod-product-compliance
Lightning Source LLC
Chambersburg PA
CBHW061451210926
51850CB00008B/1680
9783725866205